Malverd A. (Malverd Abijah) Howe

A Teatise on Arches

Designed for the Use of Engineers and Students in Technical Schools

Malverd A. (Malverd Abijah) Howe

A Teatise on Arches
Designed for the Use of Engineers and Students in Technical Schools

ISBN/EAN: 9783743687035

Printed in Europe, USA, Canada, Australia, Japan

Cover: Foto ©berggeist007 / pixelio.de

A TREATISE ON ARCHES.

DESIGNED FOR THE USE OF ENGINEERS AND STUDENTS IN TECHNICAL SCHOOLS.

BY

MALVERD A. HOWE, C.E.,

Professor of Civil Engineering, Rose Polytechnic Institute;
Member of American Society of Civil Engineers.

FIRST EDITION

FIRST THOUSAND.

NEW YORK:
JOHN WILEY & SONS.
LONDON: CHAPMAN & HALL, LIMITED.
1897.

69487

ROBERT DRUMMOND, ELECTROTYPER AND PRINTER, NEW YORK.

PREFACE.

THE theory of the elastic arch as developed in the following pages is based upon four fundamental equations demonstrated by Weyrauch in 1879. From these equations have been deduced formulas similar to those commonly given in American text-books, but in a simplified form for practical use. In addition to these a large number of general formulas have been introduced, many of which are new.

In Chapter V an attempt has been made to give a set of general formulas which can be applied to any symmetrical arch either fixed or hinged, and subjected to either vertical or horizontal loads. These formulas readily reduce to the common forms, and can be applied in their integral form to any symmetrical arch when the equation of the axis and the law of variation of the moments of inertia of the cross-sections are known. In many cases the reduction of the integrals to a simple form for a given case would be complicated and perhaps impossible; for such cases these formulas are given in their summation form when they apply to *any symmetrical arch subjected to any loading.*

The effect of the axial stress, which is usually neglected by American authors, is thoroughly discussed, exact as well as approximate formulas being given for all cases likely to occur in practice. It is shown that in flat arches fixed at the ends the effect of this stress should not be neglected if economy of material is considered.

Formulas for vertical and horizontal loads are deduced for each case considered, making it possible easily to treat loads making any angle with the axis of the arch. The effect of a couple is discussed, and general as well as special formulas given.

Changes of temperature and of shape have been considered, and when not too complicated, formulas for special cases are given.

Masonry arches are considered, and the many difficulties and inaccuracies of the common methods of treatment pointed out. With a little good judgment it is easy to so design a masonry arch that the stresses will practically follow the laws demonstrated for the elastic arch. This has been experimentally shown by the "Austrian Experiments" and by many large arches designed and erected by European engineers.

Alexander and Thomson's method for designing segmental masonry arches has been given as being the most consistent of the many methods which assume all loading due to material to act as vertical forces upon the arch.

It is hoped that the practising engineer, who has, as a rule, little time to study mathematical demonstrations or to search through several pages of transformations for a desired formula, will appreciate the collection in simple form (Chapter II) of all of the necessary formulas likely to be needed in practice, and also the ease and celerity with which they can be applied, with the aid of the tables, to the case in hand. A fair trial of the summation formulas given in the same chapter will, it is believed, lead to the adoption of metal arches more artistic in form than the usual American type.

These summation formulas are readily applied in the designing of masonry arches.

Nearly all of the formulas given have been deduced for this treatise by two radically different methods. Many of these formulas are old, and while it was desired to give full credit in every particular, it was not found either expedient or possible to do so for each form.

The tables were carefully computed, and when possible by the method of differences, each tenth value being checked by direct computation.

The demonstrations are believed to be sufficiently simple to be easily followed by senior students in Technical schools. With the aid of the tables, class problems can be solved which otherwise would be impossible on account of the time required where direct computation of the various terms must be resorted to.

The author will esteem it a favor if any errors that may be found are at once brought to his notice.

M. A. H.

TERRE HAUTE, May 1897.

TABLE OF CONTENTS.

CHAPTER I.

GENERAL PRELIMINARY FORMULAS.

CHAPTER II.

FORMULAS FOR PRACTICAL USE.

CHAPTER III.

PARABOLIC ARCHES WITH THE MOMENTS OF INERTIA VARYING ACCORDING TO THE RELATION $E\theta \cos \phi$ = A CONSTANT.

CHAPTER IV.

CIRCULAR ARCHES HAVING $\frac{2E\theta}{R}$ = A CONSTANT.

CHAPTER V.

SYMMETRICAL ARCHES HAVING A VARIABLE MOMENT OF INERTIA.

CHAPTER VI.

COMPARISON OF FOUR TYPES OF ARCHES.

CHAPTER VII.

APPLICATIONS.

CHAPTER VIII.

APPLICATION OF GENERAL SUMMATION FORMULAS TO ARCHES HAVING A HINGE AT EACH SUPPORT.

CHAPTER IX.

APPLICATION OF GENERAL SUMMATION FORMULAS TO ARCHES WITHOUT HINGES.

CHAPTER X.

THE ST. LOUIS ARCH.

CHAPTER XI.

THE SPANDREL-BRACED ARCH.

CHAPTER XII.

THE MASONRY ARCH.

CHAPTER XIII.

ALEXANDER AND THOMSON'S METHOD FOR DESIGNING SEGMENTAL MASONRY ARCHES.

CHAPTER XIV.

EXAMPLES ILLUSTRATING ALEXANDER AND THOMSON'S METHOD FOR DESIGNING SEGMENTAL MASONRY ARCHES.

CHAPTER XV.

TESTS OF ARCHES.

APPENDICES.

TABLES.

INTRODUCTION.

PROBABLY the Chinese first employed the arch in the construction of bridges across small streams. No authentic information is obtainable in reference to the time of its first use. It is known, however, that bridges and other public works were executed in China 2900 B.C., and that possibly the arch may have been used at as early a date as this.

CHINESE BRIDGE.

Stone and brick arches have been found in Egypt, but the dates of their construction are not positively known.

In "Campbell's Tomb" an arch of brick composed of

CHINESE BRIDGE.

four ring-courses, the inner ring having a span of eleven feet, was found. This arch, according to Wilkinson, was built about 1540 B.C.

"CAMPBELL'S TOMB."

In Ethiopia, arched porticos were found before some of the pyramids, as well as arches over tombs, which it is claimed were erected 1540 B.C.

Five brick arches of from twelve to fourteen feet span

have been discovered in Thebes, in ancient Greece, said to have been inscribed with the name of Sesostris, a legendary monarch whose deeds are supposed to refer to those of Rameses III., who lived about 1200 B.C.

Although the arch was known to the Egyptians, yet they did not use it to any great extent in their structures, preferring a solid lintel as a covering for openings, rooms, etc.

TREASURY OF ATREUS.

A large number of apparent arches have been found, composed of masonry in horizontal layers, corbelled out over the openings and then cut to resemble arches. This method of spanning openings seems to have been almost universal, judging from ruins found in all parts of the world. The Greeks employed this method for covering quite large areas, although it is claimed that they were familiar with the true arch

To the Romans belongs the credit of first using the arch for spanning openings of considerable magnitude.

The earliest Roman arch of which we have authentic knowledge was the Cloaca Maxima sewer, constructed about 615 B.C. This arch consisted of three concentric rings of stone, the inner ring having a span of about fourteen feet. Like the majority of early Roman arches, it was circular.

CLOACA MAXIMA.

Bridges of from fifty to seventy feet in span were built oi stone by Æmilius Scaurus 120 B.C.

Trajan (about 104 A.D.) is credited with having constructed a wooden arch bridge having a *span of one hundred and seventy feet*, but some authorities doubt that such a structure ever existed.

One of the largest stone arch bridges constructed by the Romans was built by Trajan, 105 A.D., at Alcantara in Spain. The largest arch was semicircular, and had a span of one hundred and ten feet. Outside of China this is probably the oldest stone arch (of any considerable magnitude) which exists at the present time.

Many aqueducts were constructed by the Romans which were carried across valleys upon arch bridges sometimes built in three tiers one above the other.

Accurate data exist of many masonry arches constructed in the seventeenth and eighteenth centuries by the French and the English, of which the general dimensions of some of the most noted are given in Table XXX, which also contains data in reference to masonry arches built in the nineteenth century.

The greatest distance spanned by a single masonry arch is supposed to be *two hundred and fifty-one* feet, which was the span of the Trezzo Bridge constructed in 1380 and destroyed by Carmagnola in 1416.

At the present day the Cabin John aqueduct bridge has the greatest span of any masonry arch. The span is about *two hundred and twenty* feet.

Cast-iron arch bridges were first constructed in England, and the Coalbrookdale bridge, with a span of *one hundred* feet and a rise of *fifty* feet, has the distinction of being the first cast-iron arch bridge which was successfully constructed. This bridge was built by Abraham Darby, an iron-founder, in 1779.

From this time up to the introduction of wrought iron, many very artistic cast-iron bridges were constructed; and even as late as 1871 a cast-iron arch bridge of *one hundred* feet span was constructed at Nottingham. As far as known, with the exception of the Chestnut-street bridge, Philadelphia, there are no cast-iron arch bridges of any magnitude in the United States.

The maximum span of any cast-iron arch bridge is that of the Southwark bridge, built in 1819; this has a span of *two hundred and forty* feet.

The dimensions of a few cast-iron bridges are given in Table XXXI.

With but few exceptions, these bridges were arches *without hinges.*

The use of wrought iron and steel in the construction of arch *bridges* is of recent date.

The first arch bridge* with ribs practically of wrought iron was probably the Cron bridge at St. Denis, which was constructed in 1808. Wrought iron and steel have come into general use for large arch bridges since 1870.

* William H. Wahl, A.M., Ph.D., "Iconographic Encyclopædia," vol. v. p. 268.

The maximum span at the present time* is that of the Luiz I. bridge at Oporto, Portugal, which has a span of *five hundred and sixty-six* feet.

The dimensions of a few wrought-iron and steel arches are given in Table XXXII.

Wooden arches are probably not very recent. The maximum span constructed was built by Louis Wernway in 1812 in Philadelphia. The bridge crossed the Schuylkill River, and had a span of about *three hundred and forty* feet. It was burned in 1838.

Since the time of the Romans the arch in some form has been the favorite method for roof construction in stone, wood, and metal, where artistic interior effects were sought and means were obtainable for executing the work.

In the United States the arch is freely used for roofs covering large areas, as train-sheds, armories, exhibition buildings, etc. These arches are usually of the three-hinged type.

The dimensions of a few large roof-arches are given in Table XXXII.

Whether the ancients had any knowledge of the theoretical principles of the arch is not known, but it is known that they were very successful in designing arch structures which have remained until the present time. It is probable that their knowledge was purely the results of experiments, and in the case of masonry arches very little advancement has been made even now, as one may see by comparing the dimensions and details of arches constructed since 1750.

Since the time of Newton (1642–1727) volumes have been written upon the theory of arches, especially the masonry arch. The theory of the masonry arch has been and is now quite unsatisfactory from a practical point of view, since we are unable to discover the *directions* and *magnitudes* of the

---

* See Viaur Viaduct in Table XXXII.

forces caused by the spandrel-filling.* The necessary assumptions which must be made for computation according to ordinary methods have been a source of much controversy among engineers, and will probably continue unsettled for a long time.

The theory of the elastic arch and its application to metal arched ribs has been developed since 1840, and is now generally accepted as being sufficiently accurate for practical purposes.

---

* See Chapter XII for a method of construction which fixes the directions and magnitudes of the forces caused by the spandrel-filling

NOMENCLATURE.

NOMENCLATURE USED IN CHAPTERS II TO XI INCLUSIVE.

$A = E\theta \cos \phi =$ constant for *Parabolic Arches.*

$A = \frac{2E\theta}{R} =$ constant for *Circular Arches.*

$a =$ the abscissa of the point of application of any load P or Q.

a_1 and $a_2 =$ the abscissas of the extreme limits of any uniform horizontally distributed load.

$b =$ the ordinate of the point of application of any load P or Q.

$c =$ the difference in elevation of the right and left supports.

$E =$ the modulus of elasticity.

$e =$ coefficient of expansion.

$f =$ the rise of the linear arch.

$F_x =$ the area of the arch-rib at any section x.

$g =$ the abscissa of the crown of the linear arch.

$H_x =$ the horizontal thrust at any section x.

$H_1 =$ the horizontal thrust at the *left* support.

$H_2 =$ the horizontal thrust at the *right* support.

$\mathfrak{H}_1 =$ the horizontal thrust at the left support due to two equal and symmetrically placed loads.

$k = a/l$.

$k' = R - f$.

$l =$ the span of the linear arch.

$m = \left(\textit{radius of gyration}\right)^2 = \frac{\theta}{F}$ for *Parabolic Arches.*

$m = \left(\frac{radius\ of\ gyration}{R}\right)^2 = \frac{\theta}{FR^2}$ for *Circular Arches.*

M_1 = the moment at the *left* support.

M_2 = the moment at the *right* support.

M_x = the moment at any section x.

N_1 and N_2 = the normal intensities of the resultant force upon any section at the *extreme fibres.*

N_x = the normal component of any force acting upon the section x.

$n = f/l$.

P = any vertical concentrated load.

p = parameter of parabola.

p_0 = the *average* intensity of the resultant force acting upon any section.

Q = any horizontal concentrated load.

r = radius of gyration.

R_1 = the resultant of V_1 and H_1.

R_2 = the resultant of V_2 and H_2.

R = the radius of a circular linear arch.

R_x = the resultant force acting upon any section.

s = the length of the linear arch curve.

T_x = the normal shear at any point x.

t° = the number of degrees of change in temperature.

V_x = the vertical shear at any section x.

V_1 = the vertical reaction at the *left* support.

V_2 = the vertical reaction at the *right* support.

w = the load per lineal unit of span.

x = the abscissa of any point of the linear arch.

x_0 = the abscissa of the point of intersection of Q, R_1 and R_2.

x_1 = the abscissa of the point where R_1 cuts the horizontal passing through the *left* support.

x_2 = the abscissa of the point where R_2 cuts the horizontal passing through the *right* support.

y = the ordinate of any point having the abscissa x.

y_1 = the ordinate of the point where R_1 cuts the vertical through the *left* support.

y_2 = the ordinate of the point where R_2 cuts the vertical through the *right* support.

y_0 = the ordinate of the point of intersection of R_1 and R_2.

α = angular distance of the point of application of the load P or Q from the crown.

θ = moment of inertia of a normal section of the arch-rib.

θ_x = the moment of inertia at the section x.

β = the angle made by the resultant at any section with the horizontal.

ϕ = the angular distance of any point x from the crown of the arch.

ϕ_0 = the angular distance of the left support from the crown of the arch.

ϕ_l = the angular distance of the right support from the crown.

Δ_1, Δ_2, etc., = value of Δ, will be found in Tables I, II, etc., respectively.

Δl = small finite change in l.

$\Delta\phi$ = small finite change in ϕ.

$\Delta\phi_0$ = small finite change in ϕ_0.

$\Delta\phi_l$ = small finite change in ϕ_l.

Δx = a finite value of dx.

Δy = a finite value of dy.

Δs = a finite value of ds.

$\overset{x}{\Sigma}$ = algebraic sum up to the section x.

MASONRY ARCHES.

NOMENCLATURE USED IN ALEXANDER AND THOMSON'S METHOD.

b = distance of directrix to centre of described circle.

$2c$ = clear span of arch.

d = distance of directrix from soffit of arch at crown.

e = base of Naperian system of logarithms.

k = the clear rise of arch.

m = the parameter of catenary.

t_0 = depth of arch-ring at the crown.
t_s = depth of arch-ring at the skew-backs.
w = weight of a unit mass of masonry.
r = ratio of transformation = $\sqrt{s}$.
R_1 = the radius of the described circle.
R_2 = the radius of the three-point circle.
x and y = general co-ordinates.
x_1 and y_1 = co-ordinates of the nose of a two-nosed catenary.
x_2 and y_2 = the co-ordinates of the point where the two-nosed catenary is cut by the three-point circle.
y_0 = the ordinate of the two-nosed catenary at the crown.
Y_0 = the ordinate of the described circle at the crown.
$\delta_0 = y_0 - Y_0$.
δ_2 = departure of the two-nosed catenary from the described circle at the skew-backs, measured radially.
δ_1 = departure of the two-nosed catenary from the three-point circle at the noses.
ρ_0 = radius of curvature of two-nosed catenary at the crown.
ρ_1 = radius of two-nosed catenary at the nose.
ρ_2 = radius of the two-nosed catenary at the point where it cuts the three-point circle.
ϕ_1 = the angle which ρ_1 makes with the vertical.
ϕ_2 = the angle which ρ_2 makes with the vertical.
β_2 = the angle which R_2 makes with the vertical.
H_1, V_1, M_1, etc., have the same meaning in general as for elastic arches.

SOME FORMULAS CONSTANTLY REFERRED TO.

$$\Delta x = -\int \Delta\phi dy + et^\circ \int dx - \frac{1}{E}\int \frac{N_x}{F_x} dx. \quad (a)$$

$$\Delta y = \int \Delta\phi dx + et^\circ \int dy - \frac{1}{E}\int \frac{N_x}{F_x} dy. \quad (b)$$

$$\Delta s = et^\circ \int ds - \frac{1}{E}\int \frac{N_x}{F_x} ds. \quad (c)$$

$$\Delta\phi = \frac{1}{E}\int\frac{M_x}{\theta_x}ds. \quad \ldots \quad (d)$$

$$N_n = \frac{N_x}{F_x} + n\frac{M_x}{\theta_x}. \quad \ldots \quad (e)$$

$$H_x = H_1 - \overset{x}{\Sigma} Q \quad \ldots \quad x > a. \quad \ldots \quad (39)$$

$$V_x = V_1 - \overset{x}{\Sigma} P \quad \ldots \quad x > a. \quad \ldots \quad (40)$$

$$N_x = V_x \sin\phi + H_x \cos\phi. \quad \ldots \quad (42)$$

$$M_x = M_1 + V_1 x - H_1 y - \overset{x}{\Sigma} P(x - a) + \overset{x}{\Sigma} Q(y - b). \quad (41)$$

$$M_1 = M_2 - V_1 l + H_1 c_1 + \overset{l}{\Sigma} P(l - a) - \overset{l}{\Sigma} Q(c - b). \quad \ldots \quad (49)$$

$$V_1 = \frac{1}{l}\{M_2 - M_1 + H_1 c + \overset{l}{\Sigma} P(l - a) - \overset{l}{\Sigma} Q(c - b). \quad (47)$$

$$y_0 = \frac{M_1 + V_1 a}{H_1}. \quad \ldots \quad (50)$$

$$y_1 = \frac{M_1}{H_1} \quad \text{and} \quad y_2 = \frac{M_2}{H_2}. \quad \ldots \quad (51)$$

$$x_1 = \frac{M_1}{V_1} \quad \text{and} \quad x_2 = \frac{M_2}{V_2}. \quad \ldots \quad (54)$$

$$x_0 = \frac{bx_1 - x_1 y_1}{y_1}.$$

A TREATISE ON ARCHES.

CHAPTER I.

GENERAL PRELIMINARY FORMULAS.

DEFORMATION FORMULAS.*

LET Fig. 1 represent a portion of an elastic arch; then the relation between the length of any fibre between two adjacent

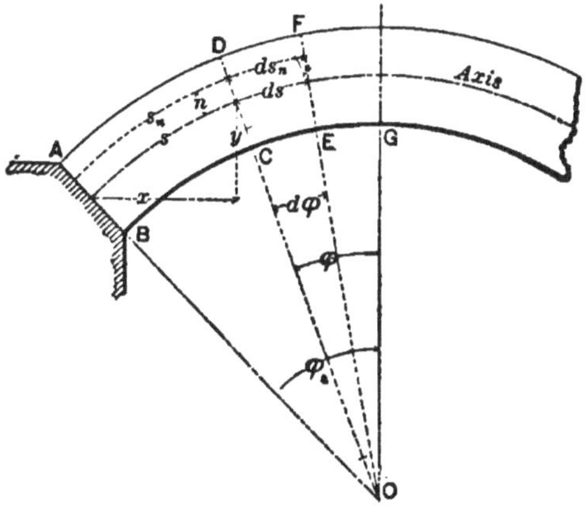

FIG. 1.

radial sections can be expressed in terms of the length of the neutral fibre (limited by the same radial sections) by the equation

$$ds_n = ds + n \sin(-d\phi) = ds - n d\phi. \quad . \quad . \quad . \quad (1)$$

* The formulas and demonstrations in this article are essentially the same as given by Prof Weyrauch in "Theorie der Elastigen Bogenträger" (München, 1879).

Now suppose some circumstance, as the application of a load, changes the lengths of these fibres, and let s become $s + \Delta s$, s_n become $s_n + \Delta s_n$, etc., as shown in Fig. 2. Then we have for the new condition

$$d(s_n + \Delta s_n) = d(s + \Delta s) - nd(\phi + \Delta\phi). \quad . \quad . \quad (2)$$

Combining (1) and (2),

$$d\Delta s_n = d\Delta s - nd\Delta\phi, \quad . \quad . \quad . \quad . \quad . \quad . \quad (3)$$

or

$$\frac{d\Delta s_n}{d\Delta s} = 1 - \frac{nd\Delta\phi}{d\Delta s}, \quad . \quad . \quad . \quad . \quad . \quad . \quad . \quad (4)$$

where $d\Delta s_n$ represents the change in magnitude of ds_n and $d\Delta s$ that of ds.

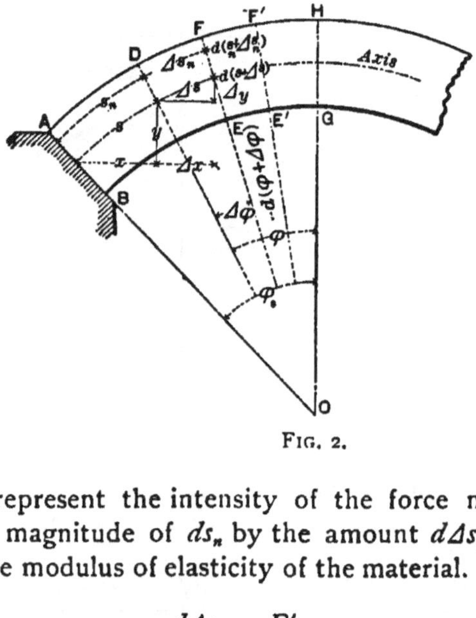

FIG. 2.

Let F' represent the intensity of the force necessary to change the magnitude of ds_n by the amount $d\Delta s_n$, and let E represent the modulus of elasticity of the material. Then

$$\frac{d\Delta s_n}{ds_n} = \frac{F'}{E}. \quad . \quad . \quad . \quad . \quad . \quad . \quad . \quad (5)$$

If the force F' is due to a change of temperature, and e represents the coefficient of expansion per degree of change, then for an *increase* of temperature of t° we have

$$\frac{d\Delta s_n}{ds_n} = et^\circ. \quad . \quad . \quad . \quad . \quad . \quad . \quad . \quad (6)$$

Let N_n be the intensity of any normal force acting upon the fibre s_n; assuming that this force acts at the same time with the change of temperature but of opposite effect, then we have from (5) and (6)

$$\frac{d\Delta s_n}{ds_n} = et_0 - \frac{N_n}{E}. \quad . \quad . \quad . \quad . \quad . \quad . \quad . \quad (7)$$

From (1) and (2),

$$\frac{d\Delta s_n}{ds_n} = \frac{d\Delta s - nd\Delta\phi}{ds - nd\phi}. \quad . \quad . \quad . \quad . \quad . \quad (8)$$

Substituting (8) in (7) and solving for N_n, we obtain

$$N_n = Eet^\circ - E\frac{d\Delta s - nd\Delta\phi}{ds - nd\phi}, \quad . \quad . \quad . \quad . \quad . \quad . \quad (9)$$

or

$$N_n = E\frac{d\Delta s - nd\Delta\phi}{ds}\frac{1}{n\frac{d\phi}{ds} - 1} + Eet^\circ. \quad . \quad . \quad (10)$$

Now $ds = R\sin(-d\phi) = -Rd\phi$, or $\frac{d\phi}{ds} = -\frac{1}{R}$; hence

$$N_n = E\left\{n\frac{d\Delta\phi}{ds} - \frac{d\Delta s}{ds}\right\}\frac{R}{n+R} + Eet^\circ. \quad . \quad . \quad . \quad (11)$$

Let f' represent the area of the fibre s_n, and N_x the resultant normal force acting upon the section. Then

$$N_x = \Sigma N_n f' = E\left\{\frac{d\Delta\phi}{ds}\Sigma\frac{f'nR}{n+R} - \frac{d\Delta s}{ds}\Sigma\frac{f'R}{n+R}\right\} + Eet^\circ\Sigma f'. (12)$$

If we take the centre of moments on the axis of the arch at the section x, the moment of the radial force upon the section will be zero.

If x_0 is the distance of the point of application of the force N_x from the axis, we have

$$N_x x_0 = \Sigma N_n f'n = M_x$$
$$= \text{the moment of the external forces acting upon the section;}$$

then from (12) we can write

$$M_x = \Sigma N_n f'n = E\left\{\frac{d\Delta\phi}{ds}\Sigma\frac{f'n^2R}{n+R} - \frac{d\Delta s}{ds}\Sigma\frac{f'nR}{n+R}\right\}$$
$$+ Eet^\circ \Sigma f'n. \quad . \quad . \quad (13)$$

Let $R\Sigma\dfrac{f'n^2}{n+R} = W$ and $\Sigma f' = F_x$; then $\Sigma f'n = 0$;

hence in (12)

$$\Sigma\frac{f'nR}{n+R} = \Sigma f'n - \Sigma\frac{f'n^2}{n+R} = -\frac{W}{R},$$

$$\Sigma\frac{f'R}{n+R} = \Sigma f' - \frac{\Sigma f'n}{R} + \Sigma\frac{f'n^2}{R(n+R)} = F_x + \frac{W}{R^2}.$$

Substituting these values in (12) and solving for $\dfrac{N_x}{E}$, we obtain

$$\frac{N_x}{E} = \frac{d\Delta\phi}{ds}\left(-\frac{W}{R}\right) - \frac{d\Delta s}{ds}\left(F_x + \frac{W}{R^2}\right) + F_x et^\circ,$$

or

$$\frac{N_x}{E} = -\left\{\frac{d\Delta\phi}{ds} + \frac{1}{R}\frac{d\Delta s}{ds}\right\}\frac{W}{R} - \left\{\frac{d\Delta s}{ds} - et^\circ\right\}F_x. \quad (14)$$

(13) reduces to

$$\frac{M_x}{E} = \left\{\frac{d\Delta\phi}{ds}R + \frac{d\Delta s}{ds}\right\}\frac{W}{R}. \quad . \quad . \quad . \quad . \quad . \quad . \quad . \quad . \quad (15)$$

From (15)

$$\left\{\frac{d\Delta\phi}{ds}+\frac{1}{R}\frac{d\Delta s}{ds}\right\}=\frac{M_x}{EW}.$$

Substituting this in (14),

$$\frac{N_x}{E}=-\frac{M_x}{ER}-\frac{d\Delta s}{ds}F_x+F_x et^\circ,$$

or

$$\frac{d\Delta s}{ds}=-\left(N_x+\frac{M_x}{R}\right)\frac{1}{EF_x}+et^\circ=Y. \quad . \quad . \quad . \quad (16)$$

Substituting (16) in (15) and reducing, we have

$$\frac{d\Delta\phi}{ds}=\left\{N_x+\frac{M_x}{R}\right\}\frac{1}{REF_x}+\frac{M_x}{WE}-\frac{et^\circ}{R}=X. \quad . \quad (17)$$

Substituting (16) and (17) in (11), and reducing, we have

$$N_n=\left\{\frac{N_x}{F_x}+\frac{M_x}{F_xR}+\frac{M_x}{W}\frac{nR}{n+R}\right\}. \quad . \quad . \quad . \quad . \quad . \quad (18)$$

From Fig. 2,

$$d(x+\Delta x)=d(s+\Delta s)\cos(\phi+\Delta\phi);$$

$$d(y+\Delta y)=d(s+\Delta s)\sin(\phi+\Delta\phi);$$

but

$$\cos(\phi+\Delta\phi)=\cos\phi\cos\Delta\phi-\sin\phi\sin\Delta\phi=\frac{dx}{ds}-\Delta\phi\frac{dy}{ds}; \quad (19)$$

$$\sin(\phi+\Delta\phi)=\sin\phi\cos\Delta\phi+\cos\phi\sin\Delta\phi=\frac{dy}{ds}+\Delta\phi\frac{dx}{ds}. \quad (20)$$

Substituting (19) and (20) in the two expressions above, and integrating, we have

$$\Delta x=-\int\Delta\phi dy+\int^{u}\frac{\Delta s}{ds}(dx-dy\Delta\phi);$$

or, from (16),

$$\Delta x=-\int\Delta\phi dy+\int Ydx-\int Y\Delta\phi dy. \quad . \quad . \quad (21)$$

But $\int Y\Delta\phi dy$ can be neglected in comparison with the terms preceding. Hence

$$\Delta x = -\int \Delta\phi dy + \int Y dx. \quad \ldots \ldots \quad (22)$$

In a similar manner,

$$\Delta y = \int \Delta\phi dx + \int Y dy. \quad \ldots \ldots \quad (23)$$

From (16) and (17),

$$\Delta s = \int Y ds, \quad \ldots \ldots \quad (24)$$

and

$$\Delta\phi = \int X ds. \quad \ldots \ldots \quad (25)$$

These four equations, (22), (23), (24), and (25), completely determine the effect of any change of position of any point in the axis of the arch when X, Y, and the equation of the axis of the arch are known.

The expressions for X, Y, and N_n can be simplified by replacing W by $\theta = \Sigma f'n^2 =$ *the moment of inertia of the cross-section.* Since in the expression $W = R\Sigma\frac{f'n^2}{n+R}$, R is usually very large in comparison with n, no material error results from the change.

In (16) $\frac{1}{REF_x}$ is a very small quantity, and consequently we can neglect $\frac{M_x}{REF_x}$.

In (17) $\frac{M_x}{R^2EF_x}$ and $\frac{et^\circ}{R}$ can be omitted.

In (18) $\frac{M_x}{RF_x}$ can be neglected and $\frac{nR}{n+R}$ be assumed to equal n.

Making these modifications and collecting our formulas, we have finally

$$\Delta x = -\int \Delta\phi dy + et^\circ \int dx - \frac{1}{E}\int \frac{N_x}{F_x} dx ; \quad . \quad . \quad (a)$$

$$\Delta y = \int \Delta\phi dx + et^\circ \int dy - \frac{1}{E}\int \frac{N_x}{F_x} dy ; \quad . \quad . \quad . \quad (b)$$

$$\Delta s = et^\circ \int ds - \frac{1}{E}\int \frac{N_x}{F_x} ds ; \quad . \quad . \quad . \quad . \quad . \quad . \quad . \quad . \quad (c)$$

$$\Delta\phi = +\frac{1}{E}\int \frac{M_x}{\theta} ds ; \quad . \quad . \quad . \quad . \quad . \quad . \quad . \quad . \quad . \quad . \quad . \quad (d)$$

$$N_n = \frac{N_x}{F_x} + n\frac{M_x}{\theta}. \quad . \quad . \quad . \quad . \quad . \quad . \quad . \quad . \quad . \quad . \quad . \quad (e)$$

The term containing N_x shows the effect of the *axial stress*, which is usually neglected in the common investigation of the problem. In many cases the influence of this stress is of little or no importance, but in very *flat arches* it should *not* be neglected.

Omitting the terms containing N_x greatly simplifies the deduction of equations for special forms of arches, and also the solution of these equations in the determination of the reactions, bending-moments, etc.

Omitting the terms containing N_x, we have

$$\Delta x = -\int \Delta\phi dy + et^\circ \int dx ; \quad . \quad . \quad . \quad . \quad (aa)$$

$$\Delta y = \int \Delta\phi dx + et^\circ \int dy ; \quad . \quad . \quad . \quad . \quad . \quad (bb)$$

$$\Delta s = et^\circ \int ds ; \quad . \quad . \quad . \quad . \quad . \quad . \quad . \quad . \quad . \quad . \quad (cc)$$

$$\Delta\phi = +\frac{1}{E}\int \frac{M_x}{\theta} ds ; \quad . \quad . \quad . \quad . \quad . \quad . \quad . \quad . \quad (dd)$$

$$N_n = n\frac{M_x}{\theta}. \quad . \quad . \quad . \quad . \quad . \quad . \quad . \quad . \quad . \quad . \quad . \quad . \quad (ee)$$

THE DISTRIBUTION OF STRESS UPON ANY RADIAL SECTION x OF THE ELASTIC ARCH.

In Fig. 3 let AC represent any radial section of the arch, and N_x the resultant normal force applied to the section at a distance x_0 from the axis of the arch passing through the centre of gravity of the section; then

$$N_x x_0 = \Sigma N_n f'n = M_x. \quad . \quad . \quad . \quad . \quad . \quad (29)$$

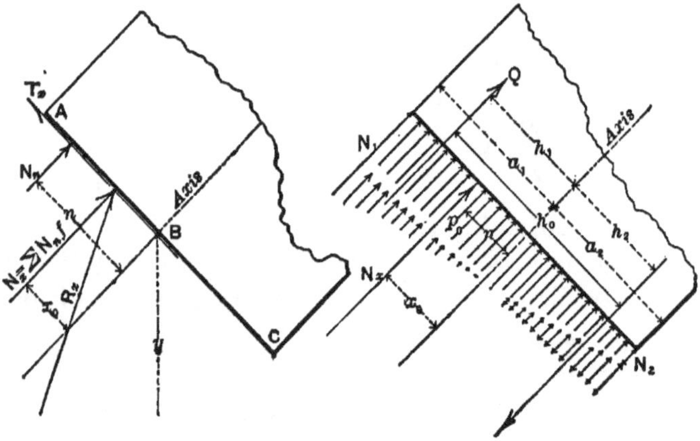

FIG. 3.

From (e), after substituting (29),

$$N_n = \frac{N_x}{F_x} + \frac{n}{\theta} N_x x_0,$$

or

$$N_n = \frac{N_x}{F_x} \left\{ 1 + \frac{nF_x}{\theta} x_0 \right\}. \quad . \quad . \quad . \quad . \quad (30)$$

Now $\frac{N_x}{F_x}$ represents the average intensity of the pressure on the section, and may be represented by p_0 for convenience; and $\frac{N_x F_x x_0}{F_x \theta} n$ represents an intensity which varies directly with n; hence N_n is composed of the algebraic sum of an average intensity and a uniformly varying intensity.

Replacing $\frac{N_x}{F_x}$ by p_0, and remembering that $\frac{F_x}{\theta} = \frac{1}{r^2}$, where r represents the *radius of gyration*, (30) becomes

$$N_n = p_0\left(1 + \frac{n}{r^2}x_0\right), \quad . \quad . \quad . \quad . \quad . \quad . \quad (f)$$

from which the intensity of pressure at any point of the section can be determined.

Let $\frac{r^2}{a_1} = k_2$ and $\frac{r^2}{a_2} = k_1$. (Fig. 3.)

Making $n = a_1$ in (f), we have

$$N_1 = p_0\left(1 + \frac{x_0}{k_2}\right). \quad . \quad . \quad . \quad . \quad . \quad . \quad (31)$$

Making $n = a_2$ in (f), we have

$$N_2 = p_0\left(1 - \frac{x_0}{k_1}\right). \quad . \quad . \quad . \quad . \quad . \quad . \quad (32)$$

(31) and (32) determine the intensities upon the extreme fibres of the section.

When $x_0 = -k_2$, $N_1 = 0$.
" $x_0 = +k_1$, $N_2 = 0$.
" $x_0 > -k_2$, N_1 and N_x have the same sign.
" $x_0 < +k_1$, N_2 and N_x " " " "

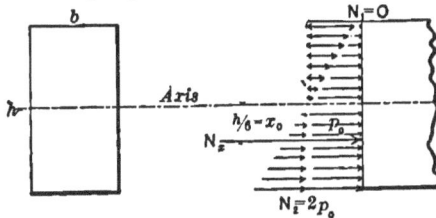

FIG. 4.

Hence when $x_0 > -k_2$ and $< k_1$ the entire section is subjected to the same kind of stress.

To illustrate: Suppose the section to be rectangular; then $F = bh$ and $\theta = \frac{1}{12}bh^3$, and

$$N_1 = p_0\left(1 + \frac{6x_0}{h}\right).$$

$N_1 = 0$ when $1 + \frac{6x_0}{h} = 0$, or $x_0 = -\frac{h}{6} = -k_2$; or the resultant stress acting upon the section must cut the section at a point distant from the axis one sixth the depth of the section and below the axis (see Fig. 4). Evidently, if N_x were applied above the axis and $x_0 = \frac{h}{6}$, $N_2 = 0$ and $N_1 = 2p_0$. Hence, in order that all parts of a rectangular section be subjected to the same kind of stress, the resultant stress N_x must be applied within the *middle third of the section.*

Adding (31) and (32),

$$N_1 + N_2 = 2p_0 + p_0\left(\frac{x_0}{k_2} - \frac{x_0}{k_1}\right),$$

or

$$N_1 + N_2 = 2p_0 + p_0 x_0(a_1 - a_2). \quad . \quad . \quad . \quad (33)$$

If $a_1 = a_2$,

$$\frac{N_1 + N_2}{2} = p_0. \quad . \quad . \quad . \quad . \quad . \quad . \quad . \quad . \quad (34)$$

Returning to (e),

$$N_n = p_0 + n\frac{M_x}{\theta} = p_0 + p_n';$$

$$p_n' = n\frac{M_x}{\theta}, \quad p_n'f' = \frac{M_x}{\theta}nf', \quad \text{and} \quad \Sigma p_n'f' = \frac{M_x}{\theta}\Sigma nf' = 0.$$

From Fig. 3, letting Q represent the force whose intensity is uniformly varying, we have

$$Q = \frac{M_x}{\theta}\sum_0^{a_1} f'n = \frac{M_x}{\theta}\sum_0^{a_2} f'n. \quad . \quad . \quad . \quad . \quad . \quad (35)$$

But $M_x = N_x x_0 = Qh_0$; therefore

$$h_0 = \frac{\theta}{\sum_0^{a_1} f'n} = \frac{\theta}{\sum_0^{a_2} f'n}, \quad . \quad . \quad . \quad . \quad . \quad . \quad . \quad . \quad (36)$$

which completely determines the arm of the couple whose moment is Qh_0. Now since the intensities of the force Q vary directly with n, the intensity at the axis of the arch must be zero, and the application of Q be $\frac{1}{2}h_0$ from the axis as indicated in Fig. 3.

ARCHES HAVING TWO FLANGES OR CHORDS.

In case the arch is composed of two flanges connected by a thin web or by struts and ties, it is customary to consider the material of each flange concentrated at its center of gravity, and that the flanges resist all stresses excepting radial stresses T_x.

From Fig. 5,

$$N_x(x_0 + h_2) = Q'h_0,$$

or

$$Q' = \frac{N_x x_0 + N_x h_2}{h_0} = \frac{h_2 N_x + M_x}{h_0}. \quad . \quad . \quad . \quad (37)$$

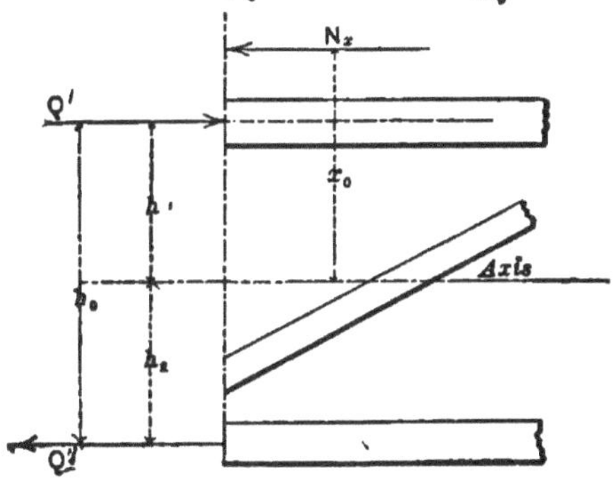

FIG. 5.

Also

$$N_x(h_1 - x_0) = Q''h_0,$$

and

$$Q'' = \frac{N_x h_1 - x_0 N_x}{h_0} = \frac{h_1 N_x - M_x}{h_0}. \quad . \quad . \quad . \quad (38)$$

Q' and Q'' will be of the same kind as long as $+ x_0$ is less than h_1 and $- x_0$ less than h_2, or N_x must lie between Q' and Q''.

VALUES OF x_0 FOR VARIOUS SECTIONS.

The following table contains the maximum values which x_0 can have when the entire cross-section is subjected to the same kind of stress for the various sections shown.

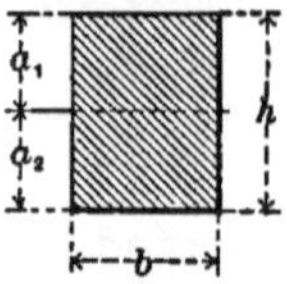

$$F = bh, \qquad a_1 = a_2 = \tfrac{1}{2}h, \qquad r^2 = \tfrac{1}{12}h^2.$$
$$\theta = \tfrac{1}{12}bh^3, \qquad x_0 = \pm \tfrac{1}{6}h,$$

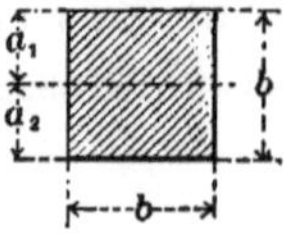

$$F = b^2, \qquad a_1 = a_2 = \tfrac{1}{2}b, \qquad r^2 = \tfrac{1}{12}b^2.$$
$$\theta = \tfrac{1}{12}b^4, \qquad x_0 = \pm \tfrac{1}{6}b,$$

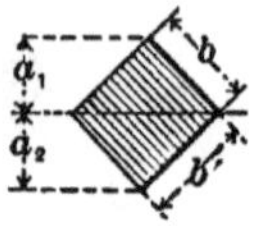

$$F = b^2, \qquad a_1 = a_2 = \frac{b}{1.414}, \qquad r^2 = \tfrac{1}{12}b^2$$
$$\theta = \frac{b^4}{12}, \qquad x_0 = \pm\, 0.1178b,$$

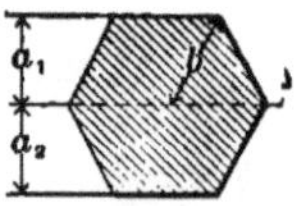

$$F = 2.598b^2, \qquad a_1 = a_2 = 0.866b,$$
$$\theta = 0.5413b^4, \qquad r^2 = 0.2083b^2,$$
$$x_0 = \pm\, 0.240b.$$

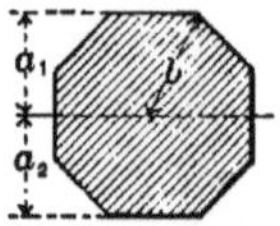

$$F = 2.598b^2, \qquad a_1 = a_2 = b,$$
$$\theta = 0.5413b^4, \qquad r^2 = 0.2083b^2,$$
$$x_0 = \pm\, 0.2083b.$$

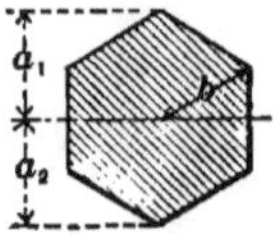

$$F = 2.828b^2, \qquad a_1 = a_2 = 0.924b,$$
$$\theta = 0.638b^4, \qquad r^2 = 0.2256b^2,$$
$$x_0 = \pm\, 0.244b.$$

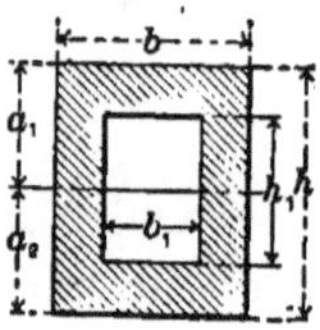

$$F = bh - b_1h_1, \qquad a_1 = a_2 = \frac{h}{2},$$
$$\theta = \tfrac{1}{12}(bh^3 - b_1h_1^3), \quad r^2 = \frac{1}{12}\,\frac{bh^3 - b_1h_1^3}{bh - b_1h_1},$$
$$x_0 = \pm\, \frac{1}{6h}\,\frac{bh^3 - b_1h_1^3}{bh - b_1h_1}.$$

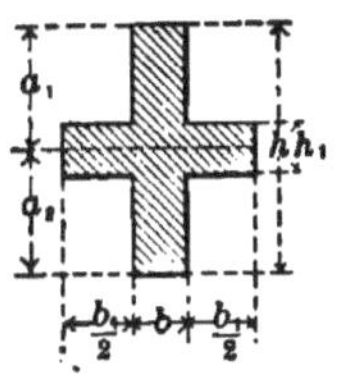	$F = bh + b_1h_1, \qquad a_1 = a_2 = \frac{h}{2},$ $\theta = \frac{1}{12}(bh^3 + b_1h_1^3), \qquad r^2 = \frac{1}{12}\frac{bh^3 + b_1h_1^3}{bh + b_1h_1},$ $x_0 = \pm\frac{1}{6h}\frac{bh^3 + b_1h_1^3}{bh + b_1h_1}.$
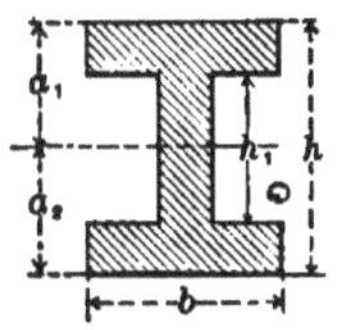	$F = bh - (b - b_1)h_1, \qquad a_1 = a_2 = \frac{h}{2},$ $\theta = \frac{1}{12}[bh^3 - (b - b_1)h_1^3],$ $r_2 = \frac{1}{12}\frac{bh^3 - (b - b_1)h_1^3}{bh - (b - b_1)h_1},$ $x_0 = \pm\frac{1}{6h}\frac{bh^3 - (b - b_1)h_1^3}{bh - (b - b_1)h_1}.$
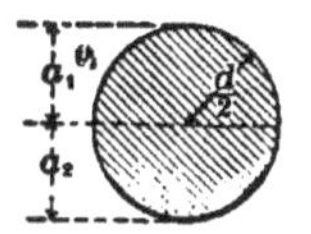	$F = \frac{\pi}{4}d^2 = 0.7854d^2, \qquad a_1 = a_2 = \frac{d}{2},$ $\theta = 0.0491d^4, \qquad r^2 = 0.0625d^2,$ $x_0 = \pm 0.125d = \pm\frac{1}{4}$ radius.
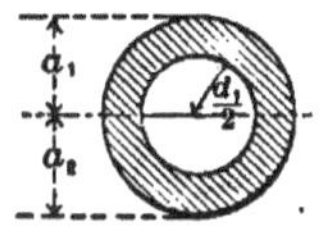	$F = 0.7854(d^2 - d_1^2), \qquad a_1 = a_2 = \frac{d}{2},$ $\theta = 0.0491(d^4 - d_1^4), \qquad r^2 = 0.0625(d^2 + d_1^2),$ $x_0 = \pm 0.125\left(\frac{d^2 + d_1^2}{d}\right).$
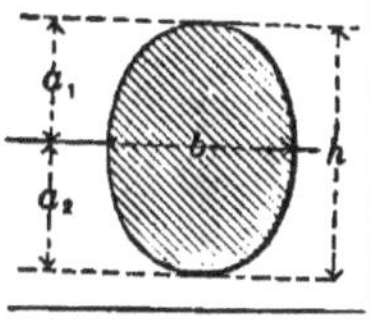	$F = 0.7854bh, \qquad a_1 = a_2 = \frac{h}{2},$ $\theta = 0.0491bh^3, \qquad r^2 = 0.0625h^2,$ $x_0 = \pm 0.125h.$
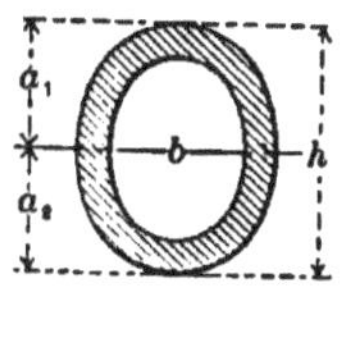	$F = 0.7854(bh - b_1h_1), \quad a_1 = a_2 = \frac{h}{2},$ $\theta = 0.0491(bh^3 - b_1h_1^3), \quad r^2 = 0.0625\frac{bh^3 - b_1h_1^3}{bh - b_1h_1},$ $x_0 = \pm\frac{0.125}{h}\frac{bh^3 - b_1h_1^3}{bh - b_1h_1}$

GENERAL RELATIONS BETWEEN THE EXTERNAL FORCES.

In Fig. 6 let ABC represent the axis of any elastic arch, and P and Q the vertical and horizontal components respec-

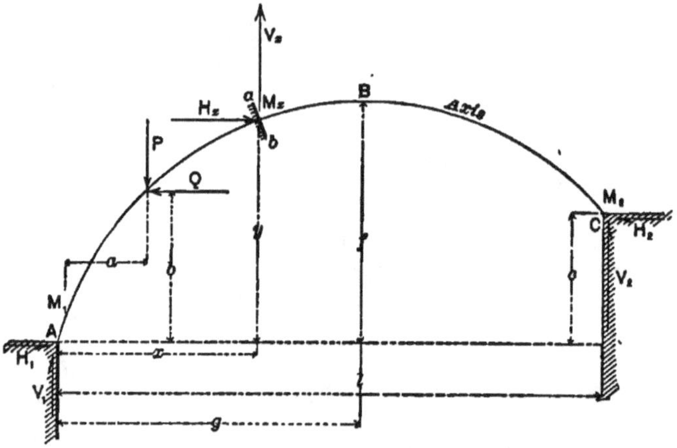

FIG. 6.

tively, of any load applied at a point having the co-ordinates a and b; then for equilibrium we have

$$H_x = H_1 - \sum^x Q; \quad x > a. \quad . \quad . \quad . \quad . \quad . \quad (39)$$

$$V_x = V_1 - \sum^x P; \quad x > a. \quad . \quad . \quad . \quad . \quad . \quad (40)$$

$$M_x = M_1 + V_1 x - H_1 y - \sum^x P(x - a) + \sum^x Q(y - b). \quad . \quad (41)$$

Referring to Fig. 7,

$$N_x = V_x \sin \phi + H_x \cos \phi ; \quad . \quad . \quad . \quad . \quad (42)$$

$$T_x = V_x \cos \phi - H_x \sin \phi ; \quad . \quad . \quad . \quad . \quad (43)$$

$$\tan \beta = \frac{V_x}{H_x}; \quad . \quad . \quad . \quad . \quad . \quad . \quad (44)$$

$$R_x = \frac{H_x}{\cos \beta} = \frac{V_x}{\sin \beta} = \sqrt{H_x^2 + V_x^2} = \sqrt{N_x^2 + T_x^2}. \quad (45)$$

Differentiating (41),

$$\frac{dM_x}{dx} = V_1 - \frac{H_1 dy}{dx} - \sum^x P + \sum^x Q \frac{dy}{dx}$$

$$= V_1 - H_1 \tan \phi - \sum^x P + \sum^x Q \tan \phi.$$

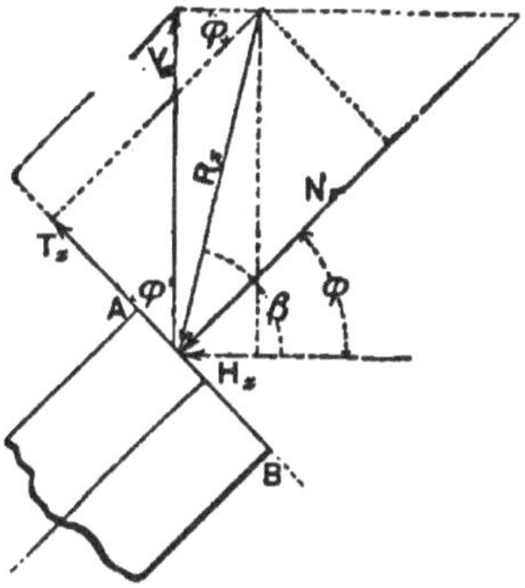

FIG. 7.

Since $dx = ds \cos \phi$, we have from (43)

$$T_x = V_x \frac{dx}{ds} - H_x \frac{dx}{ds} \tan \phi.$$

Also, from (40), $V_x = V_1 - \sum^x P.$
Hence

$$\frac{dM_x}{dx} = V_x - H_1 \tan \phi + \sum^x Q \tan \phi$$

$$= T_x \frac{ds}{dx} + H_x \tan \phi - (H_1 - \sum^x Q) \tan \phi$$

$$= T_x \frac{ds}{dx}.$$

Therefore

$$\frac{dM_x}{ds} = T_x. \quad . \quad . \quad . \quad . \quad . \quad . \quad . \quad . \quad (46)$$

Making $x = l$ and $y = c$ in (41), M_x becomes M_2, and we have

$$M_2 = M_1 + V_1 l - H_1 c - \sum^l P(l - a) + \sum^l Q(c - b).$$

Solving this for V_1, we obtain

$$V_1 = \frac{1}{l}\{M_2 - M_1 + H_1c + \overset{l}{\Sigma}P(l-a) - \overset{l}{\Sigma}Q(c-b)\}. \quad (47)$$

Making $x = l$ in (40), and combining with (47), we have

$$V_2 = \frac{1}{l}\{M_2 - M_1 + H_1c - \overset{l}{\Sigma}Pa - \overset{l}{\Sigma}Q(c-b)\}. \quad . \quad . \quad (48)$$

Collecting the equations which will be employed in the investigation of special cases, we have

$$H_x = H_1 - \overset{x}{\Sigma}Q; \quad x > a. \quad . \quad . \quad . \quad . \quad . \quad . \quad . \quad . \quad . \quad (39)$$

$$V_x = V_1 - \overset{x}{\Sigma}P; \quad x > a. \quad . \quad . \quad . \quad . \quad . \quad . \quad . \quad . \quad . \quad (40)$$

$$N_x = V_x \sin\phi + H_x \cos\phi. \quad . \quad . \quad . \quad . \quad . \quad . \quad . \quad . \quad . \quad (42)$$

$$M_x = M_1 + V_1x - H_1y - \overset{x}{\Sigma}P(x-a) + \overset{x}{\Sigma}Q(y-b); \quad (41)$$

$$M_1 = M_2 - V_1l + H_1c + \overset{l}{\Sigma}P(l-a) - \overset{l}{\Sigma}Q(c-b); \quad (49)$$

$$V_1 = \frac{1}{l}\{M_2 - M_1 + H_1c + \overset{l}{\Sigma}P(l-a) - \overset{l}{\Sigma}Q(c-b)\}; \quad (47)$$

$$V_2 = \frac{1}{l}\{M_2 - M_1 + H_1c - \overset{l}{\Sigma}Pa - \overset{l}{\Sigma}Q(c-b)\}. \quad . \quad (48)$$

ORDINATES LOCATING THE EQUILIBRIUM POLYGONS.

(a) *Vertical Components.*

In Fig. 8 let ABC represent the axis of any elastic arch

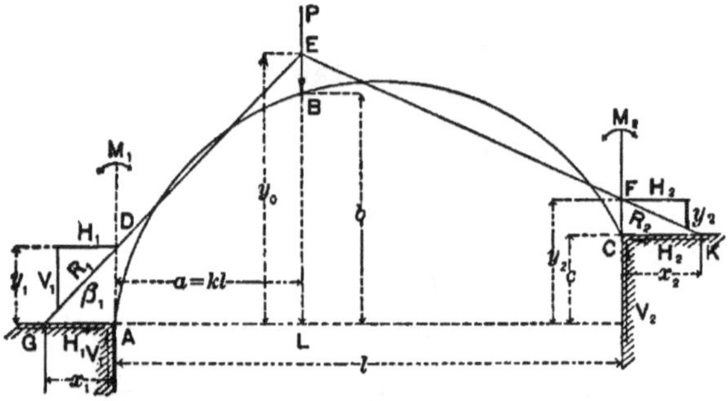

FIG. 8.

and let a single vertical load (corresponding to the vertical component of any load) be applied at B. This load causes the reactions R_1 and R_2 and the moments M_1 and M_2 at A and C respectively. This condition can be represented graphically by the equilibrium polygon GEK, which must be so situated that $H_1y_1 = M_1$, $H_2y_2 = M_2$, $\tan\beta_1 = \frac{V_1}{H_1}$, and $\tan\beta_2 = \frac{V_2}{H_2}$.

From Fig. 8, taking moments about E,

$$M_1 + V_1a - H_1y_0 = 0,$$

or

$$y_0 = \frac{M_1 + V_1a}{H_1}, \quad \ldots \ldots \quad (50)$$

which locates the point E when M_1, V_1, and H_1 are known.

Taking moments about D,

$$M_1 - H_1y_1 = 0, \quad \text{or} \quad y_1 = \frac{M_1}{H_1}. \quad \ldots \quad (51)$$

Taking moments about F,

$$M_2 - H_2(y_2 - c) = 0 \quad \text{or} \quad y_2 = c + \frac{M_2}{H_2}. \quad \ldots \quad (52)$$

From the triangles DGA and CKF,

$$\tan\beta_1 = \frac{V_1}{H_1} \quad \text{and} \quad \tan\beta_2 = \frac{V_2}{H_2}. \quad \ldots \quad (53)$$

From the triangles GAD and GLE,

$$x_1 : x_1 + a :: y_1 : y_0,$$

and

$$x_1y_0 = x_1y_1 + ay_1,$$

or

$$x_1 = \frac{a}{y_0 - y_1}y_1 = \frac{M_1}{V_1}. \quad \ldots \ldots \quad (54)$$

We also have

$$x_2 = \frac{M_2}{V_2}. \quad \ldots \ldots \ldots \quad (55)$$

The above equations completely determine the locations of $GDEF$ and K, and hence the equilibrium polygon GEK

can be drawn in its true position and the values of R_1 and R_2 at once determined.

Having determined R_1 and R_2 in magnitude and position, the distribution of pressure over the section at A can be found, and then the stresses in other portions of the arch determined. When the arch is solid in section the stresses are best determined by equations (39), (40), etc. If, however, the arch is composed of two flanges connected by a thin web or by bracing, the graphical method is the more expeditious.

The methods of determining the stresses, etc., at different points of the arch will be fully illustrated by problems, but a brief outline of one method of procedure after R_1 has been determined will be given here.

In Fig. 9 let AB be any radial section of a solid elastic arch. Suppose 1, 2, . . . 5 represent the positions and

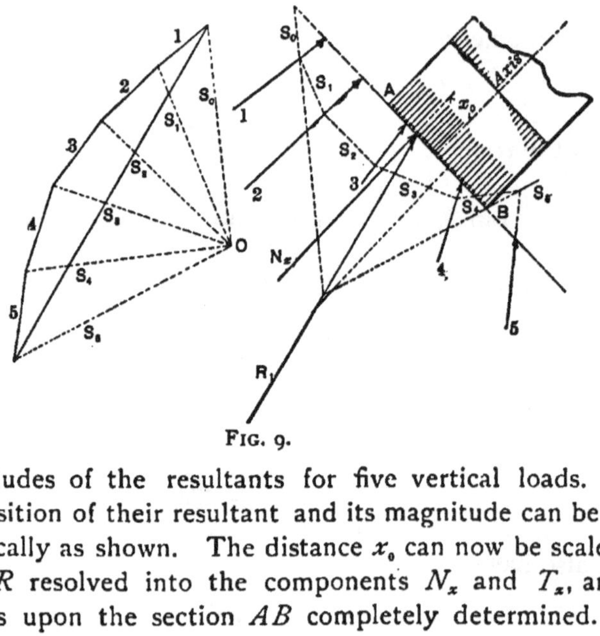

FIG. 9.

magnitudes of the resultants for five vertical loads. Then the position of their resultant and its magnitude can be found graphically as shown. The distance x_0 can now be scaled, the force R resolved into the components N_x and T_x, and the stresses upon the section AB completely determined. (See page 8.)

If the arch is composed of flanges, the method is practically the same, with the exception that each flange is assumed to have a uniform stress over its entire section, as explained above. (See page 11.)

(b) *Horizontal Components.*

An examination of Fig. 10 shows that we can locate the equilibrium polygon *GEK* for the horizontal load *Q* in a

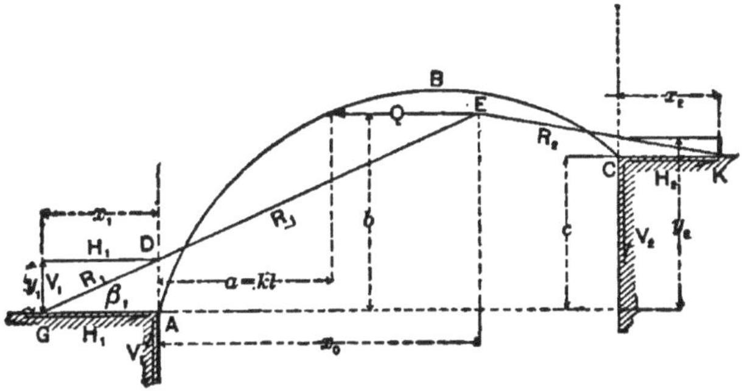

FIG. 10.

manner similar to that employed for the vertical component; in fact, all of the equations will be the same, with the exception of that for y_0, which in this case becomes x_0.

We have then

$$y_1 = \frac{M_1}{H_1} \quad \text{and} \quad y_2 - c = \frac{M_2}{H_2}. \quad . \quad . \quad . \quad . \qquad (56)$$

Also,

$$x_1 = \frac{M_1}{V_1} \quad \text{and} \quad x_2 = \frac{M_2}{V_2}. \quad . \quad . \quad . \quad . \quad . \qquad (57)$$

From the figure

$$x_1 : y_1 :: x_1 + x_2 : b,$$

or

$$x_2 = \frac{bx_1 - x_1y_1}{y_1}. \quad . \quad . \quad . \quad . \quad . \quad . \quad . \quad . \quad . \qquad (58)$$

CHAPTER II.

FORMULAS FOR PRACTICAL USE.

IN this chapter all of the important formulas for parabolic and circular arches have been collected and arranged for ready application. The demonstrations of these formulas are given in chapters which follow.

(A) SYMMETRICAL PARABOLIC ARCHES.

$A = E\theta \cos \phi$ = a constant, or θ varies inversely as $\cos \phi$.

$$E\theta \cos \phi = A = \text{a constant}; \quad p(59^*). \quad (59)$$

$$m = \frac{\theta}{F} = (\text{radius of gyration})^2; \quad p(60) \quad (60)$$

$$p = \text{parameter of parabola} = \frac{l^2}{8f}; \quad (61)$$

$$b = 4k(1 - k)f = y \quad \text{for} \quad x = a = kl; \quad (62)$$

$$\tan \phi = \frac{8(\frac{1}{2}l - x)}{l^2}f. \quad (63)$$

Δ_1 = function given in Table I,

Δ_2 = function given in Table II,

Δ_3 = etc. etc.

ARCH WITH TWO HINGES, ONE AT EACH SUPPORT.

(a) Vertical Loads, with Effect of Axial Stress neglected—Common Method.

$$H_1 = \frac{5}{8}\frac{l}{f}\sum_x^l P[k(1 - 2k^2 + k^3)] \quad . \ . \ p(91) \quad (64)$$

* $p(59)$ indicates that this equation is taken from the chapter on Parabolic Arches (Chap. III), its number in that place being $p(59)$.

or

$$H_1 = \frac{5}{8}\frac{l}{f}\sum_{0}^{l} P\Delta_1 (\Delta_1 = \text{function given in Table I}). \quad . \ (64a)$$

$$V_1 = \sum_{0}^{l} P(1 - k). \quad . \ . \ . \ . \ . \ . \ . \ . \ . \quad p(93) \quad . \ . \quad (65)$$

$$y_0 = \frac{8}{5}\frac{1}{1 + k - k^2}f \quad . \ . \ . \ . \ . \ . \ . \ . \quad p(95) \quad . \ . \quad (66)$$

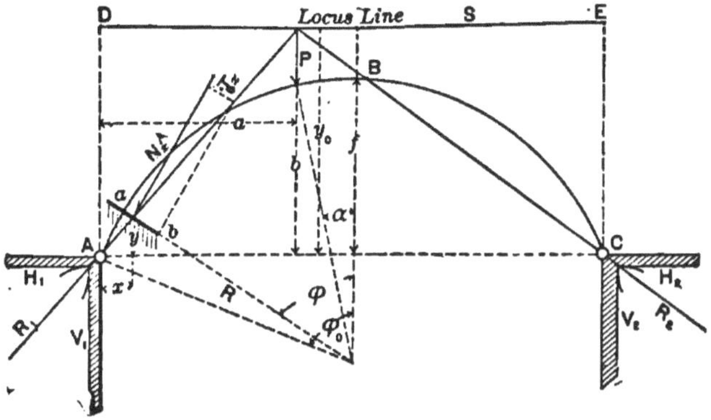

FIG. 11.

or

$$y_0 = f\Delta_2; \quad . \ . \ . \ . \ . \ . \ . \ . \ . \ . \ . \ . \ . \ . \ (66a)$$

$$T_x = (V_1 - \sum^x P)\cos\phi - H_1 \sin\phi. \qquad p(96) \quad . \ . \quad (67)$$

From (42),

$$N_x = (V_1 - \sum^x P)\sin\phi + H_1\cos\phi. \qquad (42) \quad . \ . \quad (68)$$

From (41),

$$M_x = V_1 x - H_1 y - \sum^x P(x - a). \quad . \ . \ . \quad p(97) \quad . \ . \quad (69)$$

The application of the above formulas to either the solid or open arch rib is quite simple. The formulas are exact, of course, for the solid rib alone, and then only when the depth of the rib is small and the loading is applied upon the centre line; yet for practical purposes they can be applied to open ribs.

SOLID RIB.

For the solid rib we compute the values of H_1 and V_1, and then determine the values of M_x, N_x, and T_x for each section of the arch, the sections being taken at convenient distances apart.

The values of M_x, T_x, and N_x can be found from a graphical construction as shown in Fig. 12.

Draw the *locus line* S after computing y_0 (formula 66), and then draw FA and FC for the load being considered. By resolution of forces R_1, R_2, H_1, H_2, V_1, and V_2 can be determined.

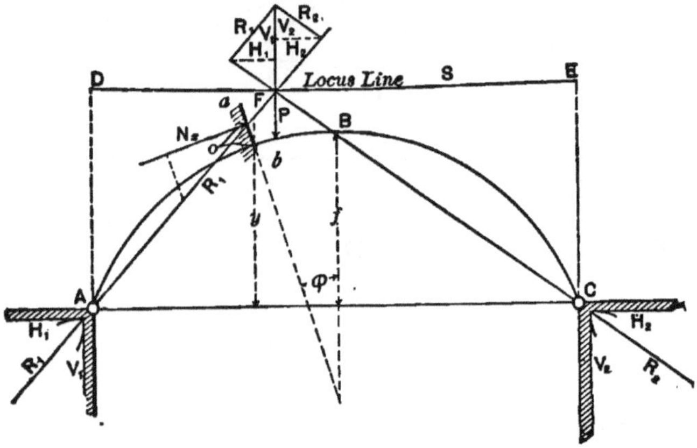

FIG. 12.

Let ab be any radial section where M_x, N_x, and T_x are desired. Then N_x equals the normal component of R_1 upon ab, T_x equals the tangential component of R_1 upon ab, and M_x equals the ordinate (o) multiplied by N_x.

Maximum Value of M_x.—Let ab, Fig. 14, be any section where the maximum moment is desired. Draw the lines Ao and Co until they cut the *locus line* S. Then since the loads at these points produce no moment at the section ab, these points separate the loadings which cause moments of different signs. The shaded sections in the figure clearly indicate the loadings which cause maximum $\pm M_x$.

Maximum Value of T_x.—At any section ab, Fig. 13, draw AD perpendicular to ab. Then the loading causing positive and negative shear is distributed as shown by the shaded portions in the figure; or, *for maximum upward shear the arch must be loaded on the left up to the section ab and on the right between D and E.*

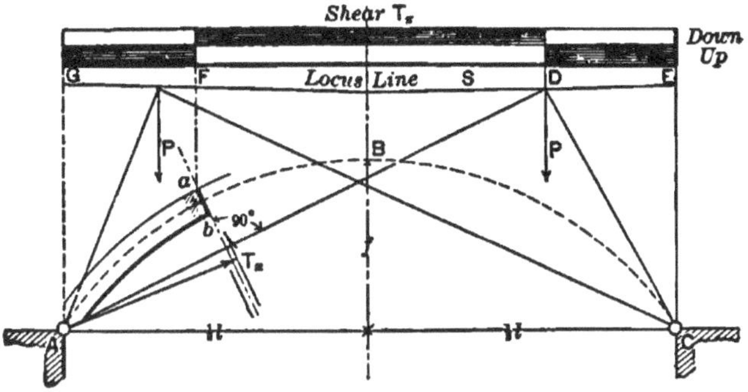

FIG. 13.

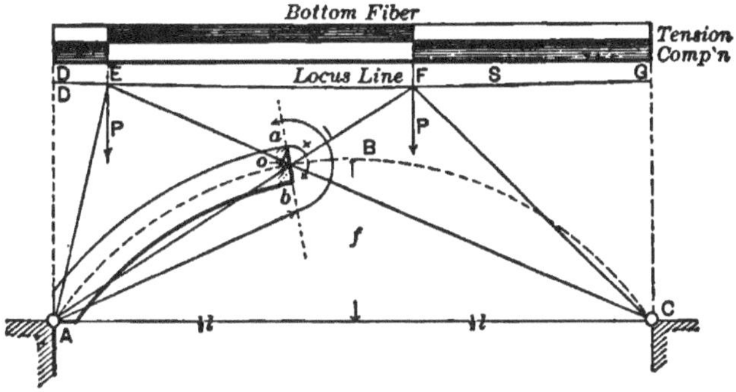

FIG. 14.

Uniform Loading.—Thus far only concentrations have been considered. If, however, the loading is uniformly distributed *horizontally*, we have

$$H_1 = \frac{1}{16n} wl \Big[_{k''}^{k'} k^2(5 - 5k^2 + 2k^3). \quad . \quad p(132) \qquad (70)$$

$n = f/l$ and w = load per unit length of span.

$$V_1 = \frac{wl}{2} \Big[_{k''}^{k'} k(2 - k). \quad . \quad . \quad . \quad . \quad . \quad . \quad . \quad . \quad . \quad p(133) \qquad (71)$$

$$M_x = \frac{wl}{2}\Big[_{k''}^{k'} \left\{ xk(2-k) - \frac{y}{8m}k^2(5-5k^2+2k^3) \right\}$$

$$-\frac{wl}{2}\{(x-a')^2-(x-a'')^2\}\ x \gtreqless a' \text{ and } a''. \quad p(134) \qquad (72)$$

$$T_x = \frac{wl}{2}\Big[_{k''}^{k'} k(2-k)\cos\phi - \frac{1}{16n}wl\Big[_{k''}^{k'} k^2(5-5k^2+2k^3)\sin\phi$$

$$-(wl(k'-k'')\cos\phi, \text{ where } k' \gtreqless \frac{x}{l}). \quad . \quad p(135) \qquad (73)$$

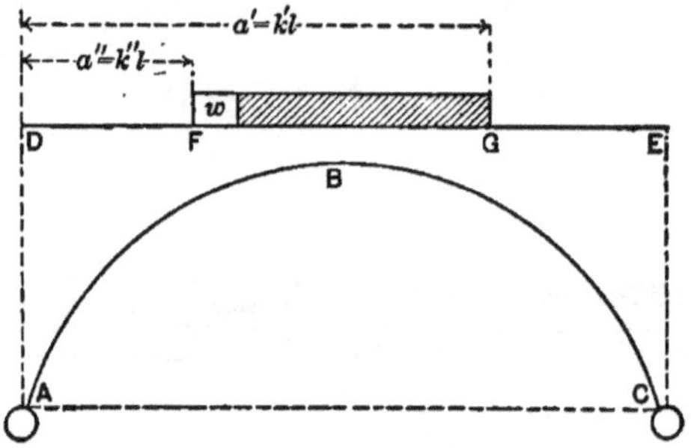

FIG. 15.

OPEN RIB.

The values of H_1 and V_1 are found by the formulas given for the solid rib. The internal stresses can then be found by Clerke-Maxwell's method of graphics.

The loadings causing maximum values of M_x and T_x are clearly defined in Figs. 16 to 18 inclusive.

Fig. 18 is strictly true only when *cd* and *ab* are parallel.

PLATE-GIRDER RIB.

In plate-girder ribs the flanges usually are assumed to resist the bending-moment, and the web the shear or T_x; hence we may treat them the same as the open rib.

EQUILIBRIUM POLYGON.

If in any of the above cases it is desired to construct an equilibrium polygon for any given loading, it can be done as

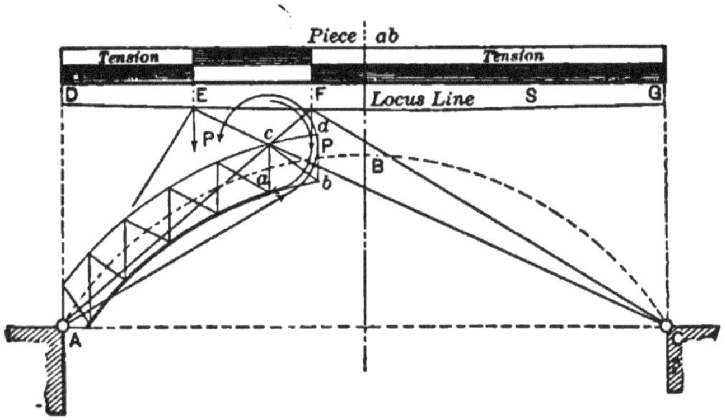

FIG. 16.

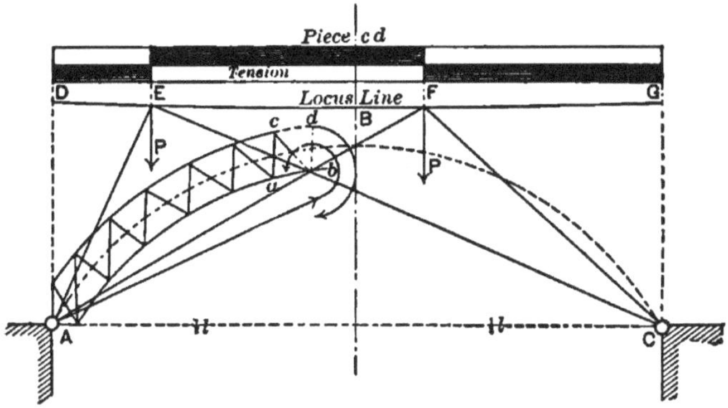

FIG. 17.

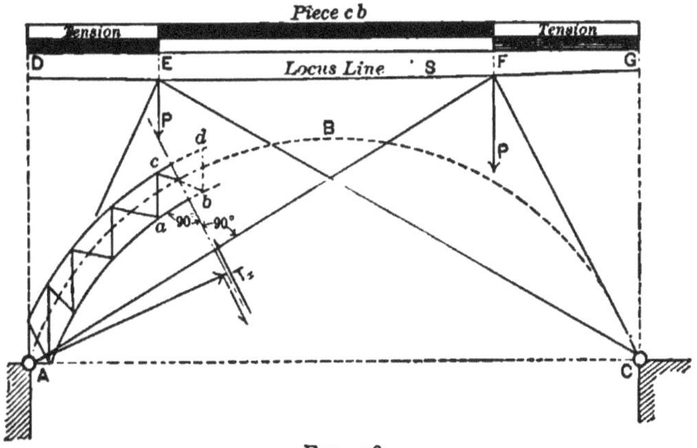

FIG. 18.

follows (Fig. 19): Construct the resultants R_1 for each concentration, and find the resultant of the system, also the corre-

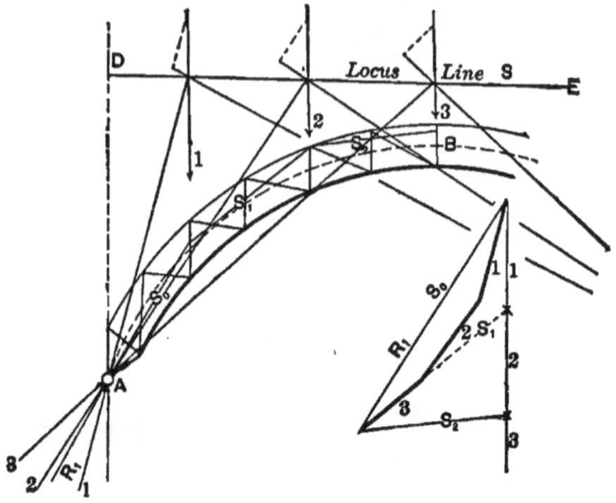

FIG. 19.

sponding values of H_1 and V_1; then the polygon can be constructed by the usual methods.

(b) *Vertical Loads, with Effect of Axial Stress included.*

$$H_1 = \frac{15}{8lf^2 + 30mp\phi_0}\left\{\frac{8lf^2}{15}\mathfrak{H} - \frac{ml^2}{2(p+2f)}\sum^{l} Pk(1-k)\right\}, \quad (74)$$

$p(103)$

where $\mathfrak{H}$ is the value found from (64*a*) for H_1, or approximately

$$* H_1 = \mathfrak{H}(1 - \epsilon). \quad . \quad . \quad . \quad . \quad . \quad . \quad (75)$$

(Tables I and V.)

The axial stress affects only the value of H_1; hence to include the effect of the axial stress we have merely to compute H_1 by (74), and then proceed as already outlined for the case where the axial stress is neglected.

* See Appendix C.

If the *locus line* S is determined by computing the ordinates y_0 they must be deduced from the formula

$$y_0 = \frac{V_1}{H_1} a, \quad . \quad . \quad . \quad . \quad . \quad (50) \qquad (76)$$

where H_1 is to be found from (74).

(c) Horizontal Loads, with Effect of Axial Stress neglected—Common Method.

$$H_1 = \sum^{l} Q \left\{ 1 - \frac{k}{2}[5(1-k-2k^2+4k^3)-8k^4] \right\} \qquad p(111) \quad . \quad (77)$$

or

$$H_1 = \sum^{l} Q\Delta_3 . \quad . \quad . \quad . \quad . \quad . \quad . \quad . \quad . \quad (77a)$$

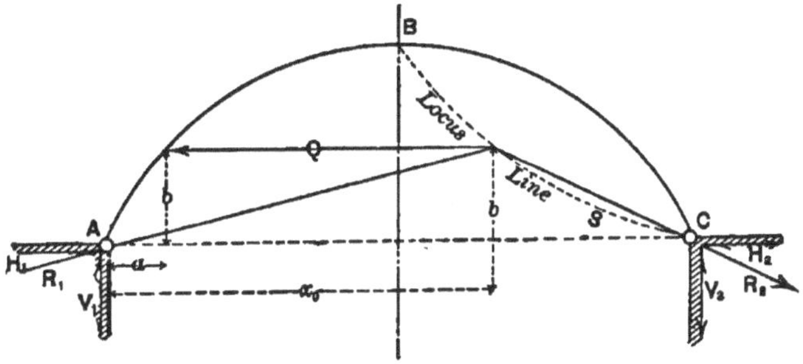

FIG. 20.

$$V_1 = 4k(1-k)nQ \quad . \quad . \quad . \quad . \quad . \quad . \quad p(113) \quad . \quad (78)$$

or

$$V_1 = 4n\Delta_6 Q . \quad . \quad . \quad . \quad . \quad . \quad . \quad . \quad . \quad (78a)$$

$$x_0 = \left\{ 1 - \frac{k}{2}[5(1-k-2k^2+4k^3)-8k^4] \right\} l, \qquad p(116) \quad . \quad (79)$$

or

$$x_0 = l\Delta_3 . \quad . \quad . \quad . \quad . \quad . \quad . \quad . \quad . \quad (79a)$$

From (39), (40), and (43),

$$T_x = V_1 \cos\phi - (H_1 - \sum^{x} Q) \sin\phi . \quad . \quad . \quad . \quad . \quad (43) \quad . \quad . \quad (80)$$

From (39), (40), and (42),

$$N_x = V_1 \sin \phi + (H_1 - \overset{x}{\Sigma} Q) \cos \phi. \quad . \quad . \quad . \quad . \quad (42) \quad . \quad . \quad (81)$$

From (41),

$$M_x = V_1 x - H_1 y + \overset{x}{\Sigma} Q(y - b). \quad . \quad . \quad . \quad . \quad . \quad p(119) \quad . \quad (82)$$

The application of the above formulas to either the solid or open arch rib is quite simple. After the *locus line S* has been located by means of (79), the reactions can be drawn as shown in Fig. 20, and from these the values of H and V determined.

Horizontal loads are usually caused by wind; hence the ordinary case to consider is half the arch covered with a steady load.

(*d*) *Horizontal Loads, with Effect of Axial Stress included.*

$$H_1 = \frac{4lf^2}{15B} \overset{l}{\Sigma} Q \left\{ 2\left[1 - \frac{k}{2}(5[1 - k - 2k^2 + 4k^3] - 8k^4) \right] \right\}$$

$$+ \overset{l}{\Sigma} Q \frac{mp(\alpha + \phi_0)}{B}, \quad . \quad . \quad . \quad . \quad . \quad p(125) \quad . \quad (83)$$

where

$$\frac{1}{B} = \frac{15}{8lf^2 + 30mp\phi_0}. \quad . \quad . \quad . \quad . \quad . \quad . \quad . \quad . \quad . \quad . \quad p(126) \quad . \quad (84)$$

$$x_0 = \frac{H_1}{V_1} b = \frac{H_1}{Q} l. \quad . \quad . \quad . \quad . \quad . \quad . \quad . \quad . \quad . \quad . \quad . \quad . \quad (58) \quad . \quad . \quad (85)$$

The axial stress does not affect the other equations excepting where they contain H_1 or x_0, the values of which must be found from the above expressions.

After the values of H_1, V_1, etc., have been determined for horizontal loads, the stresses can be found in the manner outlined for vertical loads.

(e) Temperature.

$$H_1 = \frac{15A}{8f^2} et^\circ, \quad . \quad . \quad . \quad . \quad . \quad p(128) \quad . \quad (86)$$

neglecting the effect of the ~~actual~~ axial stress; or

$$H_1 = \frac{60Af}{32f^2 + 15ml\phi_0} et^\circ, \quad p(129) \quad . \quad (87)$$

including the effect of the axial stress.

A rise in temperature causes a horizontal thrust similar in character to that produced by vertical loads acting downward.

$$V_1 = 0. \quad . \quad . \quad . \quad . \quad . \quad . \quad . \quad . \quad (47) \quad . \quad . \quad (88)$$

$$T_x = - H_1 \sin \phi. \quad . \quad . \quad . \quad . \quad (43) \quad . \quad . \quad (89)$$

$$M_x = - H_1 y. \quad . \quad . \quad . \quad . \quad . \quad . \quad (41) \quad . \quad . \quad (90)$$

In case of the open rib arch the stresses in the individual members can be found by graphics after H_1 has been determined.

(f) Change of Length in Span.

By replacing et° by $\frac{\Delta l}{l}$ in the above equations they may be applied to any change in length of span.

ARCH WITHOUT HINGES.

(a) Vertical Loads, neglecting Effect of Axial Stress— Common Method.

$$H_1 = \frac{15}{4n} \sum^{l} Pk^2(1 - k)^2 \quad . \quad . \quad . \quad . \quad . \quad . \quad . \quad . \quad . \quad . \quad . \quad p(143) \quad (91)$$

or

$$H_1 = \frac{15}{4n} \sum^{l} P\Delta_{11}, \text{ where } n = f/l. \quad . \quad . \quad . \quad . \quad . \quad . \quad . \quad . \quad . \quad . \quad (91a)$$

$$M_1 = \frac{l}{2}\overset{l}{\Sigma} Pk(1-k)^2(5k-2) \quad . \quad . \quad . \quad . \quad . \quad . \quad . \quad . \quad p(148) \qquad (92)$$

or

$$M_1 = \frac{l}{2}\overset{l}{\Sigma} P\Delta_6, \text{ reading } (1-k) \text{ for } k. \quad . \quad . \quad . \quad . \quad . \quad . \quad . \quad (92a)$$

$$M_2 = \frac{l}{2}\overset{l}{\Sigma} P\Delta_6. \quad . \quad . \quad . \quad . \quad . \quad . \quad . \quad . \quad . \quad . \quad . \quad . \quad (92b)$$

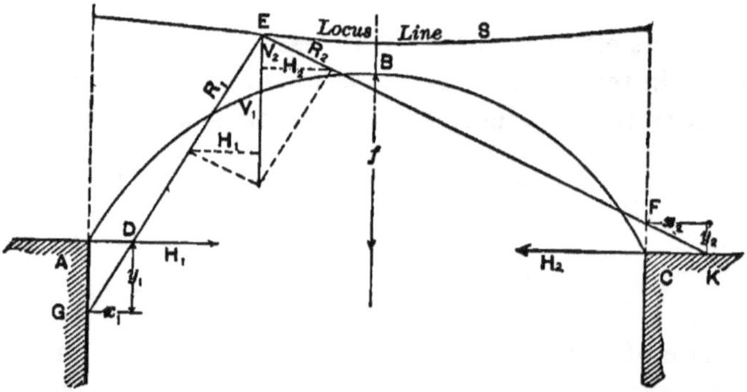

FIG. 21.

$$V_1 = \overset{l}{\Sigma} P(1-k)^2(1+2k) \quad . \quad . \quad . \quad . \quad . \quad . \quad . \quad . \quad . \quad p(150) \qquad (93)$$

or

$$V_1 = \overset{l}{\Sigma} P\Delta_7. \quad . \quad . \quad . \quad . \quad . \quad . \quad . \quad . \quad . \quad . \quad . \quad . \quad . \quad (93a)$$

$$y_0 = \frac{6}{5}f \quad \text{(positive upwards).} \quad . \quad . \quad . \quad . \quad . \quad . \quad p(151) \qquad (94)$$

$$y_1 = \frac{6}{5}\,\frac{5k-2}{9k}f = f\Delta_8. \quad . \quad . \quad . \quad . \quad . \quad . \quad . \quad . \quad p(152) \qquad (95)$$

$$y_2 = \frac{6}{5}\,\frac{3-5k}{9(1-k)}f = f\Delta_8 \text{ reading } (1-k) \text{ for } k. \quad p(153) \qquad (96)$$

$$x_1 = -\frac{k(5k-2)}{2(1+2k)}l \quad . \quad . \quad . \quad . \quad . \quad . \quad . \quad . \quad . \quad . \quad p(154) \qquad (97)$$

or

$$x_1 = -\frac{l}{10}\Delta_9. \quad . \quad . \quad . \quad . \quad . \quad . \quad . \quad . \quad . \quad . \quad . \quad . \quad (97a)$$

$$x_2 = -\frac{(1-k)(3-5k)}{2(3-2k)}l \quad . \quad . \quad . \quad . \quad . \quad . \quad . \quad . \quad p(155) \quad . \quad (98)$$

or

$$x_2 = -\frac{l}{10}\Delta_2, \text{ reading } (1-k) \text{ for } k. \quad . \quad . \quad . \quad . \quad . \quad . \quad . \quad . \quad (98a)$$

$$T_x = (V_1 - \overset{x}{\Sigma}P)\cos\phi - H_1 \sin\phi. \quad . \quad . \quad . \quad . \quad (43) \quad . \quad (99)$$

$$M_x = M_1 + V_1 x - H_1 y - \overset{x}{\Sigma}P(x-a). \quad . \quad . \quad . \quad (41) \quad . \quad (100)$$

As in the case of the two-hinged arch, it is necessary only to compute H_1, V_1, and y_0 to determine all the outer forces acting upon the arch, and then the stresses; but as a check it is advisable to compute x_1, x_2, y_1, and y_2.

The methods of determining the fields of loading which cause maximum values of M_x and T_x are the same as for the two-hinged arch, only the resultants R_1 and R_2 do not necessarily pass through the supports, but must have their locations fixed by the ordinates x_1, y_1, x_2, and y_2.

(b) Vertical Loads, with Effect of Axial Stress included.

$$H_1 = C\left\{ l\overset{l}{\Sigma}Pk^2(1-k)^2 - \frac{3lm}{2f(p+2f)}\overset{l}{\Sigma}Pk(1-k)\right\}, \quad . \quad (101)$$

$p(162)$

where

$$C = \frac{15lf}{4lf^2 + 90mp\phi_0}. \quad . \quad . \quad . \quad . \quad . \quad . \quad . \quad . \quad . \quad . \quad . \quad . \quad (102)$$

Approximately,

$$^*H_1 = \mathfrak{H}(1-\epsilon'), \quad . \quad . \quad . \quad . \quad . \quad (103)$$

where $\mathfrak{H} = H_1$ in (91).

$$k(1-k) = \Delta_2. \quad . \quad . \quad . \quad . \quad . \quad . \quad . \quad (105)$$

$$k^2(1-k)^2 = \Delta_{11}. \quad . \quad . \quad . \quad . \quad . \quad . \quad . \quad (106)$$

---

* See Appendix C.

Note that all quantities in the above equations excepting those given by the tables are constant for any given arch.

$$M_2(2D - \tfrac{3}{2}) = H_1\left\{\frac{8D}{5}f - f + \frac{6mp\phi_0}{lf}D\right\}$$

$$- lD\sum^{l} Pk(1 - 2k^2 + k^3) + \frac{l}{2}\sum^{l} P(2k - 3k^2 + k^3)$$

$$\frac{3mDl}{2f(p + 2f)}\sum^{l} P(1 - k)k$$

$$- \frac{3mp}{l^2}\sum^{l} P\{\phi_0(2k - 1) + \alpha\}, \quad . \quad . \quad . \quad . \quad p(164) \quad . \quad (107)$$

where H_1 is to be found from (101).

$$D = 1 + \frac{3m}{l^2} - \frac{6mp\phi_0}{l^3}. \quad . \quad . \quad . \quad . \quad . \quad (108)$$

$$k(1 - 2k^2 + k^3) = \Delta_1. \quad . \quad . \quad . \quad . \quad . \quad . \quad (109)$$

$$2k - 3k^2 + k^3 = \Delta_{10}. \quad . \quad . \quad . \quad . \quad . \quad . \quad (110)$$

$$k(1 - k) = \Delta_6. \quad . \quad . \quad . \quad . \quad . \quad . \quad (111)$$

* To determine M_1 replace k by $(1 - k)$ in (107), or compute M_2 from (107) and substitute the value in

$$M_1 + M_2 = H_1\frac{8f}{5} - l\sum^{l} Pk(1 - 2k^2 + k^3)$$

$$+ H_1\frac{6mp\phi_0}{fl} + \frac{3ml}{2f(p + 2f)}\sum^{l} Pk(1 - k). \quad . \quad p(161) \qquad (112)$$

$$V_1 = \frac{1}{l}\{M_2 - M_1 + \sum^{l} Pl(1 - k)\}. \quad . \quad p(149) \qquad (113)$$

$$y_1 = \frac{M_1}{H_1}. \quad . \quad . \quad . \quad . \quad . \quad . \quad . \quad . \quad . \quad . \quad . \quad . \quad (51) \qquad (114)$$

---

* Note that only the terms containing $2k - 3k^2 + k^3$ and $2k - 1$ change in magnitude when $1 - k$ is used in place of k.

Having computed H_1, M_1, M_2, V_1, and y_1, all the other outside forces can be found as follows (Fig. 22): Lay off H_1 and V_1 at A and complete the parallelogram of forces, thereby determining the direction and magnitude of R_1. Then lay off y_1

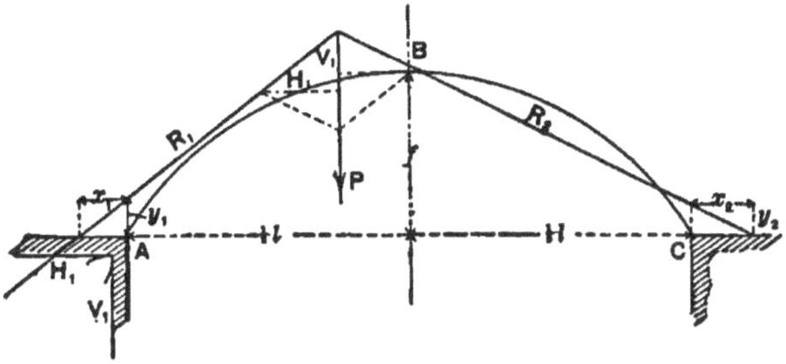

FIG. 22.

at A above or below, according to the sign, and draw R_1 in its proper position, extending its direction until it cuts the load being considered. By parallelogram of forces V_2 and R_2 are readily found. As a check, H_2, V_2, and y_2 should be computed.

(*c*) *Horizontal Loads, with Effect of Axial Stress neglected—Common Method.*

$$H_1 = \sum^{l} Q\{1+k^2(-15+50k-60k^2+24k^3)\} \quad . \quad p(168) \quad . \quad (115)$$

or

$$H_1 = \sum^{l} Q\Delta_{12}. \quad . \quad . \quad . \quad . \quad . \quad . \quad . \quad (116)$$

$$M_1 = +f\sum^{l} Q\{2k(1-k)^2(2-7k+8k^2)\} \quad . \quad p(171) \quad . \quad (117)$$

or

$$M_1 = +f\Sigma Q\Delta_{13}. \quad . \quad . \quad . \quad . \quad . \quad . \quad . \quad (118)$$

$$M_2 = -f\sum^{l} Q\{2k^2(1-k)(3-9k+8k^2)\} \quad . \quad p(172) \quad . \quad (119)$$

or

$M_2 = -f\Sigma Q\Delta_{13}$, entering table with $1-k$. . . . . . (120)

$$V_1 = 12n\sum^{l} Qk^2(1-k)^2 \quad . \; . \; . \; . \; . \; . \; . \; . \; . \; . \quad p(176) \qquad (121)$$

or

$$V_1 = 12n\sum^{l} Q\Delta_{11}. \quad . \; . \; . \; . \; . \; . \; . \; . \; . \; . \; . \; . \; . \; . \qquad (121a)$$

$$y_0 = b. \quad . \; . \; . \; . \; . \; . \; . \; . \; . \; . \; . \; . \; . \; . \quad p(177) \qquad (122)$$

$$y_1 = \frac{2k(1-k)^2(2-7k+8k^2)}{1+k^2[-15+50k-60k^2+24k^3]}f \quad . \quad p(178) \qquad (123)$$

or

$$y_1 = f\Delta_{14}. \quad . \; . \; . \; . \; . \; . \; . \; . \; . \; . \; . \; . \; . \; . \; . \; . \; . \qquad (123a)$$

$$y_2 = \frac{2(1-k)^2(3-9k+8k^2)}{15-50k+60k^2-24k^3}f \quad . \; . \; . \; . \; . \quad p(179) \qquad (124)$$

or

$y_2 = f\Delta_{14}$, reading $1-k$ for k. . . . . (124a)

$$x_1 = \frac{2-7k+8k^2}{6k}l \quad . \; . \; . \; . \; . \; . \; . \; . \; . \; . \quad p(180) \qquad (125)$$

or

$$x_1 = l\Delta_{15}. \quad . \; . \; . \; . \; . \; . \; . \; . \; . \; . \; . \; . \; . \; . \; . \; . \; . \qquad (125a)$$

$$x_2 = \frac{3-9k+8k^2}{6(1-k)}l \quad . \; . \; . \; . \; . \; . \; . \; . \; . \quad p(181) \qquad (126)$$

or

$x_2 = l\Delta_{15}$, reading $1-k$ for k. . . . . (126a)

$$x_0 = \frac{l}{2}(3-12k+24k^2-16k^3) \quad . \; . \; . \; . \quad p(183) \qquad (127)$$

or

$$x_0 = \frac{l}{2}\Delta_{16}. \quad . \; . \; . \; . \; . \; . \; . \; . \; . \; . \; . \; . \; . \; . \; . \; . \qquad (127a)$$

(*d*) *Horizontal Loads, with Effect of Axial Stress included.*

$$H_1 = C\left\{\frac{4f}{15}\overset{l}{\Sigma} Q(1 + k^2[-15 + 50k - 60k^2 + 24k^3])\right.$$

$$\left. + \frac{3mp}{lf}\overset{l}{\Sigma} Q(\alpha + \phi_0)\right\}, \qquad p(187) \qquad (128)$$

where

$$C = \frac{15lf}{4lf^2 + 90mp\phi_0}. \quad . \quad . \quad . \quad . \quad . \quad (129)$$

$$H_1 = C\left\{\frac{4f}{15}\overset{l}{\Sigma} Q\Delta_{12} + \frac{3mp}{lf}\Sigma Q(\alpha + \phi_0)\right. . \quad . \quad . \quad . \quad . \quad . \quad . \quad (130)$$

$$1 + k^2(-15 + 50k - 60k^2 + 24k^3) = \Delta_{12}. \quad . \quad (131)$$

$$M_2(2D - \tfrac{3}{2}) = H_1\left\{\frac{8D}{5}f - f + \frac{6mp\phi_0}{fl}D\right\}$$

$$-\frac{3m}{2f}\left(\frac{p}{p + 2f} + 1 - \frac{2p\phi_0}{l}\right)\overset{l}{\Sigma} Qk(1 - k)$$

$$+ f\overset{l}{\Sigma} Q\{1 - 2k(2 - 5k + 5k^2) + 3k^4\}$$

$$-\frac{8}{5}fD\overset{l}{\Sigma} Q\left\{1 - \frac{k}{2}[5(1 - k - 2k^2 + 4k^3) - 8k^4]\right\}$$

$$-\frac{3mpD}{lf}\overset{l}{\Sigma} Q(\alpha + \phi_0), \quad . \quad . \quad . \quad . \quad . \quad . \quad . \quad p(189) \quad . \quad (132)$$

where H_1 is given by (128).

$$D = 1 + \frac{3m}{l^2} - \frac{6mp\phi_0}{l^3}. \quad . \quad . \quad . \quad . \quad . \quad . \quad (133)$$

$$1 - 2k(2 - 5k + 5k^2) + 3k^4 = \Delta_4. \quad . \quad (134)$$

$$1 - \frac{k}{2}\{5(1 - k - 2k^2 + 4k^3) - 8k^4\} = \Delta_5. \quad . \quad (135)$$

$$k(1 - k) = \Delta_1. \quad . \quad . \quad . \quad . \quad . \quad . \quad . \quad . \quad . \quad (136)$$

For M_1 read $(1 - k)$ for k, ~~and replace H_1 by H_2~~ in (132).

$$V_1 = \frac{1}{l}(M_2 - M_1 + \overset{l}{\Sigma} Qb) \quad . \quad (47) \quad . \quad (137)$$

$$y_1 = \frac{M_1}{H_1}. \quad . \quad . \quad . \quad . \quad . \quad . \quad . \quad (51) \quad . \quad (138)$$

Having the values of H_1, M_1, V_1, and y_1, the remaining outside forces are readily determined in the manner outlined for vertical loads on page 33.

(e) Temperature.

$$H_1 = \frac{15}{4lf^2 + 90mp\phi_0} 3Aet^\circ l. \quad . \quad . \quad p(190) \quad . \quad . \quad (139)$$

when the effect of the axial stress is included, and

$$H_1 = \frac{45A}{4f^2} et^\circ \quad . \quad . \quad . \quad . \quad . \quad . \quad . \quad p(191) \quad . \quad . \quad (140)$$

when the axial stress is neglected.

$$M_2\left(2D - \frac{3}{2}\right) = \frac{45Alet^\circ}{4lf^2 + 90mp\phi_0} \left\{ \frac{8D}{5} f - f + \frac{6mp\phi_0}{fl} D \right\}$$

$$- \frac{3AD}{f} et^\circ \quad . \quad . \quad . \quad . \quad . \quad p(192) \quad . \quad . \quad (141)$$

when the effect of the axial stress is considered, and

$$M_2 = \frac{15A}{2f} et^\circ = M_1 \quad . \quad . \quad . \quad . \quad p(193) \quad . \quad . \quad (142)$$

when the axial stress is neglected.

$$D = 1 + \frac{3m}{l^2} - \frac{6m\phi_0}{l^2}. \quad . \quad . \quad . \quad . \quad . \quad . \quad . \quad . \quad . \quad (143)$$

From $p(193)$ and $p(191)$,

$$M_2 = M_1 = H_1 \frac{2}{3} f; \quad \ldots \ldots \quad (144)$$

$$y_1 = y_2 = \frac{2}{3} f; \quad \ldots \quad p(195) \quad . \quad . \quad (145)$$

$$V_1 = V_2 = 0. \quad \ldots \quad p(196) \quad . \quad . \quad (146$$

(*f*) *Effect of a Change of Δl in the Length of the Span.*

If et° be replaced by $\frac{\Delta l}{l}$ in the equations for temperature, they apply to this case.

(*g*) *Uniform Loading.*

Let w be the uniform load per unit length of span. Then

$$H_1 = \frac{wl}{8n} \Big[k^3(10 - 15k + 6k^2) \Big]_{k''}^{k'}; \quad \ldots \quad p(206) \quad . \quad . \quad (147)$$

$$M_1 = \frac{wl^2}{2} \Big[k^2(-1 + 3k - 3k^2 + k^3) \Big]_{k''}^{k'}; \quad . \quad p(208) \quad . \quad . \quad (148)$$

$$M_2 = \frac{wl^2}{2} \Big[k^3(1 - 2k + k^2) \Big]_{k''}^{k'}; \quad \ldots \quad p(207) \quad . \quad . \quad (149)$$

$$V_1 = \frac{wl}{2} \Big[k(2 - 2k^2 + k^3) \Big]_{k''}^{k'}; \quad \ldots \quad p(209) \quad . \quad . \quad (150)$$

$$M_x = M_1 + V_1 x - H_1 y - \frac{w}{2} \Big[(x - a)^2 \Big]_{a''}^{a' \gtreqless x/l}. \quad p(210) \quad . \quad . \quad (151)$$

In some of the equations the quantities A and m appear. For A an average value of $E\theta \cos \phi$ may be taken, and for m an average value of θ/F.

In case it is desired to take advantage of the effect of the axial stress, it will be advisable to first proportion the arch members with the effect of the axial stress neglected; this makes it possible to take nearly correct values for A and m in a second computation which includes the effect of the axial stress.

* In parabolic arches which have a large rise in comparison with the length of the span the effect of the axial stress upon the values of M_1 and H_1 is very small,—in fact smaller than errors which are likely to be made in graphical solutions.

In flat arches the axial stress may be of considerable magnitude (about 20 per cent for H_1 in the case of the St. Louis arch); yet this is greatly reduced when the stresses are determined, and considering that a safety factor of from four to six is employed in proportioning the members, the degree of danger is exceedingly small in omitting the effect entirely.

B. Symmetrical Circular Arches.

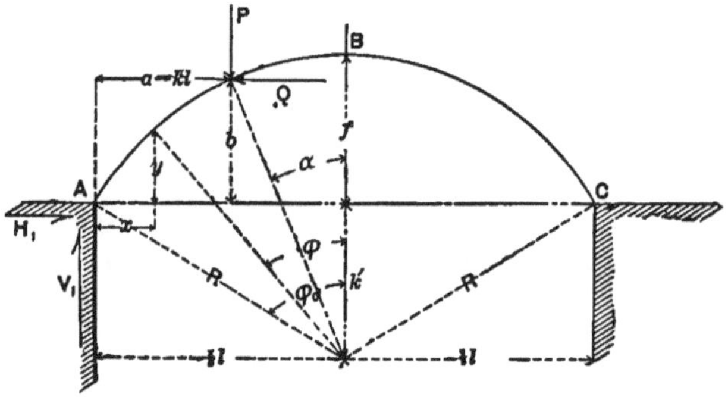

Fig. 23.

$$A = \frac{2E}{R}\theta = \text{a constant.} \quad . \quad . \qquad c(59) \quad . \quad (152)$$

* See Appendix C

$$m = \frac{\theta}{FR^2} = \left(\frac{\text{radius of gyration}}{R}\right)^2. \quad c(60) \quad (153)$$

$$k' = R - f. \quad \dots \quad c(61) \quad . \quad (154)$$

$$x = R(\sin\phi_0 - \sin\phi). \quad \dots \quad c(62) \quad . \quad (155)$$

$$y = R(\cos\phi - \cos\phi_0). \quad . \quad c(63) \quad . \quad (156)$$

$$R^2 = \left(\frac{l}{2} - x\right)^2 + (k' + y)^2. \quad . \quad c(64) \quad . \quad (157)$$

$$\sin\phi = \frac{\frac{1}{2}l - x}{R}; \quad \cos\phi = \frac{k' + y}{R}. \quad c(66) \quad . \quad (158)$$

$$\tan\phi = \frac{\frac{1}{2}l - x}{k' + y}. \quad \dots \quad c(67) \quad . \quad (159)$$

Since the general method of treating circular arches is the same as for parabolic arches, it will be necessary to only give the equations.

ARCHES HAVING TWO HINGES, ONE AT EACH ABUTMENT.

(*a*) *Vertical Loads, with Effect of Axial Stress neglected—Common Method.*

$$H = \sum^{l} P \left\{ \frac{\frac{1}{2}(\sin^2\phi_0 - \sin^2\alpha) + \cos\phi_0(\cos\alpha + \alpha\sin\alpha - \cos\phi_0 - \phi_0\sin\phi_0)}{2\phi_0\cos^2\phi_0 - 3\sin\phi_0\cos\phi_0 + \phi_0} \right\}$$

$$c(108) \quad (160)$$

or

$$H_1 = \sum^{l} P\frac{A}{B} = \sum^{l} P\Delta_{11}. \quad \dots \quad (160a)$$

$$V_1 = \sum^{l} P(1 - k) \quad \text{where} \quad k = a/l. \quad \dots \quad c(111) \quad (161)$$

$$y_0 = \frac{V_1}{H_1}a \quad . \quad . \quad . \quad . \quad c(113) \quad (163)$$

or

$$y_0 = \frac{B}{A}k(1-k)l \quad . \quad c(114) \quad (163)$$

or

$$y_0 = \frac{\Delta_6}{\Delta_{17}}l. \quad . \quad . \quad . \quad . \quad . \quad . \quad . \quad (163a)$$

(*b*) *Vertical Loads, with Effect of Axial Stress included.*

$$H_1 = \mathfrak{H}\frac{1 - \frac{m}{2A}(\sin^2\phi_0 - \sin^2\alpha)}{1 + \frac{m}{B}(\phi_0 + \sin\phi_0\cos\phi_0)} \quad c(117) \quad (164)$$

where

$$\mathfrak{H} = H_1 \text{ in } (160) = \sum^{l} P\Delta_{17}, \quad . \quad . \quad . \quad . \quad . \quad (165)$$

$$B = \Delta_{18} = 2\phi_0\cos^2\phi_0 - 3\sin\phi_0\cos\phi_0 + \phi_0. \quad . \quad . \quad (166)$$

$$A = \tfrac{1}{2}(\sin^2\phi_0 - \sin^2\alpha)$$
$$+ \cos\phi_0(\cos\alpha + \alpha\sin\alpha - \cos\phi_0 - \phi_0\sin\phi_0) \quad (167)$$

or

$$A = \Delta_{17}\Delta_{18}. \quad . \quad . \quad . \quad . \quad . \quad . \quad . \quad . \quad . \quad . \quad . \quad . \quad . \quad . \quad (168)$$

$$\phi_0 + \sin\phi_0\cos\phi_0 = \Delta_{19}. \quad . \quad . \quad . \quad . \quad . \quad (169)$$

$$V_1 = \sum^{l} P(1-k). \quad . \quad . \quad . \quad . \quad c(111) \quad (170)$$

$$y_0 = \frac{V_1}{H_1}a. \quad . \quad . \quad . \quad . \quad . \quad . \quad c(118) \quad (171)$$

(*c*) *Horizontal Loads, with Effect of Axial Stress neglected.*

$$H_1 = \frac{1}{2}\overset{l}{\Sigma} Q \left\{ 1 + \frac{\begin{array}{c}\alpha - \sin\alpha\cos\alpha \\ -2\cos\phi_0(\sin\alpha - \alpha\cos\alpha)\end{array}}{\phi_0 - 3\sin\phi_0\cos\phi_0 + 2\phi_0\cos^2\phi_0} \right\}. \quad c(120) \quad (172)$$

$$\alpha - \sin\alpha\cos\alpha = \beta_{10}. \quad . \quad . \quad . \quad . \quad . \quad (173)$$

$$\sin\alpha - \alpha\cos\alpha = \Delta\Delta_{10}. \quad . \quad . \quad . \quad . \quad (174)$$

$$\phi_0 - 3\sin\phi_0\cos\phi_0 + 2\phi_0\cos^2\phi_0 = \Delta_{10}. \quad . \quad . \quad (175)$$

$$V = -V_2 = \overset{l}{\Sigma} Q\frac{b}{l}. \quad . \quad . \quad . \quad . \quad . \quad (176)$$

$$x_0 = \frac{H_1}{V}b \quad . \quad . \quad . \quad . \quad . \quad . \quad . \quad . \quad . \quad . \quad . \quad . \quad . \quad . \quad . \quad . \quad (177)$$

or

$$x_0 = \sin\phi_0 \left\{ 1 + \frac{\begin{array}{c}\alpha - \sin\alpha\cos\alpha \\ -2\cos\phi_0(\sin\alpha - \alpha\cos\alpha)\end{array}}{\begin{array}{c}\phi_0 - 3\sin\phi_0\cos\phi_0 \\ + 2\phi_0\cos^2\phi_0\end{array}} \right\} R \quad c(123) \quad (178)$$

or

$$x_0 = \sin\phi_0\left(1 + \frac{\beta_{10} - 2\cos\phi_0(\Delta\Delta_{10})}{\Delta_{10}}\right)R. \quad . \quad . \quad . \quad . \quad . \quad (179)$$

(*d*) *Horizontal Loads, including Effect of Axial Stress.*

$$H_1 = +\overset{l}{\Sigma} Q \left\{ \frac{\begin{array}{c}\phi_0 - 3\sin\phi_0\cos\phi_0 + 2\phi_0\cos^2\phi_0 + \alpha \\ -\sin\alpha\cos\alpha - 2\cos\phi_0(\sin\alpha - \alpha\cos\alpha) \\ + m(\phi_0 + \sin\phi_0\cos\phi_0 + \alpha + \sin\alpha\cos\alpha)\end{array}}{\begin{array}{c}2(\phi_0 - 3\sin\phi_0\cos\phi_0 + 2\phi_0\cos^2\phi_0) \\ + 2m(\phi_0 + \sin\phi_0\cos\phi_0)\end{array}} \right\} \quad c(125) \quad (180)$$

or

$$H_1 = + \overset{l}{\Sigma} Q \left\{ \frac{\Delta_{18} + \beta_{19} - 2\cos\phi_0(\Delta\Delta_{19}) + m(\Delta_{19} + \Delta_{19})}{2(\Delta_{18}) + 2m(\Delta_{19})} \right\}. \quad . \quad (181)$$

$$V_1 = \overset{l}{\Sigma} Q \frac{b}{l}. \quad . \quad . \quad . \quad . \quad . \quad . \quad . \quad . \quad . \quad . \quad . \quad . \quad (182)$$

$$x_0 = \frac{H_1}{V_1} b. \quad . \quad . \quad . \quad . \quad . \quad . \quad . \quad . \quad . \quad . \quad c(127) \quad (183)$$

(*e*) *Temperature.*

$$H_1 = \frac{et^\circ A}{R} \frac{\sin\phi_0}{\phi_0 + 2\phi_0\cos^2\phi_0 - 3\sin\phi_0\cos\phi_0 + m(\phi_0 + \sin\phi_0\cos\phi_0)} \qquad c(128) \quad (184)$$

when the effect of the axial stress is included, and

$$H_1 = \frac{et^\circ A}{R} \frac{\sin\phi_0}{B}, \quad . \quad . \quad . \quad . \quad . \quad . \quad . \quad . \quad . \quad . \quad . \quad c(129) \quad (185)$$

when the effect of the axial stress is neglected.

$$B = \phi_0 + 2\phi_0\cos^2\phi_0 - 3\sin\phi_0\cos\phi_0 = \Delta_{18}. \quad . \quad (186)$$

$$\phi_0 + \sin\phi_0\cos\phi_0 = \Delta_{19}. \quad . \quad . \quad . \quad . \quad (187)$$

Change in the Length of the Span.

$$H_1 = \frac{-A}{2R^2(B + m(\phi_0 + \sin\phi_0\cos\phi_0))}\Delta l \qquad c(130) \quad (188)$$

when the effect of the axial stress is included, and

$$H_1 = \frac{-A}{2R^2 B}\Delta l \quad . \quad . \quad . \quad . \quad . \quad . \quad . \quad . \quad . \quad . \quad c(131) \quad (189)$$

when the axial stress is neglected, Δl being the actual change in the length of the span.

$$B = \phi_0 - 3 \sin \phi_0 \cos \phi_0 + 2\phi_0 \cos^2 \phi_0 = \Delta_{18}. \quad . \quad (190)$$

$$\phi_0 + \sin \phi_0 \cos \phi_0 = \Delta_{19}. \quad . \quad . \quad . \quad . \quad . \quad (191)$$

SYMMETRICAL CIRCULAR ARCH WITHOUT HINGES.

(*a*) *Vertical Loads, with Effect of Axial Stress neglected—Common Method.*

$$H_1 = \frac{1}{2}\overset{l}{\Sigma} P \left\{ \frac{2 \sin \phi_0[\cos \alpha + \alpha \sin \alpha] - \sin \phi_0[2 \cos \phi_0 + \phi_0 \sin \phi_0] - \phi_0 \sin^2 \alpha}{\phi_0^2 + \phi_0 \sin \phi_0 \cos \phi_0 - 2 \sin^2 \phi_0} \right\}. \quad \text{c(113)} \quad (192)$$

(C 133.)

$$\cos \alpha + \alpha \sin \alpha = \Delta_{22}. \quad . \quad . \quad (193)$$

$$\sin \phi_0[2 \cos \phi_0 + \phi_0 \sin \phi_0] = \Delta_{21}. \quad . \quad . \quad (194)$$

$$\phi_0^2 + \phi_0 \sin \phi_0 \cos \phi_0 - 2 \sin^2 \phi_0 = \Delta_{20}. \quad . \quad . \quad (195)$$

$$M_1 = \frac{H_1 R}{\phi_0}(\sin \phi_0 - \phi_0 \cos \phi_0)$$

$$+ \frac{\overset{l}{\Sigma} PR}{2\phi_0(\sin \phi_0 \cos \phi_0 - \phi_0)} \Big\{ \sin \alpha \phi_0(\cos \alpha \sin \phi_0 - \cos \phi_0 \sin \phi_0 - \phi_0)$$

$$+ \alpha \phi_0 \sin \phi_0$$

$$+ (\sin \phi_0 \cos \phi_0 - \phi_0)[\cos \alpha + \alpha \sin \alpha - \cos \phi_0 - \phi_0 \sin \phi_0] \Big\}. \quad \text{c(134)} \quad (196)$$

$$\sin \phi_0 - \phi_0 \cos \phi_0 = \Delta\Delta_{19}. \quad . \quad . \quad . \quad (197)$$

$$\cos \alpha + \alpha \sin \alpha = \Delta_{22}. \quad . \quad . \quad . \quad (198)$$

$$- \cos \phi_0 - \phi_0 \sin \phi_0 = \Delta_{21}. \quad . \quad . \quad . \quad (199)$$

$$- (\phi_0 - \sin \phi_0 \cos \phi_0) = \beta_{19}. \quad . \quad . \quad . \quad (200)$$

$$- (\phi_0^2 - \phi_0 \sin \phi_0 \cos \phi_0) = \Delta_{23}. \quad . \quad . \quad . \quad (201)$$

$$(\phi_0^2 + \phi_0 \sin \phi_0 \cos \phi_0) = \Delta_{24}. \quad . \quad . \quad . \quad (202)$$

The value of M_2 can be found from (196) by assuming a load so that a will become $l - a$.

$$V_1 = \frac{1}{l}\left(M_2 - M_1 + \sum^{l} Pl(1 - k)\right). \quad c(135) \quad (203)$$

$$y_0 = \frac{M_1 + V_1 a}{H_1} \quad . \ . \ . \ . \ . \ . \ . \ . \quad (50) \quad . \quad (204)$$

$$y_1 = \frac{M_1}{H_1} \quad . \ . \ . \ . \ . \ . \ . \ . \ . \ . \quad (51) \quad . \quad (205)$$

$$y_2 = \frac{M_2}{H_2} \quad . \ . \ . \ . \ . \ . \ . \ . \ . \ . \quad (52) \quad . \quad (206)$$

Independent equations for y_0, y_1, and y_2 are given in Chapter IV.

(b) Horizontal Loads, with Effect of Axial Stress neglected.

$$H_1 = + \frac{\sum^{l} Q}{2}\left\{1 + \frac{\phi_0(\sin\alpha\cos\alpha - \alpha) + 2\sin\phi_0(\sin\alpha - \alpha\cos\alpha)}{2\sin^2\phi_0 - \phi_0\sin\phi_0\cos\phi_0 - \phi_0^2}\right\}. \quad c(141) \quad (207)$$

$$\sin\alpha\cos\alpha - \alpha = -\beta_{19}. \quad . \ . \ . \ . \quad (208)$$

$$\sin\alpha - \alpha\cos\alpha = \Delta\Delta_{19}. \quad . \ . \ . \ . \ . \quad (209)$$

$$\sin^2\phi_0 - \phi_0\sin\phi_0\cos\phi_0 - \phi_0^2 = -\Delta_{20}. \quad . \quad (210)$$

$$M_1 = \frac{H_1 R}{\phi_0}(\sin\phi_0 - \phi_0\cos\phi_0) \quad . \ . \ . \ . \ . \ . \ . \ . \ . \ . \ . \ . \ . \ . \quad (211)$$

$$+ \frac{\sum^{l} QR}{2(\sin\phi_0\cos\phi_0 - \phi_0)}\left\{\begin{matrix}(\cos\alpha - \cos\phi_0)(\sin\phi_0\cos\phi_0 - \phi_0 + 2\cos\alpha) \\ - \sin\phi_0(\sin^2\phi_0 - \sin^2\alpha)\end{matrix}\right\}$$

$$- \frac{\sum^{l} QR}{2\phi_0}\left\{\sin\phi_0 - \phi_0\cos\phi_0 + \sin\alpha - \alpha\cos\alpha\right\}. \quad . \ . \ . \ . \quad c(143) \quad (212)$$

$$\sin\phi_0 - \phi_0\cos\phi_0 = \Delta\Delta_{19}. \quad . \ . \ . \ . \ . \quad (213)$$

$$\sin \phi_0 \cos \phi_0 - \phi_0 = \beta_{19}. \quad . \quad . \quad . \quad . \quad (214)$$

$$\sin \phi_0 - \phi_0 \cos \phi_0 = \Delta\Delta_{19}. \quad . \quad . \quad . \quad . \quad (215)$$

$$\sin \alpha - \alpha \cos \alpha = \Delta\Delta_{19}. \quad . \quad . \quad . \quad . \quad (216)$$

The magnitudes of H_2 and M_2 can be found from (207) and (212) by replacing a by $l - a$, etc.

$$V_1 = \frac{1}{l}(M_2 - M_1 + \overset{l}{\Sigma} Qb). \qquad c(144) \qquad (217)$$

(c) *Temperature, with Effect of Axial Stress neglected.*

$$H_1 = \frac{2A\phi_0(let^\circ)}{4R^2(\phi_0^2 + \phi_0 \sin \phi_0 \cos \phi_0 - 2 \sin^2 \phi_0)}. \qquad c(146) \qquad (218)$$

$$\phi_0^2 + \phi_0 \sin \phi_0 \cos \phi_0 - 2 \sin^2 \phi_0 = \Delta_{21}. \quad . \quad (219)$$

It will be noticed that all of our tables, with one exception, have been computed for whole degrees. In case the loads do not fall at even-degree points, it will be found advisable to make all computations for H_1, M_1, V_1, etc., for the even-degree points, and then obtain the values corresponding to the true positions of the loads by reading their values from a diagram constructed from the calculations thus made.

The effect of the axial stress has been omitted here, as the equations are long; these are given complete in Chapter IV.

When the rise of the circular arch is not greater than two tenths the span, the formulas for parabolic arches can be applied in the determination of the external forces without sensible error.

Another approximate method may also be used for arches where $f/l > 0.30$, viz.: Substitute a parabolic arch of the same span which has an area equal to the area of the given circular arch and determine the external forces, and then apply these forces to the given circular arch.

For arches approaching a semicircle this method is but a few per cent in error.

C. Summation Formulas for Symmetrical Arches of any Regular Shape and any Cross-section.

ARCH WITHOUT HINGES.

(a) *Vertical Loads, with Effect of Axial Stress considered.*

$$\mathfrak{H}_1 = \frac{\sum_0^l \frac{K'y\Delta s}{\theta_x} + \sum_0^l \frac{N_x\Delta x}{F_x} - \frac{\sum_0^l \frac{K'\Delta s}{\theta_x}}{\sum_0^l \frac{\Delta s}{\theta_x}} \sum_0^l \frac{y\Delta s}{\theta_x}}{\sum^l \frac{y^2\Delta s}{\theta_x} + \sum_0^l \frac{\Delta x}{F_x}\cos\phi - \frac{\left(\sum_0^l \frac{y\Delta s}{\theta_x}\right)^2}{\sum_0^l \frac{\Delta s}{\theta_x}}}, \quad g(87) \qquad (220)$$

where $\mathfrak{H}_1$ is the horizontal thrust due to *two equal and symmetrically placed loads.*

For a single load

$$H_1 = \frac{\sum_0^{\frac{1}{2}l} \frac{K'y\Delta s}{\theta_x} + \sum_0^{\frac{1}{2}l} \frac{N_x\Delta x}{F_x} - \frac{\sum_0^{\frac{1}{2}l} \frac{K'\Delta s}{\theta_x}}{\sum_0^{\frac{1}{2}l} \frac{\Delta s}{\theta_x}} \sum_0^{\frac{1}{2}l} \frac{y\Delta s}{\theta_x}}{2\left(\sum_0^{\frac{1}{2}l} \frac{y^2\Delta s}{\theta_x} + \sum_0^{\frac{1}{2}l} \frac{\Delta x}{F_x}\cos\phi - \frac{\left(\sum_0^{\frac{1}{2}l} \frac{y\Delta s}{\theta_x}\right)^2}{\sum_0^{\frac{1}{2}l} \frac{\Delta s}{\theta_x}}\right)}, \quad . \qquad (221)$$

where

$$\sum^{\frac{1}{2}l} \frac{K'y\Delta s}{\theta_x} = \left[\sum_0^{\frac{1}{2}l} xy\frac{\Delta s}{\theta_x} - \left(\sum_a^{\frac{1}{2}l} \frac{xy\Delta s}{\theta_x} - a\sum_a^{\frac{1}{2}l} \frac{y\Delta s}{\theta_x}\right)\right]P, \quad . \qquad (222)$$

$$\sum_0^{\frac{1}{2}l} \frac{K'\Delta s}{\theta_x} = \left[\sum_0^{\frac{1}{2}l} x\frac{\Delta s}{\theta_x} - \left(\sum_a^{\frac{1}{2}l} \frac{x\Delta s}{\theta_x} - a\sum_a^{\frac{1}{2}l} \frac{\Delta s}{\theta_x}\right)\right]P, \quad . \qquad (223)$$

and

$$\sum_0 \frac{N_x\Delta x}{F_x} = +P\sum_0^a \frac{\Delta x}{F_x}\sin\phi. \quad . \quad . \quad . \quad . \qquad (224)$$

$$M_1 = \frac{\left\{ \sum_0^l \frac{Kx\Delta s}{\theta_x} - \sum_0^l \frac{N_x \Delta y}{F_x} \right\} \sum_0^l \frac{x\Delta s}{\theta_x} - \sum_0^l \frac{K\Delta s}{\theta_x} \sum_0^l \frac{x^2 \Delta s}{\theta_x}}{\sum_0^l \frac{\Delta s}{\theta_x} \sum_0^l \frac{x^2 \Delta s}{\theta_x} - \left(\sum_0^l \frac{x\Delta s}{\theta_x} \right)^2}, \text{(g71)} \quad (225)$$

where

$$\sum_0^l \frac{Kx\Delta s}{\theta_x} = - H_1 \sum_0^l \frac{xy\Delta s}{\theta_x} - \left(\sum_a^l \frac{x^2 \Delta s}{\theta_x} - a \sum_a^l \frac{x\Delta s}{\theta_x} \right) P, \quad (226)$$

$$\sum_0^l \frac{K\Delta s}{\theta_x} = - H_1 \sum_0^l \frac{y\Delta s}{\theta_x} - \left(\sum_a^l \frac{x\Delta s}{\theta_x} - a \sum_a^l \frac{\Delta s}{\theta_x} \right) P, \quad . \quad (227)$$

$$\sum_0^l \frac{N_x \Delta y}{F_x} = H_1 \sum_0^l \frac{\Delta y}{F_x} \cos \phi \text{ (approximately)}, \quad . \quad . \quad (228)$$

and H_1 is given by (221).

$$V_1 = \frac{1}{l} (M_2 - M_1 + P(1 - k)l). \quad . \quad . \quad . \quad (229)$$

$$y_1 = \frac{M_1}{H_1}. \quad . \quad . \quad . \quad . \quad . \quad . \quad . \quad . \quad . \quad . \quad . \quad . \quad (230)$$

Having H_1, M_1, V_1, and y_1, the remaining outside forces can be found by the method explained on page 22.

(b) Vertical Loads, with Effect of Axial Stress neglected.

If the effect of the axial stress is to be neglected, we have merely to drop the terms containing N_x and F_x in (221) and (225), and proceed as before.

(c) Horizontal Loads, with Effect of Axial Stress included.

$$H_1 = \tfrac{1}{2}(\mathfrak{H}_1 + Q). \quad . \quad . \quad . \quad . \quad . \quad . \quad . \quad (231)$$

$$\mathfrak{H}_1 = \frac{\sum\limits_0^{\frac{1}{2}l}\frac{K'y\Delta s}{\theta_x} + \sum\limits_0^{\frac{1}{2}l}\frac{N_x\Delta x}{\theta_x} - \frac{\sum\limits_0^{\frac{1}{2}l}\frac{K'\Delta s}{\theta_x}}{\sum\limits_0^{\frac{1}{2}l}\frac{\Delta s}{\theta_x}}\sum\limits_0^{\frac{1}{2}l}\frac{y\Delta s}{\theta_x}}{\sum\limits_0^{\frac{1}{2}l}\frac{y^2\Delta s}{\theta_x} + \sum\limits_0^{\frac{1}{2}l}\frac{\Delta x}{F_x}\cos\phi - \frac{\left(\sum\limits_0^{\frac{1}{2}l}\frac{y\Delta s}{\theta_x}\right)^2}{\sum\limits_0^{\frac{1}{2}l}\frac{\Delta s}{\theta_x}}}, \qquad g(87) \qquad (232)$$

where

$$\sum_0^{\frac{1}{2}l}\frac{K'y\Delta s}{\theta_x} = \left[\sum_a^{\frac{1}{2}l}\frac{y^2\Delta s}{\theta_x} - b\sum_a^{\frac{1}{2}l}\frac{y\Delta s}{\theta_x}\right]Q, \quad . \quad . \quad . \quad (233)$$

$$\sum_0^{\frac{1}{2}l}\frac{K'\Delta s}{\theta_x} = \left[\sum_a^{\frac{1}{2}l}\frac{y\Delta s}{\theta_x} - b\sum_a^{\frac{1}{2}l}\frac{\Delta s}{\theta_x}\right]Q, \quad . \quad . \quad . \quad . \quad (234)$$

$$\sum_0 \frac{N_x\Delta x}{\theta_x} = -\sum_{a_2}^{a_1}\frac{Q\Delta x}{\theta_x}\cos\phi. \qquad a_2 = l - a_1. \qquad (235)$$

$$M_1 = \frac{\begin{matrix} +\left\{-H_1\sum\limits_0^l\frac{xy\Delta s}{\theta_x} + Q\sum\limits_a^l\frac{xy\Delta s}{\theta_x} - Qb\sum\limits_a^l\frac{x\Delta s}{\theta_x}\right\}\sum\limits_0^l\frac{x\Delta s}{\theta_x} \\ +\left\{-H_1\sum\limits_0^l\frac{\Delta y}{\theta_x}\cos\phi + Q\sum\limits_a^l\frac{\Delta y}{\theta_x}\cos\phi\right\}\sum\limits_0^l\frac{x\Delta s}{\theta_x}. \\ +\left\{+H_1\sum\limits_0^l\frac{y\Delta s}{\theta_x} - Q\sum\limits_a^l\frac{y\Delta s}{\theta_x} + Qb\sum\limits_a^l\frac{\Delta s}{\theta_x}\right\}\sum\limits_0^l\frac{x^2\Delta s}{\theta_x} \end{matrix}}{\sum\limits_0^l\frac{\Delta s}{\theta_x}\sum\limits_0^l\frac{x^2\Delta s}{\theta_x} - \left(\sum\limits_0^l\frac{x\Delta s}{\theta_x}\right)^2 \qquad g(106)}. \qquad (236)$$

M_2 for a load situated a distance $\underline{a}$ from the origin $= M_1$ for a load situated $(l - a)$ from the origin.

$$V_1 = \frac{1}{l}\{M_2 - M_1 + Qb\}. \quad . \quad . \quad . \quad . \quad (237)$$

$$y_1 = \frac{M_1}{H_1}. \quad . \quad . \quad . \quad . \quad . \quad . \quad . \quad . \quad . \quad . \quad (238)$$

Having determined H_1, M_1, M_2, V_1, and y_1, the other outer forces are readily found, as explained on page 22.

(d) Horizontal Loads, with Effect of Axial Stress neglected.

If the effect of the axial stress is to be neglected, we have merely to omit all the terms which contain ϕ in (236) and (232), and proceed as before.

(e) Temperature.

$$H_t = \frac{Eet^\circ l}{\sum_0^l \frac{y^2 \Delta s}{\theta_x} + \sum_0^l \frac{\Delta x}{F_x}\cos\phi - \frac{\left(\sum_0^l \frac{y\Delta s}{\theta_x}\right)^2}{\sum_0^l \frac{\Delta s}{\theta_x}}}. \quad g(145) \qquad (239)$$

$$M_1 = -\frac{Eet^\circ l \sum_0^l \frac{y\Delta s}{\theta_x}}{\left\{\sum_0^l \frac{y^2 \Delta s}{\theta_x} + \sum_0^l \frac{\Delta x}{F_x}\cos\phi - \frac{\left(\sum_0^l \frac{y\Delta s}{\theta_x}\right)^2}{\sum_0^l \frac{\Delta s}{\theta_x}}\right\}\sum_0^l \frac{\Delta s}{\theta_x}}. \qquad (240)$$

$$V_1 = 0. \quad . \quad . \quad . \quad . \quad . \quad . \quad . \quad . \qquad (241)$$

If the effect of the axial stress is to be neglected, omit the terms containing ϕ in (239) and (240).

ARCH WITH A HINGE AT EACH SUPPORT.

(a) Vertical Loads, with Effect of Axial Stress included.

$$H_1 = \frac{\left[\begin{array}{l} + \left\{\sum_0^l xy\frac{\Delta s}{\theta_x} - \sum_a^l xy\frac{\Delta s}{\theta_x} + a\sum_a^l y\frac{\Delta s}{\theta_x}\right\}P \\ + \left\{\sum_0^l \frac{\Delta x}{F_x}\sin\phi - \sum_a^l \frac{\Delta x}{F_x}\sin\phi\right\}P \end{array}\right]}{2\left\{\sum_0^l y^2\frac{\Delta s}{\theta_x} + \sum_0^l \frac{\Delta x}{F_x}\cos\phi\right\}}. \quad g(131) \qquad (242)$$

$$V_1 = P(1 - k). \quad . \quad . \quad . \quad . \quad . \quad . \quad . \quad . \quad . \quad . \quad . \quad . \quad . \qquad (243)$$

If the axial stress is neglected, omit the terms containing F_x in the above equations.

(*b*) *Horizontal Loads, with Effect of Axial Stress included.*

$$H_1 = Q\left\{\frac{1}{2}+\frac{1}{2}\,\frac{\sum\limits_{a_1}^{a_2} y^2\frac{\Delta s}{\theta_x} - b\sum\limits_{a_1}^{a_2} y\frac{\Delta s}{\theta_x} - \sum\limits_{a_1}^{a_2}\frac{\Delta x}{F_x}\cos\phi}{\sum\limits_0^l y^2\frac{\Delta s}{\theta_x}+\sum\limits_0^l\frac{\Delta x}{F_x}\cos\phi}\right\}, \quad g(140) \quad (244)$$

where

$$a_2 = l - a_1.$$

$$V_1 = Q\frac{b}{l}. \quad . \quad . \quad . \quad . \quad . \quad . \quad . \quad . \quad (245)$$

(*c*) *Temperature.*

$$H_t = \frac{Eet^\circ l}{\sum\limits_0^l \frac{y^2\Delta s}{\theta_x}+\sum\limits_0^l\frac{\Delta s}{F_x}}. \quad . \quad . \quad g(148) \quad (246)$$

When the effect of the axial stress is neglected the terms containing F_x are to be omitted.

DEFLECTION OF ARCH.

In considering the deflection we will assume that the axial stress is neglected and that no change takes place in the relative positions of the several points of the arch other than that produced by the loading.

ARCH WITHOUT HINGES OR WITH TWO HINGES—(SYMMETRICAL LOADING).

$$\delta x = -\frac{1}{E}\sum_0^x M_x\frac{y\Delta s}{\theta_x} + et^\circ x. \quad . \quad g(60) \quad (247)$$

$$\delta y = \frac{1}{E}\sum_0^x M_x\frac{x\Delta s}{\theta_x} + et^\circ y. \quad . \quad . \quad g(61) \quad (248)$$

The above summation formulas are sufficiently accurate for practical purposes, and are quite simple in their application. They apply to any regular arch figure, such as circular, parabolic, oval, elliptic, gothic, spandrel-braced, etc. They are especially useful in the solution of the spandrel-braced arch, and all arches which have variable or constant moments of inertia not following the laws upon which the formulas of Chapters III and IV are based.

CHAPTER III.

PARABOLIC ARCHES, WITH THE MOMENTS OF INERTIA VARYING ACCORDING TO THE RELATION $A = E\theta \cos \phi =$ a constant.

GENERAL RELATIONS.

In large arches it is convenient often to arrange the sections so that their moments of inertia vary according to the relation $A = E\theta \cos \phi =$ a constant. This assumption enables us to deduce quite simple formulas for the determination of the reactions, bending-moments, etc.

The nomenclature used in this chapter will be the same as heretofore employed, and any new symbols appearing will be found clearly represented in Fig. 24.

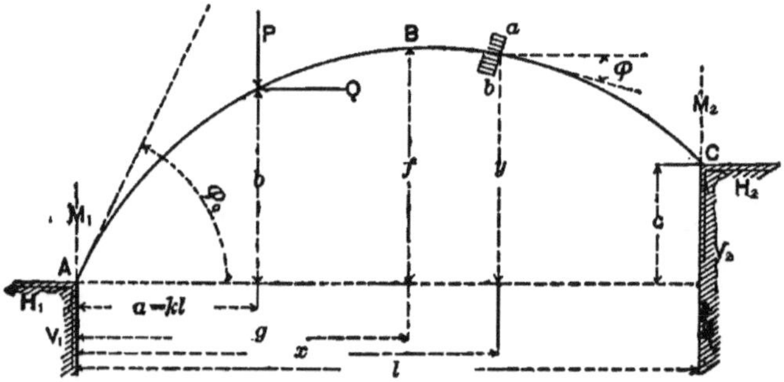

FIG. 24.

We have then

$$A = E\theta \cos \phi = \text{a constant.} \quad . \quad . \quad . \quad . \quad p(59)$$

Let

$$m = \frac{\theta}{F} = \text{(radius of gyration)}^2 \quad . \quad . \quad . \quad p(60)$$

and

$$p = \text{the parameter of the parabola.}$$

The equation of the parabolic curve referred to its vertex is

$$(g - x)^2 = 2p(f - y) \quad . \quad . \quad . \quad . \quad . \quad . \quad . \qquad p(61)$$

or

$$g^2 - 2gx + x^2 = 2pf - 2py. \quad . \quad . \quad . \qquad p(62)$$

For $x = 0$, $y = 0$, and $g^2 = 2pf$. . . . . $p(63)$

From $p(62)$,

$$y = \frac{2pf - g^2 + 2gx - x^2}{2p}; \quad \text{but} \quad g^2 = 2pf;$$

therefore

$$y = \frac{2g - x}{2p}x \quad . \quad . \quad . \quad . \quad . \quad . \quad . \quad . \quad . \quad . \quad . \quad . \qquad p(64)$$

and

$$dy = \frac{g - x}{p}dx \quad \text{or} \quad \frac{dx}{dy} = \frac{g - x}{p} = \tan \phi. \quad . \quad . \qquad p(65)$$

From (d),

$$\Delta\phi = \frac{1}{E}\int \frac{M_x}{\theta}ds. \quad . \quad . \quad . \quad . \quad . \quad . \quad . \quad . \quad . \quad . \quad . \qquad (d)$$

If $\Delta\phi_0$ represents any change in ϕ_0, the corresponding change up to any section x will be represented by

$$\Delta\phi = \Delta\phi_0 + \frac{1}{E}\int_0^x \frac{M_x}{\theta}ds. \quad . \quad . \quad . \quad . \quad . \quad . \quad . \quad . \qquad p(66)$$

But $\dfrac{ds}{dx} = \dfrac{1}{\cos \phi}$; hence

$$\Delta\phi = \Delta\phi_0 + \frac{1}{E}\int_0^x \frac{M_x}{\theta}\,\frac{dx}{\cos \phi}$$

or, since $E \cos \phi = \frac{A}{\theta}$ (see $p(59)$),

$$\Delta\phi = \Delta\phi_0 + \frac{1}{A}\int_0^x M_x dx. \quad . \quad . \quad . \quad . \quad p(67)$$

From (41),

$$M_x = M_1 + V_1 x - H_1 y - \overset{x}{\Sigma} P(x - a) + \overset{x}{\Sigma} Q(y - b). \quad (41)$$

Substituting (41) in $p(67)$, remembering that $y = \frac{2g - x}{2p}x$ (see $p(64)$), we have

$$\Delta\phi = \Delta\phi_0 + \frac{1}{A}\int_0^x \left\{ M_1 + V_1 x - H_1 \frac{2g - x}{2p} x \right\} dx$$

$$- \frac{1}{A}\overset{x}{\Sigma}\left\{ P\int_a^x (x - a)dx \right\} + \frac{1}{A}\overset{x}{\Sigma}\left\{ Q\int_a^x (y - b)dx \right\}. \quad p(68)$$

Performing the integrations indicated, factoring, and collecting, we obtain

$$\Delta\phi = \Delta\phi_0 + \frac{1}{2A}\left\{ 2M_1 x + V_1 x^2 - H_1 \frac{3g - x}{3p} x^2 - \overset{x}{\Sigma} P(x - a)^2 \right.$$

$$\left. + \overset{x}{\Sigma} Q \frac{1}{3p}(3g(x^2 - a^2) - x^3 + a^3 - 6pb(x - a)) \right\}. \quad . \quad p(69)$$

From (a) the expression for Δx from 0 up to the section x becomes

$$\Delta x = -\int_0^x \Delta\phi dy + et^\circ \int_0^x dx - \frac{1}{E}\int_0^x \frac{N_x}{F_x} dx. \quad . \quad . \quad p(70)$$

Substituting the value of $\Delta\phi$ obtained above, integrating and reducing, $p(70)$ becomes (see Appendix A)

$$\Delta x = et^\circ x - y\Delta\phi_0 - \frac{x^2}{6Ap}\left\{ M_1(3g - 2x) + V_1 x\left(g - \frac{3x}{4}\right) \right.$$

$$- H_1\frac{x}{p}\left(g^2 - gx + \frac{x^2}{5}\right) - \sum^x P\frac{1}{4x^2}[(4g - 3x - a)(x - a)^3]$$

$$+ \sum^x Q\frac{1}{10x^2p}\Big[2x^5 - 10gx^4 + 10(g^2 + 2pb)x^3$$

$$+ (15ga^2 - 30pba - 5a^3)x^2$$

$$+ 10(ga^3 - 3g^2a^2)x - 30pbg(x - a)^2$$

$$+ 3a^5 - 15a^4g + (20g^2 + 10pb)a^3\Big]\Big\}$$

$$- \frac{m}{A}\frac{p}{p+2f}\Big\{V_1y + H_1(p+2f)(\phi_0 - \phi) - \sum^x P(y - b)$$

$$- \sum^x Q(p + 2f)(\alpha - \phi)\Big\}. \quad . \quad . \quad . \quad . \quad . \quad . \quad . \quad . \quad . \quad p(79)$$

From (b) the expression for Δy from 0 up to the section x becomes

$$\Delta y = \int_0^x \Delta\phi dx + et^\circ \int_0^x dy - \frac{1}{E}\int_0^x \frac{N_x}{F_x}dy, \quad . \quad . \quad . \quad p(80)$$

which reduces to (see Appendix B)

$$\Delta y = et^\circ y + x\Delta\phi_0 + \frac{x^2}{6A}\Big\{3M_1 + V_1x - H_1\frac{x}{p}\left(g - \frac{x}{4}\right)$$

$$- \frac{1}{x^2}\sum^x P(x - a)^3 + \frac{1}{px^2}\sum^x Q\Big[-\frac{x^4}{4} + gx^3 + (a^3 - 3ga^2)x$$

$$- 3pb(x - a)^2 - \frac{3a^4}{4} + 2ga^3\Big]\Big\} - \frac{pm}{A}\Big\{V_1\left(\frac{x}{p} - \phi_0 + \phi\right)$$

$$+ H_1\frac{y}{p+2f} - \sum^x P\left(\frac{x-a}{p} - \alpha + \phi\right)$$

$$- \sum^x Q\frac{y - b}{p + 2f}\Big\}. \quad . \quad . \quad . \quad . \quad . \quad . \quad . \quad . \quad . \quad . \quad . \quad p(84)$$

Equations $p(69)$, $p(79)$, and $p(84)$ are general expressions for the elastic parabolic arch, symmetrical or non-symmetrical, when $E\theta \cos \phi$ is constant. They can be employed in the determination of temperature stresses and stresses caused by concentrated loads acting in any direction in the plane of the arch. *Those terms containing the factor m show the influence of the "axial stress."*

SYMMETRICAL ARCHES—GENERAL FORMULAS.

For symmetrical arches these equations become somewhat more simple, as

$$g = \tfrac{1}{2}l \quad \text{and} \quad \phi_l = -\phi_0$$

Let $x = l$, and assume the arch to be symmetrical; then $\phi = \phi_l = -\phi_0$ and $y = c = 0$.

Making these changes in $p(69)$, we have

$$\Delta\phi_l = \Delta\phi_0 + \frac{1}{2A}\left\{2M_1 l + V_1 l^2 - H_1\frac{l^3}{6p} - \sum^{l} P(l-a)^2 \right.$$
$$\left. + \sum^{l} Q\frac{1}{3p}\left[\tfrac{3}{2}l(l^2 - a^2) - l^3 + a^3 - 6pb(l-a)\right]\right\}. \quad p(85)$$

From (47),

$$V_1 l^2 = M_2 l - M_1 l + \sum^{l} P(l-a)l + \sum^{l} Qbl. \quad . \quad p(86)$$

Substituting $p(86)$ in $p(85)$ and reducing,

$$\Delta\phi_l = \Delta\phi_0 + \frac{l}{2A}\left\{M_1 + M_2 - H\frac{4}{3}f + \frac{1}{l}\sum^{l} P(l-a)a \right.$$
$$\left. + \frac{4f}{l^2}\sum^{l} Q(l-a)a + \frac{8f}{6l^3}\sum^{l} Q[l^3 + 9a^2 l - 6al^2 - 4a^3]\right\}. \quad p(87)$$

From $p(79)$ we obtain

$$\Delta l = ct^\circ l + \frac{lf}{3A}\left\{M_1 + M_2 - \frac{8}{5}H_1 f + \frac{1}{l^3}\sum^{l} P(l-a)(l^2 + al - a^2)a\right.$$

$$+ \sum^{l} Q\frac{4(l-a)a}{l^2}f - \frac{8}{5}\frac{f}{l^5}\sum^{l} Q[-l^5 + 5al^4 - 5a^2l^3$$

$$- 5a^3l^2 + 10a^4l - 4a^5]\Big\} - \frac{m}{A}\Big\{2H_1p\phi_0 + \frac{\sum^{l} P(l-a)a}{2(p+2f)}$$

$$- \sum^{l} Qp(\alpha + \phi_0)\Big\} \quad . \quad . \quad . \quad . \quad . \quad . \quad . \quad . \quad . \quad . \quad . \quad p(88)$$

Equation $p(84)$ reduces to

$$\Delta c = l\Delta\phi_0 + \frac{l^2}{3A}\Big\{M_1 + \frac{1}{2}M_2 - H_1f + \frac{1}{2l^3}\sum^{l} P[a(l-a)(2l-a)]$$

$$+ \sum^{l} Q\frac{2(l-a)a}{l^2}f + \frac{f}{l^4}\sum^{l} Q[l^4 - 6al^3$$

$$+ 12a^2l^2 - 10a^3l + 3a^4]\Big\} - \frac{m}{A}\Big\{(M_2 - M_1)\Big(1 - 2\phi_0\frac{p}{l}\Big)$$

$$+ p\sum^{l} P\Big(2\frac{a\phi_0}{l} + \alpha - \phi_0\Big)$$

$$+ \sum^{l} Q\frac{a(l-a)}{2p}\Big[\frac{p}{p+2f} + 1 - \frac{2p\phi_0}{l}\Big]\Big\} \quad . \quad . \quad . \quad p(89)$$

SYMMETRICAL PARABOLIC ARCH WITH A HINGE AT EACH ABUTMENT.

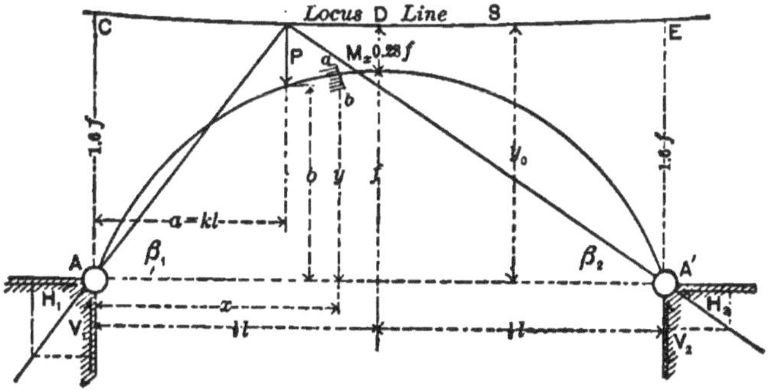

FIG. 25.

In Fig. 25 let ABA' represent a symmetrical parabolic arch having a hinge at A and A'; then there can be no bending-moments at these points; hence M_1 and $M_2 = 0$, and the resultants R_1 and R_2 will pass through the hinges.

For convenience, the effects of vertical loads, horizontal loads, and a change in temperature, with and without omitting the effect of N_x, will be considered independently.

In all that follows, $k = \frac{a}{l}$.

(a) *Vertical Loads, with the Effect of N_x omitted—Common Method.*

Assuming that l remains constant, $\Delta l = 0$, and by remembering that M_1 and $M_2 = 0$, and also that all terms containing Q and m do not appear, we have at once from $p(88)$, by solving for H_1,

$$\Sigma_1 = \frac{5}{8fl^3} \overset{l}{\Sigma} P(l^3 - 2a^2l + a^3)a \quad . \quad . \quad . \quad p(90)$$

or

$$H_1 = \frac{5}{8} \frac{l}{f} \overset{l}{\Sigma} P[k(1 - 2k^2 + k^3)]. \quad . \quad . \quad p(91)$$

Values of $k(1 - 2k^2 + k^3)$ are given in Table I.

Since all loads are vertical, the horizontal thrust is the same throughout the arch.

From (39),

$$H_x = H_1, \quad . \quad . \quad . \quad . \quad . \quad . \quad . \quad . \quad (39)$$

making $x = l$, $H_x = H_2$, and we have $H_1 - H_2 = 0$, or H_1 and H_2 are equal in magnitude, but act in opposite directions.

From $p(91)$ the values of H_1 for each load can be very quickly found with the aid of Table I, which gives the values of the expression $k(1 - 2k^2 + k^3)$ for values of k from 0 to 1.00.

From (47),

$$V_1 = \overset{l}{\Sigma} P \frac{l - a}{l} \quad . \quad . \quad . \quad . \quad . \quad . \quad . \quad p(92)$$

or

$$V_1 = \sum^{l} P(1 - k), \quad . \quad . \quad . \quad . \quad . \quad p(93)$$

from which the value of V_1 for each vertical load is readily obtained. The value of $1 - k$ can be taken directly from Tables I or V, for $k = 0$ to $k = 1.0$.

Having H_1 and V_1, the direction of R_1 for any particular load is found from

$$\tan \beta_1 = \frac{V_1}{H_1} = \frac{8}{5} \frac{f}{l} \frac{1}{k(1 + k - k^2)}. \quad . \quad . \quad p(94)$$

When the stresses are to be determined by graphics, we need only to use $p(94)$ and determine $\tan \beta_1$ for each load; then since R_1 for each load must pass through the left hinge, R_1 can be drawn in its proper position at once. Since R_1, the load, and R_2 meet in a point, R_2 must pass through the point of intersection of R_1 and the vertical force (load); it must also pass through the right hinge, and hence its direction is completely determined. The values of H_1, H_2, V_1, and V_2 can now be found by simple resolution of forces. The intermediate stresses can be found by Clerk Maxwell's method of graphics when the arch is trussed.

To facilitate the calculation of $\tan \beta_1$ the values of $\frac{8}{5} \frac{1}{(1 + k - k^2)}$ have been tabulated in Table II.

From (50) we have for each load

$$y_0 = \frac{8}{5} \frac{1}{1 + k - k^2} f, \quad . \quad . \quad . \quad . \quad . \quad . \quad . \quad p(95)$$

which locates the point of intersection of R_1 and R_2, making the application of graphics still easier than the method using $p(94)$.

The values of $\frac{8}{5} \frac{1}{1 + k - k^2}$ are given in Table II for values of k from 0 to 1.00.

From (39), (40), and (43) we obtain

$$T_x = (V_1 - \overset{x}{\Sigma}P)\cos\phi - H_1 \sin\phi. \quad . \quad . \quad p(96)$$

From (41),

$$M_x = V_1 x - H_1 y - \overset{x}{\Sigma}P(x-a). \quad . \quad . \quad . \quad . \quad p(97)$$

By means of $p(91)$, $p(93)$, $p(96)$, and $p(97)$ the stresses at any point of the arch can be completely determined by computation.

Change in Shape Due to the Action of Vertical Loads (N_x omitted).

From $p(89)$,

$$\Delta\phi_0 = \frac{\Delta c}{l} + \frac{1}{3A}\left\{H_1 lf - \frac{1}{2l}\overset{l}{\Sigma}Pa(l-a)(2l-a)\right\}. \quad p(98)$$

The term $\frac{\Delta c}{l}$ shows the effect of any change in the elevations of the hinges. This does not mean a slight difference of level in the hinges before the arch is in place, but any change which may take place afterwards.

In construction an attempt is made to so design the abutments, etc., that Δc will be zero.

$p(98)$ may be written (Δc assumed zero)

$$\Delta\phi_0 = \frac{1}{3A}\left\{H_1 lf - \frac{l^2}{2}\overset{l}{\Sigma}P(2k - 3k^2 + k^3)\right\}. \quad . \quad p(99)$$

From $p(69)$, remembering that $g = \frac{1}{2}l$,

$$\Delta\phi = \Delta\phi_0 + \frac{1}{2A}\left\{V_1 x^2 - H_1\frac{3l-2x}{6P}x^2 - \overset{x}{\Sigma}P(x-a)^2\right\}. p(100)$$

From $p(79)$,

$$\Delta x = -y\Delta\phi_0 - \frac{x^2}{6Ap}\left\{V_1\frac{x}{4}(2l-3x) - H_1\frac{x}{p}\left(\frac{l^2}{4} - \frac{lx}{2} + \frac{x^2}{5}\right)\right.$$
$$\left. - \overset{x}{\Sigma}P\frac{1}{4x^2}((2l-3x-a)(x-a)^3)\right\}. \quad . \quad . \quad p(101)$$

From $p(84)$

$$\Delta y = x\Delta\phi_0 + \frac{x^2}{6A}\left\{ V_1 x - H_1\frac{x}{4p}(2l-x) - \frac{1}{x^2}\sum^{x} P(x-a)^3 \right\}, \quad p(102)$$

in which

$$H_1 = \frac{5}{8}\frac{l}{f}\sum^{l} Pk(1 - 2k^2 + k^3) \quad . \; . \; . \; . \; . \; . \; . \; . \; . \; . \; . \quad p(91)$$

and

$$V_1 = \sum^{l} P(1 - k). \; . \; . \; . \; . \; . \; . \; . \; . \; . \; . \; . \; . \; . \quad p(93)$$

(*b*) *Vertical Loads, Effect of the Axial Stress included.*

From $p(88)$,

$$H_1 = \frac{15}{8lf^2 + 30mp\phi_0}\left\{ \frac{l^2 f}{3}\sum^{l} Pk(1-2k^2+k^3) - \sum^{l} P\frac{a(l-a)}{2(p+2f)}m \right\}$$

or

$$H_1 = \frac{15}{8lf^2 + 30mp\phi_0}\left\{ \frac{8lf^2}{15}\mathfrak{H} - \frac{ml^2}{2(p+2f)}\sum^{l} Pk(1-k) \right\}, \quad p(103)$$

in which $\mathfrak{H}$ is the value of H_1 given by $p(91)$.

Let $f = nl$; then

$$p = \frac{1}{8n}l. \; . \; . \; . \; . \; . \; . \; . \quad p(104)$$

Substituting $p(104)$ in $p(103)$,

$$H_1 = \frac{60n}{32n^3l^2 + 15m\phi_0 l}\left\{ \frac{8n^2l^2}{15}\mathfrak{H} - \frac{4n\,ml}{1+16n^2}\sum^{l} Pk(1-k) \right\}. \quad p(105)$$

The values of ϕ_0 are given in Table XXV.

$$^* H_1 = \mathfrak{H}(1 - \epsilon), \quad (\textit{approximately}) \quad p(106)$$

where $\mathfrak{H} = H_1$ as given by $p(91)$.

---

* See Appendix C.

For a brief discussion of the effect of the axial stress, see Appendix C.

The expression for V_1 is not affected by the axial stress; hence

$$V_1 = \sum^{l} P(1 - k). \quad \ldots \ldots \ldots \quad p(93)$$

For any load, from (50) we have

$$y_0 = \frac{V_1}{H_1} a = \frac{V_1}{H_1} kl. \quad \ldots \ldots \ldots \quad p(107)$$

From (39),

$$H_x = H_1, \quad \ldots \ldots \ldots \quad p(108)$$

$$V_x = V_1 - \sum^{x} P, \quad \ldots \ldots \ldots \quad (40)$$

$$N_x = V_x \sin \phi + H_x \cos \phi, \quad \ldots \ldots \quad (42)$$

$$M_x = V_1 x - H_1 y - \sum^{x} P(x - a). \quad \ldots \quad p(109)$$

(c) *Horizontal Loads* (N_x *omitted*).

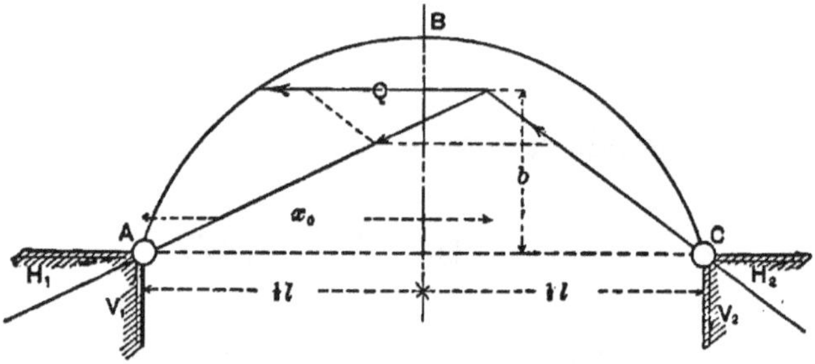

FIG. 26

In Fig. 26 is represented a single horizontal load Q acting from the right to the left, which produces a horizontal reaction H_1 similar in character to that produced by a vertical load acting downward.

From $p(88)$, for any number of horizontal loads,

$$H_1 = \sum^{l} Q \left\{ 1 - \frac{1}{2l^5}[5al^4 - 5a^2l^3 - 10a^3l^2 + 20a^4l - 8a^5] \right\} \quad p(110)$$

or

$$H_1 = \sum^{l} Q \left\{ 1 - \frac{k}{2}[5(1 - k - 2k^2 + 4k^3) - 8k^4] \right\} \ldots \quad p(111)$$

The values of the quantity in brackets,

$$\left\{ 1 - \frac{k}{2}[5(1 - k - 2k^2 + 4k^3) - 8k^4] \right\},$$

are given in Table III for values of k from 0 to 1.00.

For any load Q,

$$H_2 = Q - H_1 \ldots \quad p(112)$$

From (47),

$$V_1 = \sum^{l} Q\left(\frac{4k(1-k)}{l}f = 4k(1-k)n\right) \ldots \quad p(113)$$

$$V_2 = -V_1 \ldots \quad p(114)$$

From Fig. 26, for a single load,

$$V_1 x_0 = H_1 b \quad \text{or} \quad x_0 = \frac{H_1}{V_1} b \ldots \quad p(115)$$

From (47), for a single load,

$$V_1 = Q\frac{b}{l}.$$

Therefore

$$x_0 = \frac{H_1}{Q} l$$

or

$$x_0 = \left\{ 1 - \frac{k}{2}[5(1 - k - 2k^2 + 4k^3) - 8k^4] \right\} l. \quad p(116)$$

The coefficient of l is given in Table III.

Having the value of x_0 for any load Q, the values of H_1, V_1, R_1, etc., are readily determined by graphics.

From (39),

$$H_x = H_1 - \overset{x}{\Sigma} Q. \quad . \quad . \quad . \quad . \quad . \quad . \quad . \quad . \quad . \quad p(117)$$

From (40),

$$V_x = V_1. \quad . \quad . \quad . \quad . \quad . \quad . \quad . \quad . \quad . \quad . \quad p(118)$$

From (41),

$$M_x = V_1 x - H_1 y + \overset{x}{\Sigma} Q(y - b). \quad . \quad . \quad . \quad p(119)$$

From (43),

$$T_x = V_x \cos\phi - H_x \sin\phi ; \quad . \quad . \quad . \quad . \quad . \quad (43)$$

$$\tan\beta_1 = \frac{V_1}{H_1}. \quad . \quad . \quad . \quad . \quad . \quad . \quad . \quad . \quad . \quad . \quad p(120)$$

(a′) Change of Shape Due to Horizontal Loads
(N_x neglected).

From p(89, we have

$$\Delta\phi_0 = \frac{\Delta c}{l} + \frac{lf}{3A}\{H_1 - \overset{l}{\Sigma} Q[1 - 2k(2 - 5k + 5k^2) + 3k^4]\}. \quad p(121)$$

The coefficient of $\overset{l}{\Sigma} Q$ is tabulated in Table IV.

From $p(79)$,

$$\Delta x = - y\Delta\phi_0 - \frac{x^2}{6Ap}\left\{ V_1\frac{x}{4}(2l - 3x) - H_1\frac{x}{p}\left(\frac{l^2}{4} - \frac{lx}{2} + \frac{x^2}{5}\right) \right.$$
$$+ \overset{x}{\Sigma} Q\frac{1}{10x^2p}\left[2x^5 - 5lx^4 + 10\left(\frac{l^2}{4} + 2pb\right)x^3 + 10\left(\frac{la^3}{2} - \frac{3a^2l^2}{4}\right)x\right.$$
$$\left.\left. - 15pbl(x - a)^2 + 3a^5 - \tfrac{15}{2}a^4l + 10(\tfrac{1}{3}l^2 + pb)a^3\right]\right\}. \quad p(122)$$

$+ 15\left(\frac{a^2l}{2}\right.$ — $30p$ … ^{2}

From $p(84)$,

$$\Delta y = x\Delta\phi^0 + \frac{x^2}{6A}\Big\{ V_1 x - H_1\frac{x}{4p}(2l - x)$$

$$+\frac{1}{px^2}\sum^{x} Q\Big[-\frac{x^4}{4}+\frac{lx^3}{2}+(a^3 - \tfrac{3}{2}a^2l)x - 3pb(x-a)^2$$

$$-\frac{3a^4}{4}+a^3l\Big]\Big\}, \quad . \quad . \quad . \quad . \quad . \quad . \quad . \quad . \quad . \quad . \quad . \quad p(123)$$

in which

$$H_1 = \sum^{l} Q\Big\{1 - \frac{k}{2}[5(1-k-2k^2+4k^3) - 8k^4]\Big\} \quad p(111)$$

and

$$V_1 = 4k(1-k)n. \quad . \quad . \quad . \quad . \quad . \quad . \quad . \quad . \quad . \quad . \quad . \quad . \quad p(113)$$

(*e*) *Horizontal Loads, Effect of the Axial Stress included.*

From $p(88)$,

$$H_1 = \frac{\sum^{l} Q}{B}\Big\{\frac{4}{3}\,\frac{a(l-a)}{l}f^2 - \frac{8}{15}\frac{f^2}{l^4}\Big[\begin{matrix}-l^5+5al^4-5a^2l^3-5a^3l^2\\ +10a^4l-4a^5\end{matrix}\Big]$$

$$+\,mp(\alpha+\phi_0)\Big\} \quad . \quad . \quad . \quad . \quad . \quad . \quad . \quad . \quad . \quad . \quad . \quad . \quad p(124)$$

or

$$H_1 = \frac{4lf^2}{15B}\sum^{l} Q\{2-5k(1-k-2k^2+4k^3)+8k^4\} + \sum^{l} Q\frac{mp(\alpha+\phi_0)}{B} \quad p(125)$$

where

$$\frac{1}{B} = \frac{15}{8lf^2 + 30mp\phi_0}. \quad . \quad . \quad . \quad . \quad . \quad p(126)$$

Here we see that the effect of the axial stress is small. If $30mp\phi_0$ is neglected in $\frac{1}{B}$, the first term of the second member of $p(125)$ at once reduces to $p(111)$.

The expression $\{2 - 5k(1 - k - 2k^2 + 4k^3) + 8k^4\}$ may be written $2\left(1 - \frac{k}{2}[5(1 - k - 2k^2 + 4k^3) - 8k^4]\right)$, and hence its value quickly determined from Table III.

The value of V_1 is not affected by N_x; hence

$$V_1 = 4k(1 - k)n. \quad . \quad . \quad . \quad . \quad . \quad p(113)$$

For any particular load, from (58),

$$x_0 = \frac{H_1}{V_1}b. \quad . \quad . \quad . \quad . \quad . \quad . \quad . \quad . \quad . \quad . \quad (58)$$

The values of H_x, V_x, N_x, and T_x are given by (39), (40), (42), and (43).

$$M_x = V_1x - H_1y + \overset{x}{\Sigma}Q(y - b). \quad . \quad . \quad p(127)$$

(*f*) *Change of Shape due to the Action of Horizontal Loads, with Effect of Axial Stress included.*

The values of $\Delta\phi_0$, Δx, and Δy can be found from $p(89)$, $p(79)$, and $p(84)$ respectively.

(*g*) *Temperature.*

A change in temperature is equivalent to applying a certain horizontal load at the hinges; or, from $p(88)$,

$$H_1 = \frac{15A}{8f^2}et^\circ \quad . \quad . \quad . \quad . \quad . \quad . \quad . \quad . \quad . \quad p(128)$$

if the axial stress is neglected, and

$$H_1 = \frac{60Af}{32f^2 + 15ml\phi_0}et^\circ \quad . \quad . \quad . \quad . \quad p(129)$$

if the effect of the axial stress is included.

A rise in temperature creates a reaction H_1 acting from the left towards the right.

The values of H_x, V_x, M_x, N_x, and T_x can be found from (39), (40), (41), (42), and (43).

(*h*) *Change of Length in the Span.*

From p(88),

$$H_1 = \frac{15A}{8f^2}\frac{\Delta l}{l}, \quad . \quad . \quad . \quad . \quad . \quad . \quad . \quad . \quad . \qquad p(130)$$

neglecting the axial stress; or

$$H_1 = \frac{60A}{32f^2 + 15ml\phi_0}\frac{f}{l}\Delta l \quad . \quad . \quad . \quad . \quad . \qquad p(131)$$

if the axial stress is included.

If the span is shortened, H_1 acts from the left towards the right.

The values of H_x, V_x, etc., can be found from (39), (40), etc.

(*i*) *Sinking of a Support.*

In case one of the supports changes its elevation after the arch is in place, a slight change in the stresses may result from the effect of the change in the length of the span; but any change likely to occur may be neglected in the calculation of stresses.

(*j*) *Uniform Loads.*

Thus far we have considered only concentrated loads. If the load is uniformly distributed (horizontally),

$$\Sigma P = \int w da = wl \int dk,$$

where w represents the load per unit length of the span.

Let Fig. 27 represent an arch having a partial uniform load; then, from $p(91)$,

$$H_1 = \frac{1}{16n} wl \Big[_{k''}^{k'} k^2(5 - 5k^2 + 2k^3). \quad . \quad . \quad . \qquad p(132)$$

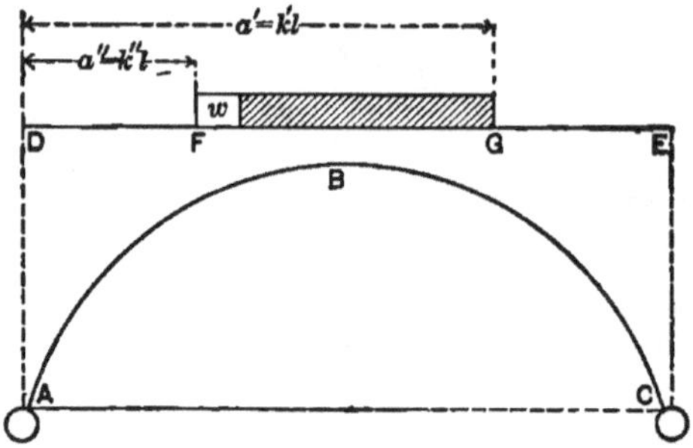

FIG. 27.

From $p(93)$,

$$V_1 = \frac{wl}{2} \Big[_{k''}^{k'} k(2 - k). \quad . \quad . \quad . \quad . \quad . \quad . \quad . \qquad p(133)$$

From $p(97)$,

$$M_x = \frac{wl}{2} \Big[_{k''}^{k'} \left\{ xk(2 - k) - \frac{y}{8n} k^2(5 - 5k^2 + 2k^3) \right\}$$

$$- \frac{w}{2} \Big[_{a'' \leqq x}^{a' \leqq x} (x - a)^2 \Big\} . \quad . \quad . \quad . \quad . \qquad p(134)$$

From $p(96)$,

$$T_x = \frac{wl}{2} \Big[_{k''}^{k'} k(2 - k) \cos \phi - \frac{1}{16n} wl \Big[_{k''}^{k'} k^2(5 - 5k^2 + 2k^3) \sin \phi$$

$$- \left(wl(k' - k'') \cos \phi, \quad \text{where} \quad k' \leqq \frac{x}{l} \right). \qquad p(135)$$

The above equations enable us to determine all the stresses in the arch when the axial stress is neglected.

(k) Uniform Load Over All.

In case the load is distributed horizontally and uniformly over the entire span, then $k'' = 0$ and $k' = 1$ or x/l, and we have, from $p(132)$,

$$H_1 = \frac{1}{8n} wl. \qquad p(136)$$

If $n = 0$, $\qquad H_1 = 0.$

From $p(133)$,

$$V = \frac{wl}{2}. \qquad p(137)$$

From $p(134)$,

$$M_x = \frac{wl}{2}x - \frac{wl}{2}\frac{y}{4n} - \frac{w}{2}x^2 = 0. \qquad p(138)$$

If $n = 0$, then $y = 0$, and

$$M_x = \frac{wl}{2}x - \frac{wx^2}{2},$$

the expression for the bending-moment in the ordinary straight girder.

From $p(135)$,

$$T_x = \frac{wl}{2} \cos \phi - \frac{1}{8n} wl \sin \phi - wx \cos \phi. \qquad p(139)$$

If $n = 0$, ϕ becomes zero, since our radius is now infinity; hence

$$T_x = \frac{wl}{2} - wx,$$

the expression for shear in the ordinary straight girder.

SYMMETRICAL PARABOLIC ARCH WITHOUT HINGES.

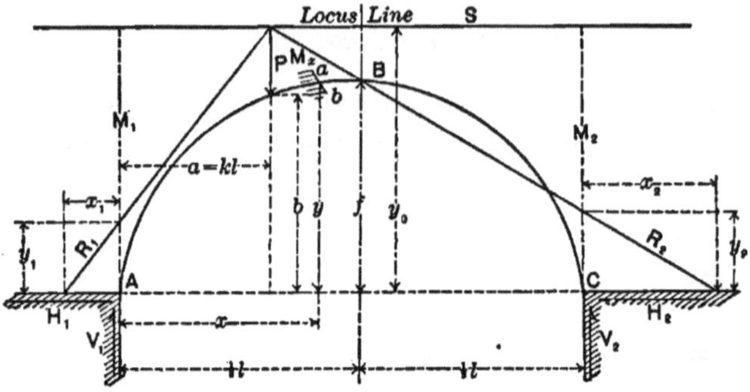

FIG. 28.

In this case we have several conditions which must be satisfied. As in the case of the arch with two hinges, we shall consider the various loadings, etc., independently.

(*a*) *Vertical Loads* (*N_x neglected*).

From $p(87)$, by transposition,

$$M_1 + M_2 = H_1 \tfrac{4}{3} f - \frac{2A}{l}(\Delta\phi_0 - \Delta\phi_l) - \frac{1}{l}\overset{l}{\Sigma} Pa(l-a); \qquad p(140)$$

and from $p(88)$,

$$M_1 + M_2 = H_1 \tfrac{8}{15} f + \frac{3A}{lf}(\Delta l - et^\circ l) - \frac{1}{l^3}\overset{l}{\Sigma} Pa(l-a)(l^2 + al - a^2). \qquad p(141)$$

Now if there are no hinges the ends of the arch must be fixed in direction, and $\phi_0 = \phi_l$ cannot change under any condition of loading.

If the length of the span be assumed unchanged and the effect of temperature omitted, we have, by combining $p(140)$ and $p(141)$ and reducing,

$$H_1 = \frac{1}{4lf^2}\left\{\frac{15f}{n}\sum^{l} Pa^2(l-a)^2\right\} \quad . \quad . \quad . \quad p(142)$$

or

$$H_1 = \frac{15}{4n}\sum Pk^2(1-k)^2 \quad . \quad . \quad . \quad . \quad . \quad . \quad . \quad . \quad p(143)$$

The values of $k^2(1-k)^2$ are given in Table XI for values of k from 0 to 1.00 inclusive.

From $p(89)$, assuming that Δc and $\Delta\phi_0$ are zero,

$$M_1 + \tfrac{1}{2}M_2 = H_1 f - \frac{1}{2l^2}\sum^{l} P(l-a)(2l-a)a. \quad . \quad p(144)$$

Combining $p(141)$ and $p(144)$, assuming Δl and $et^\circ l$ to be zero, we obtain by reduction

$$M_2 = H_1\tfrac{8}{15}f + \frac{2}{l^3}\sum^{l} P(l-a)a\left[\frac{l(2l-a)}{2} - l^2 - al + a^2\right] \quad p(145)$$

or

$$M_2 = H_1\tfrac{8}{15}nl + l\sum^{l} P(5k^3 - 3k^2 - 2k^4). \quad . \quad . \quad . \quad . \quad . \quad . \quad p(146)$$

Substituting the value of H_1 from $p(143)$, we have

$$M_2 = \frac{l}{2}\sum^{l} Pk^2(1-k)(3-5k). \quad . \quad . \quad . \quad p(147)$$

The values of $k^2(1-k)(3-5k)$ are given in Table VI for values of k from 0 to 1.00 inclusive.

Substituting $(1-k)$ for k in $p(147)$, we have

$$M_1 = \frac{l}{2}\sum^{l} Pk(1-k)^2(5k-2). \quad . \quad . \quad . \quad p(148)$$

The values of $k(1-k)^2(5k-2)$ are given in Table VI for values of k from 0 to 1.00 inclusive, reading $(1-k)$ for k.

From (47),

$$V_1 = \frac{1}{l}\left\{M_2 - M_1 + \overset{l}{\Sigma} Pl(1 - k)\right\}. \quad . \quad . \quad p(149)$$

Substituting $p(147)$ and $p(148)$ in $p(149)$, we obtain by reduction

$$V_1 = \overset{l}{\Sigma} P(1 - k)^2(1 + 2k), \quad . \quad . \quad . \quad . \quad . \quad p(150)$$

$$V_2 = \overset{l}{\Sigma} P - V_1.$$

The values of $(1 - k)^2(1 + 2k)$ are given in Table VII for values of k from 0 to 1.00 inclusive.

The above equations completely determine all of the external forces. The stresses at any section of the arch can be determined from equations (39) to (43) inclusive, remembering that the terms containing Q disappear, as we are not considering the horizontal components or loads.

The values of H_1, V_1, and R_1 can be found graphically after the ordinates y_1, y_0, and y_2 are determined.

From (50), (51), and (52),

$$y_0 = \frac{M_1 + V_1 a}{H_1}, \quad y_1 = \frac{M_1}{H_1}, \quad \text{and} \quad y_2 = \frac{M_2}{H_2}$$

Substituting the values of H_1, V_1, etc., given above, we have

$$y_0 = \frac{6}{5} f; \qquad \text{(positive upward)} \quad . \quad . \quad p(151)$$

$$y_1 = \frac{6}{5} \frac{5k - 2}{9k} f; \quad . \quad . \quad . \quad . \quad . \quad . \quad . \quad . \quad . \quad . \quad p(152)$$

$$y_2 = \frac{6}{5} \frac{3 - 5k}{9(1 - k)} f; \quad . \quad . \quad . \quad . \quad . \quad . \quad . \quad . \quad . \quad p(153)$$

which completely determines the position of the equilibrium polygon for any vertical load.

When y_1 or y_2 become very large it is more convenient to use the abscissas x_1 and x_2.

From (54) and (55),

$$x_1 = -\frac{k(5k-2)}{2(1+2k)}l. \quad . \quad . \quad . \quad . \quad . \quad p(154)$$

x_1 is negative when measured towards the left.

$$x_2 = -\frac{(1 \quad k)(3-5k)}{2(3-2k)}l. \quad . \quad . \quad . \quad p(155)$$

x_2 is negative when measured towards the right.

The coefficients in $p(152)$ and $p(153)$ are tabulated in Table VIII, and those in $p(154)$ and $p(155)$ in Table IX.

(*b*) *Change of Shape due to the Action of Vertical Loads (N$_x$ neglected).*

From $p(89)$,

$$\Delta\phi_0 = \frac{\Delta c}{l} - \frac{l}{3A}\left\{M_1 + \frac{1}{2}M_2 - H_1 f + \frac{l}{2}\overset{x}{\Sigma}Pk(2-3k+k\ )\right\},\ p(156)$$

where M_1, M_2, and H_1 are to be found from $p(148)$, $p(147)$, and $p(143)$ respectively.

The values of $k(2-3k+k^2) = 2k - 3k^2 + k^3$ are given in Table X for values of k from 0 to 1.00 inclusive.

From $p(69)$,

$$\Delta\phi = \Delta\phi_0 + \frac{1}{2A}\left\{2M_1x + V_1x^2 - H_1\frac{3l-2x}{6p}x^2 - \overset{x}{\Sigma}P(x-a)^2\right\} p(157)$$

in which M_1, V_1, and H_1 are to be found from $p(148)$, $p(150)$, and $p(143)$ respectively.

From $p(79)$,

$$\Delta x = -y\Delta\phi_0 - \frac{x^2}{6Ap}\Big\{M_1\Big(\frac{3}{2}l - 2x\Big) + V_1\frac{x}{4}(2l - 3x)$$

$$- H_1\frac{x}{p}\Big(\frac{l^2}{4} - \frac{lx}{2} + \frac{x^2}{5}\Big)$$

$$- \sum^{x} P\frac{1}{4x^2}\big((2l - 3x - a)(x - a)^3\big)\Big\}, \quad . \quad . \quad . \quad p(158)$$

where M_1, V_1, and H_1 are to be found from $p(148)$, $p(150)$, and $p(143)$ respectively.

From $p(84)$,

$$\Delta y = x\Delta\phi_0 + \frac{x^2}{6A}\Big\{3M_1 + V_1x - H_1\frac{x}{4p}(2l - x) - \frac{1}{x^2}\sum^{x} P(x - a)^3\Big\}, \qquad p(159)$$

where M_1, V_1, and H_1 are to be found from $p(148)$, $p(150)$, and $p(143)$ respectively.

These four equations completely determine the change of shape due to the action of any vertical load.

(*c*) *Vertical Loads, with Effect of Axial Stress included.*

From $p(87)$,

$$M_1 + M_2 = H_1\tfrac{4}{3}f - \frac{1}{l}\Sigma Pa(l - a). \quad . \quad . \quad . \quad . \quad . \quad . \quad . \quad . \quad p(160)$$

From $p(88)$,

$$M_1 + M_2 = H_1\tfrac{8}{15}f - \frac{1}{l^3}\sum^{l} P(l - a)(l^2 + al - a^2)a$$

$$+ H_1\frac{6mp\phi_0}{fl} + \frac{3m}{2lf(p + 2f)}\sum^{l} Pa(l - a). \quad . \quad . \quad . \quad p(161)$$

Subtracting $p(160)$ from $p(161)$ and solving for H_1, we have, by reduction,

$$H_1 = C\left\{ l\Sigma Pk^2(1-k^2) - \frac{3lm}{2f(p+2f)}\sum^{l} Pk(1-k) \right\}, \quad p(162)$$

where

$$C = \frac{15lf}{4lf^2 + 90mp\phi_0}.$$

The values of $k^2(1-k)^2$ and $k(1-k)$ are given in Tables XI and V.

We have, approximately,

$$H_1 = \mathfrak{H}(1-\epsilon'),^* \quad . \quad . \quad . \quad . \quad . \quad . \quad p(162a)$$

where $\mathfrak{H} = H_1$ in $t(143)$.

From $p(89)$,

$$M_1 + \tfrac{1}{2}M_2 - \frac{3m}{l^2}\left(1 - 2\phi_0\frac{p}{l}\right)(M_2 - M_1) = H_1 f$$

$$-\frac{1}{2l^2}\sum^{l} Pa(l-a)(2l-a) + \frac{3mp}{l^2}\sum^{l} P\left(\frac{2a\phi_0}{l} + \alpha - \phi_0\right). \quad p(163)$$

The first member of p(~~148~~ 163) may be written

$$M_1\left(1 + \frac{3m}{l^2} - \frac{6mp\phi_0}{l^3}\right) + M_2\left(\frac{1}{2} - \frac{3m}{l^2} + \frac{6mp\phi_0}{l^3}\right);$$

or if

$$1 + \frac{3m}{l^2} - \frac{6mp\phi_0}{l^3} = D, \quad . \quad . \quad . \quad . \quad . \quad . \quad (m)$$

$$M_1 D + M_2\left(\tfrac{3}{2} - D\right), \quad . \quad . \quad . \quad . \quad . \quad . \quad p(163a)$$

multiplying $p(161)$ by D, we have

---

* Appendix C.

$$M.D + M_2D = H_1\frac{8D}{5}f - \frac{D}{l^3}\overset{l}{\Sigma}P(l-a)(l^2+al-a^2)a$$

$$+ H_1\frac{6mp\phi_0}{fl}D + \frac{3mD}{2lf(p+2f)}\overset{l}{\Sigma}Pa(l-a). \quad . \quad p(163b)$$

Eliminating M_1 from $p(163)$ and $p(163b)$,

$$M_2(2D - \tfrac{3}{2}) = H_1\left\{\frac{8D}{5}f - f + \frac{6mp\phi_0}{fl}D\right\}$$

$$- lD\overset{l}{\Sigma}Pk(1-2k^2+k^3) + \frac{l}{2}\overset{l}{\Sigma}P(2k-3k^2+k^3)$$

$$+ \frac{3mDl}{2f(p+2f)}\overset{l}{\Sigma}Pk(1-k) - \frac{3mp}{l^2}\overset{l}{\Sigma}P[\phi_0(2k-1)+\alpha], \quad p(164)$$

in which H_1 is to be found from $p(162)$ and D from (m). For values of $k(1-2k^2+k^3)$, see Table I; $2k-3k^2+k^3$, see Table X; and for $k(1-k)$ see Table V.

It is to be noticed that D and the coefficients containing m are constant for any particular arch, hence by the aid of the tables, $p(164)$ can be evaluated very rapidly.

The value of M_1 can be obtained from $p(168)$ by taking everywhere $(1-k)$ for k, or by first computing the value of M_2 from $p(164)$ and substituting in $p(161)$.

The value of V_1 can be found from

$$V_1 = \frac{1}{l}\{M_2 - M_1 + \overset{l}{\Sigma}Pl(1-k)\}. \quad . \quad . \quad p(149)$$

The stresses at any point of the arch can now be determined from (39) to (43) inclusive, remembering that all terms containing Q *disappear*.

The values of y_0, y_1, and y_2 can be found from (50), (51), and (52), if graphics is employed in determining the intermediate moments and shears.

(*d*) *Change of Shape due to Vertical Loads, including Effect of Axial Stress.*

$\Delta\phi$, Δx, and Δy can be found from $p(69)$, $p(79)$, and $p(84)$ respectively, remembering that all terms containing Q disappear, and that Δc, Δl, and $\Delta\phi_0$ are zero.

(*e*) *Horizontal Loads* (N_x *neglected*).

From $p(87)$,

$$M_1 + M_2 = H_1 \frac{4}{3} f - \frac{4}{l^2} f \sum^{l} Qa(l - a$$

$$- \frac{4f}{3l^3} \sum^{l} Q(l^3 + 9a^2 l - 6al^2 - 4a^3). \quad . \quad . \quad . \quad . \quad p(165)$$

From $p(88)$,

$$M_1 + M_2 = H_1 \frac{8}{5} f - \frac{4f}{l^2} \sum^{l} Qa(l - a)$$

$$+ \frac{8f}{5l^5} \sum^{l} Q(-l^5 + 5al^4 - 5a^2 l^3 - 5a^3 l^2 + 10a^4 l - 4a^5). \; p(166)$$

Equating $p(165)$ and $p(166)$, and solving for H_1, we have

$$H_1 = \sum^{l} Q\left[1 + \frac{a^2}{l^5}(-15l^3 + 50al^2 - 60a^2 l + 24a^3)\right] \quad . \quad p(167)$$

or

$$H_1 = \sum^{l} Q[1 + k^2(-15 + 50k - 60k^2 + 24k^3)] \quad . \quad . \quad . \quad p(168)$$

in which the quantity in [ ] is tabulated in Table XII.

Eliminating M_2 from $p(88)$ and $p(89)$, and solving for M_1, we obtain

$$M_1 = \frac{f}{l^6}\overset{l}{\Sigma} Q\{4al^4 - 22a^2l^3 + 48a^3l^2 - 46a^4l + 16a^5\} \quad . \quad p(169)$$

or

$$M_1 = +\frac{f}{l^6}\overset{l}{\Sigma} Q\{2a(l^2 - 2al + a^2)(2l^2 - 7al + 8a^2)\} \quad . \quad . \quad p(170)$$

and

$$M_1 = +f\overset{l}{\Sigma} Q\{2k(1 - k)^2(2 - 7k + 8k^2)\}, \quad . \quad . \quad . \quad . \quad . \quad p(171)$$

where the expression in { } is tabulated in Table XIII.

Substituting $(1 - k)$ for k in $p(171)$, and changing signs

$$M_2 = -f\overset{l}{\Sigma} Q\{2k^2(1 - k)(3 - 9k + 8k^2)\}, \quad . \quad . \quad . \quad . \quad . \quad p(172)$$

where the expression in { } is tabulated in Table XIII, reading $(1 - k)$ for k.

From (47),

$$V_1 = \frac{1}{l}(M_2 - M_1 + \overset{l}{\Sigma} Qb). \quad . \quad . \quad . \quad p(173)$$

$$b = \frac{4al - 4a^2}{l^2}f. \quad . \quad . \quad . \quad . \quad . \quad . \quad p(174)$$

Substituting $p(174)$, $p(172)$, and $p(171)$ in $p(173)$,

$$V_1 = \frac{12f}{l}\overset{l}{\Sigma} Q(1 - k)^2k^2 \quad . \quad . \quad . \quad . \quad p(175)$$

or

$$V_1 = 12n\overset{l}{\Sigma} Q(k - k^2)^2. \quad . \quad . \quad . \quad . \quad . \quad p(176)$$

The values of $(k - k^2)^2$ can be found from Table XI.

The above equations completely determine the external

forces. The stresses at any point of the arch can be found with the aid of (39) to (43) inclusive.

The method of graphics may be employed after we have found the values of y_1, y_2, x_1, x_2, and x_0 in determining the external forces.

~~From (50), (51), and (52)~~,

$$y_0 = b, \quad \ldots \ldots \quad p(177)$$

from (51) and (52)

$$y_1 = + \frac{[2k(1-k)^2(2-7k+8k^2)]}{1+k^2[-15+50k-60k^2+24k^3]}f, \quad p(178)$$

and

$$y_2 = \frac{+2(1-k)^2(3-9k+8k^2)}{15-50k+60k^2-24k^3}f. \quad \ldots \ldots \quad p(179)$$

y_1 and y_2 are always measured upward.

The coefficients of f in $p(178)$ and $p(179)$ are tabulated in Table XIV.

From (54) and (55),

$$x_1 = \frac{2-7k+8k^2}{6k}l; \quad \ldots \ldots \quad p(180)$$

$$x_2 = \frac{3-9k+8k^2}{6(1-k)}l. \quad \ldots \ldots \quad p(181)$$

x_1 is always measured towards the left and x_2 towards the right.

The coefficients of l in $p(180)$ and $p(181)$ are given in Table XV.

From Fig. 29,

$$V x_0 = -M_1 + H_1 b, \quad \text{or} \quad x_0 = +\frac{H_1}{V_1}b - \frac{M_1}{V_1}. \quad p(182)$$

Substituting the values of H_1, V_1, and b from $p(168)$, $p(176)$, and $p(174)$, we have

$$x_0 = \frac{l}{2}(3 - 12k + 24k^2 - 16k^3). \quad . \quad . \quad . \qquad p(183)$$

x_0 is always measured from left to right.

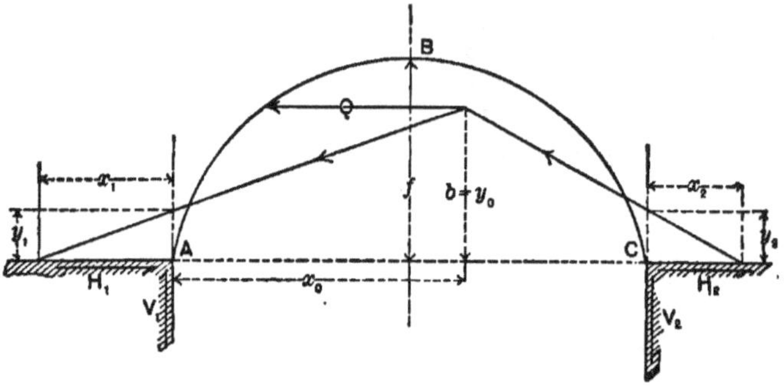

FIG. 29.

The values of $3 - 12k + 24k^2 - 16k^3$ are given in Table XVI.

(f) Change of Shape due to Horizontal Loads (N_x neglected).

From $p(69)$, since $\Delta\phi_0 = 0$,

$$\Delta\phi = \frac{1}{2A}\Big\{2M_1x + V_1x^2 - H_1\frac{3l - 2x}{6p}x^3$$

$$+ \sum^{l} Q\frac{1}{3p}\Big(\frac{3}{2}l(x^2 - a^2) - x^3 + a^3 - 6pb(x - a)\Big)\Big\}. \quad . \qquad p(184)$$

From $p(79)$,

$$\Delta x = -\frac{x^2}{6Ap}\Big\{M_1\Big(\frac{3}{2}l - 2x\Big) + V_1x\Big(\frac{l}{2} - \frac{3x}{4}\Big)$$

$$-H_1\frac{x}{p}\Big(\frac{l^2}{4} - \frac{lx}{2} + \frac{x^2}{5}\Big) + \sum^{x} Q\frac{1}{10x^2p}\Big[2x^6 - 5lx^4 + 10\Big(\frac{l^2}{4} + 2pb\Big)x^3$$

$$+\left(\frac{15}{2}la^2-30pba-5a^3\right)x^2+10\left(\frac{l}{2}a^3-\frac{3}{4}l^2a^2\right)x-15pbl(x-a)^2$$

$$+3a^5-\frac{15}{2}a^4l+(5l^2+10pb)a^3\Bigg]\Bigg\}. \quad . \quad . \quad . \quad . \quad p(185)$$

From $p(84)$,

$$\Delta y=\frac{x^2}{6A}\Bigg\{3M_1+V_1x-H_1\frac{x}{p}\left(\frac{l}{2}-\frac{x}{4}\right)+\frac{1}{px^2}\overset{x}{\Sigma}Q\Bigg[-\frac{x^4}{4}+\frac{1}{2}lx^3$$

$$+\left(a^3-\frac{3}{2}la^2\right)x-3pb(x-a)^2-\frac{3a^4}{4}+la^3\Bigg]\Bigg\}, \quad . \quad p(186)$$

where M_1, V_1, and H_1 are given by $p(171)$, $p(176)$, and $p(168)$ respectively.

(*g*) *Horizontal Loads, with Effect of Axial Stress included.*

From $p(87)$ and $p(88)$, in a manner similar to that employed for vertical loads, we have

$$H_1=C\Bigg\{\frac{4f}{15}\overset{l}{\Sigma}Q(1+k^2[-15+50k-60k^2+24k^3])$$

$$+\frac{3mp}{lf}\overset{l}{\Sigma}Q(\alpha+\phi_0)\Bigg\}, \quad . \quad . \quad . \quad . \quad . \quad . \quad . \quad . \quad . \quad p(187)$$

where

$$C=\frac{15lf}{4lf^2+90mp\phi_0}.$$

The values of

$$1+k^2(-15+50k-60k^2+24k^3) \quad . \quad . \quad p(188)$$

are given in Table XII.

From $p(88)$ and $p(89)$,

$$M_2\left(2D - \frac{3}{2}\right) = H_1\left\{\frac{8D}{5}f - f + \frac{6mp\phi_0}{fl}D\right\}$$

$$+ f\sum^{l} Q(1 - 4k + 10k^2 - 10k^3 + 3k^4)$$

$$- \frac{8}{5}fD\sum^{l} Q\left(1 - \frac{5}{2}k + \frac{5}{2}k^2 + 5k^3 - 10k^4 + 4k^5\right)$$

$$- \frac{3m}{2f}\left(\frac{p}{p + 2f} + 1 - \frac{2p\phi_0}{l}\right)\sum^{l} Qk(1 - k)$$

$$- \frac{3mpD}{lf}\sum^{l} Q(\alpha + \phi_0), \quad . \quad . \quad . \quad . \quad . \quad . \quad . \quad p(189)$$

from which the value of M_2 can be found.

$$(1 - 4k + 10k^2 - 10k^3 + 3k^4) = 1 - 2k(2 - 5k + 5k^2) + 3k^4,$$

and the values of this expression are given in Table IV.

$$\left(1 - \frac{5}{2}k + \frac{5}{2}k^2 + 5k^3 - 10k^4 + 4k^5\right)$$

$$= 1 - \frac{k}{2}[5(1 - k - 2k^2 + 4k^3) - 8k^4],$$

and the values of this expression are tabulated in Table III.

In $p(189)$,

$$D = 1 + \frac{3m}{l^2} - \frac{6mp\phi_0}{l^3}, \quad . \quad . \quad . \quad . \quad . \quad . \quad (m)$$

and H_1 is given by $p(187)$.

The values of $k(1 - k)$ are given in Table V.

It is to be noticed that D and the coefficients containing m are constant for any particular arch; hence by the aid of the tables, $p(189)$ can be readily evaluated.

The value of M_1 can be found from $p(189)$ by taking everywhere $(1 - k)$ for k.

The value of V_1 is found from

$$V_1 = \frac{1}{l}(M_2 - M_1 + \overset{l}{\Sigma} Qb). \quad . \quad . \quad . \quad p(149)$$

The stresses at any point of the arch can now be determined from (39) to (43) inclusive, remembering that all terms containing ΣP *disappear.*

The values of y_0, y_1, etc., can be found from (50), (51), and (52), if graphics is employed in determining the intermediate stresses

(h) Change of Shape due to Horizontal Loads, including Effect of Axial Stress.

$\Delta\phi$, Δx, and Δy can be found from, $p(69)$, $p(79)$ and $p(84)$ respectively, remembering that all terms containing ΣP disappear, and that Δc, Δl, and $\Delta\phi_0 = 0$.

(i) Temperature.

From $p(87)$ and $p(88)$,

$$H_t = \frac{15}{4lf^2 + 90mp\phi_0} 3Aet^\circ l, \quad . \quad . \quad . \quad p(190)$$

where the term containing m shows the effect of the axial stress.

If this be omitted,

$$H_1 = \frac{45A}{4f^2} et^\circ \quad (\textit{axial stress neglected}). \quad . \quad . \quad p(191)$$

Substituting $p(190)$ in $p(88)$ and $p(89)$, and solving for M_2,

letting $$D = 1 + \frac{3m}{l^2} - \frac{6mp\phi_0}{l^2},$$

$$M_2\left(2D - \frac{3}{2}\right) = \frac{45Alet^\circ}{4lf^2 + 90mp\phi_0}\left\{\frac{8D}{5}f - f + \frac{6mp\phi_0}{fl}D\right\}$$

$$- \frac{3AD}{f}et^\circ, \quad . \quad . \quad . \quad . \quad . \quad . \quad . \quad . \quad . \quad . \quad . \quad . \quad p(192)$$

If the axial stress be neglected, D becomes unity, and the terms containing m *disappear.*

$$M_2 = \frac{15A}{2f}et^\circ = M_1 \quad (\textit{axial stress neglected}) \quad . \quad p(193)$$

or

$$M_2 = H_1 . \tfrac{2}{3}f \quad . \quad . \quad . \quad . \quad . \quad . \quad . \quad . \quad . \quad . \quad . \quad . \quad p(194)$$

and

$$y_1 = y_2 = y_0 = \tfrac{2}{3}f. \quad . \quad . \quad . \quad . \quad . \quad . \quad . \quad . \quad . \quad . \quad . \quad p(195)$$

$$V_1 = 0 = V_2. \quad . \quad . \quad . \quad . \quad . \quad . \quad . \quad . \quad . \quad . \quad . \quad . \quad p(196)$$

The intermediate stresses, etc., can be found from the general equations (39) to (43) inclusive.

(j) *Effect of a Change Δl in the Length of the Span.*

An inspection of the general equations $p(87)$, $p(88)$, and $p(89)$ shows that $-\frac{\Delta l}{l}$ follows the same law as $+ et^\circ$. Hence

$$H_1 = -\frac{45A}{4lf^2 + 90mp\phi_0}\Delta l \quad (\textit{including axial stress}) \quad p(197)$$

and

$$H_1 = -\frac{45A}{4lf^2}\Delta l, \qquad (\textit{neglecting axial stress}) \qquad p(198)$$

also

$$M_2(2D - \tfrac{3}{2}) = -\frac{45A}{4lf^2 + 90mp\phi_0}\left\{\frac{8D}{5}f - f + \frac{6mp\phi_0}{fl}D\right\}\Delta l$$

$$+\frac{3AD}{fl}\Delta l \quad . \quad . \quad . \quad . \quad . \quad . \quad . \quad . \quad p(199)$$

or, if the axial stress be neglected,

$$M_2 = -\frac{15A}{2lf}\Delta l = M_1 = \tfrac{2}{3}H_1 f, \quad . \quad . \quad . \quad . \quad . \quad p(200)$$

where

$$D = 1 + \frac{3m}{l^2} - \frac{6mp\phi_0}{l^3}.$$

The intermediate stresses can be found from (39) to (43) inclusive.

(k) Effect of any Change in ϕ_0, ϕ_l, and the Relative Positions of the Supports in Elevation.

From $p(87)$ and $p(88)$,

$$H_1 = \frac{-15lf}{4lf^2 + 90mp\phi_0}\left\{\frac{2A}{l}(\Delta\phi_l - \Delta\phi_0)\right\}; \qquad p(201)$$

or, if the axial stress be neglected,

$$H_1 = -\frac{15A}{2lf}(\Delta\phi_l - \Delta\phi_0). \quad . \quad . \quad . \quad . \quad . \quad . \quad . \quad p(202)$$

From $p(88)$ and $p(89)$, by substituting $p(201)$,

$$M_2(2D - \tfrac{3}{2}) = -\frac{3A}{l^2}(\Delta c - l\Delta\phi_0)$$

$$+\frac{15lf}{4lf^2+90mp\phi_0}\left\{\frac{2A}{l}(\Delta\phi_l - \Delta\phi_0)\right\}\left\{\frac{8D}{5}f - f + \frac{6mp\phi_0}{fl}D\right\}. \; p(203)$$

If the effect of the axial stress be neglected,

$$M_1 = -\frac{3A}{l}\left\{-\frac{2\Delta c}{l} + 3\Delta\phi_0 - \Delta\phi_l\right\}; \quad . \quad p(204)$$

and

$$M_2 = -\frac{3A}{l}\left\{+\frac{2\Delta c}{l} - 3\Delta\phi_l + \Delta\phi_0\right\}. \quad . \quad . \quad p(205)$$

(*l*) *Uniform Loads.*

Using the same nomenclature as employed in discussing this case for the two-hinged arch, we have, from $p(143)$,

$$H_1 = \frac{wl}{8n}\left[k^2(10 - 15k + 6k^2)\right]_{k''}^{k'}. \quad . \quad . \quad . \quad . \quad p(206)$$

From $p(147)$,

$$M_2 = \frac{wl^2}{2}\left[k^2(1 - 2k + k^2)\right]_{k''}^{k'}. \quad . \quad . \quad . \quad . \quad . \quad p(207)$$

From $p(148)$,

$$M_1 = \frac{wl^2}{2}\left[k^2(-1 + 3k - 3k^2 + k^3)\right]_{k''}^{k'}. \quad . \quad p(208)$$

From $p(150)$,

$$V_1 = \frac{wl}{2}\left[k(2 - 2k^2 + k^3)\right]_{k''}^{k'}. \quad . \quad . \quad . \quad . \quad . \quad p(209)$$

From (41),

$$M_x = M_1 + V_1x - H_1y - \frac{w}{2}\left[(x-a)^2\right]_{a''}^{a' \overline{\underline{<}}\, x/l}. \quad . \quad p(210)$$

(*m*) *Uniform Load Over All.*

Here $k'' = 0$ and $k' = 1$ or x/l.
From $p(206)$,

$$H_1 = \frac{wl}{8n}. \quad . \quad . \quad . \quad . \quad . \quad . \quad . \quad p(211)$$

From $p(207)$,

$$M_2 = 0. \quad . \quad . \quad . \quad . \quad . \quad . \quad . \quad . \quad p(212)$$

From $p(208)$,

$$M_1 = 0. \quad . \quad . \quad . \quad . \quad . \quad . \quad . \quad . \quad p(213)$$

From $p(209)$,

$$V_1 = \frac{wl}{2}. \quad . \quad . \quad . \quad . \quad . \quad . \quad . \quad p(214)$$

From $p(210)$,

$$M_x = \frac{wl}{2}x - \frac{wl}{8n}y - \frac{wx^2}{2} = 0. \quad . \quad . \quad . \quad . \quad p(215)$$

CHAPTER IV.

CIRCULAR ARCHES HAVING $\frac{2E\theta}{R}$ = A CONSTANT.

GENERAL RELATIONS.

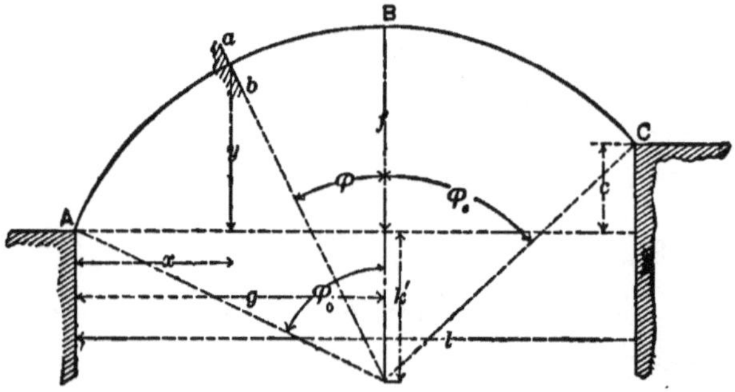

FIG. 30.

Let

$$A = \frac{2E\theta}{R} = \text{a constant}; \quad \dots \quad c(59)$$

$$m = \frac{\theta}{FR}; \quad \dots \quad c(60)$$

$$k' = R - f. \quad \dots \quad c(61)$$

Then from the equation of the circle

$$x = g - R\sin\phi = R(\sin\phi_0 - \sin\phi); \quad \dots \quad c(62)$$

$$y = R\cos\phi - k' = R(\cos\phi - \cos\phi_0); \quad \dots \quad c(63)$$

$$R^2 = (g - x)^2 + (k' + y)^2; \quad \dots \quad c(64)$$

$$g^2 - 2Rf + f^2 = 0; \quad \dots \quad c(65)$$

$$\sin\phi = \frac{g-x}{R}, \quad \cos\phi = \frac{k'+y}{R}; \quad . \quad . \quad . \quad . \quad . \qquad c(66)$$

$$\tan\phi = \frac{g-x}{k'+y}. \quad . \quad . \quad . \quad . \quad . \quad . \quad . \quad . \quad . \quad . \quad . \qquad c(67)$$

From (d), for any point x

$$\Delta\phi = \Delta\phi_0 + \int_0^x \frac{M_x}{E\theta} ds, \quad . \quad . \quad . \quad . \quad . \quad . \quad . \quad . \qquad c(68)$$

since
$$\frac{ds}{d\phi} = -R, \quad ds = -Rd\phi; \quad . \quad . \quad . \quad . \quad . \quad . \quad . \qquad c(69)$$

hence

$$\int_0^x \frac{M_x}{E\theta} ds = -\frac{2}{A}\int_0^x M_x d\phi. \quad . \quad . \quad . \qquad c(70)$$

But from (41),

$$M_x = M_1 + V_1 x - H_1 y - \overset{x}{\Sigma} P(x-a) + \overset{x}{\Sigma} Q(y-b). \qquad (41)$$

Therefore

$$\int_0^x \frac{M_x}{E\theta} ds = -\frac{2}{A}\int_0^x \left\{ \begin{array}{c} M_1 + V_1 x - H_1 y - \overset{x}{\Sigma} P(x-a) \\ + \overset{x}{\Sigma} Q(y-b) \end{array} \right\} d\phi. \quad c(71)$$

The several integrals have the following forms:

$$\int_0^x M_1 d\phi = \int_{\phi_0}^{\phi} M_1 d\phi = M_1(\phi - \phi_0); \quad . \quad . \quad . \quad . \quad . \quad . \qquad c(72)$$

$$\int_0^x V_1 x d\phi = \int_{\phi_0}^{\phi} V_1 R(\sin\phi_0 - \sin\phi) d\phi$$
$$= V_1 g(\phi - \phi_0) + V_1 y \quad . \quad . \quad . \quad . \quad . \quad . \quad . \quad . \qquad c(73).$$

$$-\int_0^x H_1 y\,d\phi = \int_{\phi_0}^{\phi} H_1 R\cos\phi_0\,d\phi - \int_{\phi_0}^{\phi} H_1 R\cos\phi\,d\phi$$

$$= H_1 k'(\phi - \phi_0) + H_1 x; \quad . \quad . \quad . \quad . \quad . \quad . \quad . \qquad c(74)$$

$$\int_0^x \overset{x}{\Sigma} P(x - a)d\phi = \int_\alpha^\phi \overset{x}{\Sigma} P(g - R\sin\phi - a)d\phi$$

$$= \overset{x}{\Sigma} P(g - a)(\phi - \alpha) + \overset{x}{\Sigma} P(y - b); \qquad c(75)$$

$$\int_0^x \overset{x}{\Sigma} Q(y - b)d\phi = \int_\alpha^\phi \overset{x}{\Sigma} QR(\cos\phi - \cos\alpha)d\phi$$

$$= -\overset{x}{\Sigma} Q\{x - a - (\alpha - \phi)(b + k')\}. \qquad c(76)$$

Substituting the above values in $c(71)$ and $c(68)$,

$$\varDelta\phi = \varDelta\phi_0 + \frac{2}{A}\left\{\begin{array}{l} (M_1 + V_1 g + H_1 k')(\phi_0 - \phi) - V_1 y - H_1 x \\ + \overset{x}{\Sigma} P(y - b) - \overset{x}{\Sigma} P(g - a)(\alpha - \phi) \\ + \overset{x}{\Sigma} Q(x - a) - \overset{x}{\Sigma} Q(b + k')(\alpha - \phi) \end{array}\right\} \cdot c(77)$$

From (a),

$$\varDelta x = -\int_0^x dy\varDelta\phi + \int_0^x et^\circ dx - \int_0^x \frac{N_x}{EF_x}dx. \quad . \quad . \quad (a)$$

After substituting the value of $\varDelta\phi$ from $c(77)$ in (a), the following integrals will aid in the reduction of the term $\int_0^x dy\varDelta\phi$:

$$\int_0^x (\phi_0 - \phi)dy = \phi_0 y + \int_{\phi_0}^{\phi} R\phi\sin\phi d\phi$$

$$= (\phi_0 - \phi)(y + k') - x; \quad . \quad . \quad . \quad . \qquad c(78)$$

$$\int_0^x H_1 x dy = \int_{\phi_0}^{\phi} H_1 R(\sin\phi_0 - \sin\phi)\sin\phi d\phi$$

Therefore

$$\int_0^x H_1 x dy = + \frac{H_1}{2}\Big\{xy + k'x + gy - R^2(\phi_0 - \phi)\Big\}; \quad c(79)$$

$$\int_0^x \sum^x P(g-a)(\alpha-\phi)dy = -\int_\alpha^\phi \sum^x P(g-a)\alpha R \sin\phi d\phi$$

$$+ \int_\alpha^\phi \sum^x P(g-a)\phi R \sin\phi d\phi$$

$$= \sum^x P(g-a)\{(\alpha-\phi)(y+k') - (x-a)\}; \quad c(80)$$

$$\int_0^x \sum^x P(y-b)dy = \sum^x P\frac{(y-b)^2}{2};$$

$$\int_0^x \sum^x Q(x-a)dy - \int_0^x \sum^x Q(b+k')(\alpha-\phi)dy$$

$$= \int_\alpha^\phi \sum^x QR^2(\sin^2\phi - \sin\alpha\sin\phi - \cos\alpha\phi\sin\phi + \alpha\cos\alpha\sin\phi)d\phi$$

$$= \sum^x Q\left[\begin{array}{l}\frac{1}{2}R^2(\phi-\alpha) - \frac{1}{2}(g-x)(k'+y) + \frac{1}{2}(g-a)(k'+b) \\ + (k'+y)\{(g-a) + (\phi-\alpha)(k'+b)\} - (g-x)(k'+b)\end{array}\right]. \quad c(81)$$

Integration of $\int_0^x \frac{N_x}{EF_x}$.

$$\int_0^x \frac{N_x}{EF_x}dx = \int_0^x N_x \frac{2m}{A}\frac{(g-x)}{\sin\phi}dx.$$

Since $E = \frac{AR}{2\theta}$ and $F = \frac{\theta}{mR^2}$, $EF = \frac{A}{2mR}$ and $\frac{1}{EF} = \frac{2mR}{A}$.

But $R = -\frac{ds}{d\phi}$; hence we have, after substituting the value of N_x as given by (42),

$$\int_0^x \frac{N_x}{EF_x}dx = \frac{2m}{A}\left\{\int_0^x V_x(g-x)dx + \int_0^x H_x(k'+y)dx\right\}. \quad c(82)$$

But from (39) and (40),

$$H_x = H_1 - \overset{x}{\Sigma} Q$$

and

$$V_x = V_1 - \overset{x}{\Sigma} P;$$

hence the second member of $c(82)$ becomes

$$\frac{2m}{A}\left\{\int_0^x V_1(g-x)dx + \int^x H_1(k'+y)dx - \int_0^x \overset{x}{\Sigma} P(g-x)dx \right.$$

$$\left. - \int_0^x \overset{x}{\Sigma} Q(k'+y)dx\right\},$$

where

$$V_1\int_0^x (g-x)dx = V_1\left(gx - \frac{x^2}{2}\right). \quad . \quad . \quad . \quad . \quad . \quad . \quad . \quad . \quad c(83)$$

$$-H_1\int_0^x (k'+y)dx = H_1\int_{\phi_0}^{\phi} R^2(\cos^2\phi)d\phi$$

$$= \tfrac{1}{2}H_1\{R^2(\phi_0 - \phi) + k'x - gy + xy\} \quad c(84)$$

$$-\int_0^x \overset{x}{\Sigma} Q(k'+y)dx = \int_{x=0}^{\phi} \overset{x}{\Sigma} QR^2 \cos^2\phi d\phi$$

$$= \tfrac{1}{2}\overset{x}{\Sigma} Q\{R^2(\phi - \alpha) + (k'+y)(g-x) - (g-a)(k'+b)\} \quad c(85)$$

$$-\int_0^x \overset{x}{\Sigma} P(g-x)dx = \int_a^{\phi} \overset{x}{\Sigma} PR^2 \sin\phi\cos\phi d\phi$$

$$= \tfrac{1}{2}\overset{x}{\Sigma} P\{(x-a)(a+x) + 2g(a-x)\}. \quad . \quad . \quad . \quad . \quad . \quad . \quad c(86)$$

Therefore (*a*) becomes

$$
\begin{aligned}
\Delta x = et^\circ x - y\Delta\phi_0 \\
-\frac{2}{A}\Big[& (M_1 + V_1 g + H_1 k')[(\phi_0 - \phi)(k' + y) - x] \\
& - \tfrac{1}{2}V_1 y^2 \\
& - \tfrac{1}{2}H_1[xy + k'x + gy - R^2(\phi_0 - \phi)] \\
& - (y + k')\sum^x P(g-a)(\alpha-\phi) + \sum^x P(g-a)(x-a) \\
& + \tfrac{1}{2}\sum^x P(y-b)^2 \\
& + \sum^x Q\big[\tfrac{1}{2}R^2(\phi - \alpha) - \tfrac{1}{2}(g - x)(k' + y) \\
& + \tfrac{1}{2}(g - a)(k' + b) + (k' + y)\{(g - a) \\
& + (\phi - \alpha)(k' + b)\} - (g - x)(k' + b)\big] \Big]
\end{aligned}
$$

$$
\begin{aligned}
-\frac{m}{A}\Big[& V_1 x(2g - x) + H_1\{xy + k'x - gy + R^2(\phi_0 - \phi)\} \\
& - \sum^x P\{(x-a)(2g - x - a)\} \\
& + \sum^x Q\{R^2(\phi - \alpha) + (k' + y)(g - x) \\
& - (g - a)(k' + b)\} \Big]. \qquad c(87)
\end{aligned}
$$

From (*b*),

$$\Delta y = \int_0^x \Delta\phi dx + \int^x et^\circ dy - \int_0^x \frac{N_x}{EF_x} dy. \quad . \quad c(88)$$

The following integrals are employed in reducing $c(88)$:

$$\int_0^x (\phi_0 - \phi)dx = -\{g(\phi_0 - \phi) - x(\phi_0 - \phi) - y\}; \quad c(89)$$

$$\int_0^x ydx = -\tfrac{1}{2}\{-xy + k'x + gy - R^2(\phi_0 - \phi)\}; \quad . \quad c(90)$$

$$\int_0^x \overset{x}{\Sigma} P(y-b)dx$$

$$= - \overset{x}{\Sigma} P\{\tfrac{1}{2}(x-a)(b+k')+\tfrac{1}{2}(y-b)(g-x)-\tfrac{1}{2}R^2(\alpha-\phi)\}; \quad c(91)$$

$$\int_0^x \overset{x}{\Sigma} P(g-a)(\alpha-\phi)dx =$$

$$= \overset{x}{\Sigma} P\{-(g-a)(g-x)(\alpha-\phi)+(g-a)(y-b)\}; \quad c(92)$$

$$\int_0^x \overset{x}{\Sigma} Q(x-a)dx - \int_0^x \overset{x}{\Sigma} Q(\alpha-\phi)(b+k')dx$$

$$= \overset{x}{\Sigma} Q \left[\begin{array}{l} -\tfrac{1}{2}\{(x-a)(2g-x-a)\} \\ -(g-a)\{(g-x)-(g-a)\} \\ -(k'+b)\{y \mp b + \phi(g-x) - \alpha(g-a)\} \\ +\alpha(k'+b)\{(g-x)-(g-a)\} \end{array} \right] \quad c(93)$$

or

$$\overset{x}{\Sigma} Q \left\{ \tfrac{1}{2}(x-a)^2 + (k'+b)\{(g-x)(\alpha-\phi)-(y \mp b)\} \right\} \quad c(94)$$

$$-\int_0^x \frac{N_x}{EF_x}dy = \frac{2m}{A}\left\{\int_0^x (V_x R^2 \sin^2\phi d\phi + H_x R^2 \sin\phi\cos\phi d\phi)\right\},$$

in which the following integrals occur after substituting the values of V_x and H_x from (39) and (40):

$$\int_{\phi_0}^{\phi} V_1 R^2 \sin^2\phi d\phi = \frac{V_1}{2}\{xy + k'x - gy - R^2(\phi_0 - \phi)\}; \quad c(95)$$

$$\int_{\phi_0}^{\phi} H_1 R^2 \sin\phi\cos\phi d\phi = -\frac{H_1}{2}\{x(2g-x)\}; \quad . \quad . \quad . \quad . \quad c(96)$$

$$\int_a^\phi \overset{x}{\Sigma} PR^2 \sin \phi d\phi =$$

$$= \tfrac{1}{2}\overset{x}{\Sigma} P\{R^2(\phi - \alpha) + k'(x - a) - g(y - b) + xy - ab\}; \quad c(97)$$

$$\int_a^\phi \overset{x}{\Sigma} QR^2 \sin \phi \cos \phi d\phi =$$

$$= -\tfrac{1}{2}\overset{x}{\Sigma} Q\{(x-a)[2g-a-x]\}. \quad . \quad . \quad . \quad . \quad . \quad . \quad . \quad . \quad c(98)$$

Using the above integrals, $c(88)$ become

$$\Delta y = ct^\circ y + x\Delta\phi$$

$$+ \frac{2}{A}\left[\begin{array}{l} (M_1 + V_1 g + H_1 k')[y - (\phi_0 - \phi)(g - x)] - \tfrac{1}{2}H_1 x^2 \\ - \tfrac{1}{2}V_1\{xy - k'x - gy + R^2(\phi_0 - \phi)\} \\ + (g - x)\overset{x}{\Sigma} P\{(g-a)(\alpha-\phi)\} - \overset{x}{\Sigma} P(g-a)(y-b) \\ - \tfrac{1}{2}\overset{x}{\Sigma} P\{(x - a)[b + k'] + (y - b)(g - x) \\ - R^2(\alpha - \phi)\} \\ + \overset{x}{\Sigma} Q\{\tfrac{1}{2}(x - a)^2 + (k' + b)[(g - x)(\alpha - \phi) \\ - (y + b)]\} \end{array}\right]$$

$$+ \frac{m}{A}\left[\begin{array}{l} V_1\{xy + k'x - gy - R^2(\phi_0 - \phi)\} - H_1 x(2g - x) \\ + \overset{x}{\Sigma} P\{(g - x)(k' + y) - (g - a)(k' + b) \\ + R^2(\alpha - \phi)\} \\ + \overset{x}{\Sigma} Q\{(x - a)[2g - a - x]\} \end{array}\right]. \quad c(99)$$

Equations $c(77)$, $c(87)$, and $c(99)$ are perfectly general for circular arches, and can be applied for any loading, either vertical or horizontal. The equations also enable us to consider arches which are not symmetrical. The equations for symmetrical arches are considerably more simple.

SYMMETRICAL CIRCULAR ARCHES.

For this case we have $g = \frac{1}{2}l$, and for $x = l$, $y = c = 0$, and $\phi = \phi_l = -\phi_0$.

$c(77)$ now becomes, remembering that

$$V_1 = \frac{1}{l}\{M_2 - M_1 + (H_1 c = 0) + \overset{l}{\Sigma} P(l-a) + \overset{l}{\Sigma} Q(+b)\}, \quad c(100)$$

$$\Delta\phi_l = \Delta\phi_0 + \frac{2}{A}\left[\begin{array}{l}(M_1 + M_2)\phi_0 - H_1(l - 2k'\phi_0) \\ - \frac{1}{2}\overset{l}{\Sigma} P\{2b - 2a\alpha - l(\phi_0 - \alpha)\} \\ + \overset{l}{\Sigma} Q\{l - a - b\alpha - k'(\alpha + \phi_0)\}\end{array}\right]. \quad c(101)$$

From $c(101)$,

$$M_1 + M_2 = (\Delta\phi_l - \Delta\phi_0)\frac{A}{2\phi_0}$$
$$+ \frac{1}{\phi_0}\left[\begin{array}{l}H_1(l - 2k'\phi_0) + \frac{1}{2}\overset{l}{\Sigma} P\{2b - 2a\alpha - l(\phi_0 - \alpha)\} \\ \qquad - \overset{l}{\Sigma} Q\{l - a - b\alpha - k'(\alpha + \phi_0)\}\end{array}\right]. \quad c(102)$$

From $c(87)$ we have, when $x = l$,

$$\Delta l = et^\circ l + \frac{1}{A}\left[\begin{array}{l}(M_1 + M_2)(l - 2k'\phi_0) \\ \qquad - H_1(4k'^2\phi_0 - 3k'l + 2R^2\phi_0) \\ + \overset{l}{\Sigma} P\{a(l - a - 2k'\alpha) - k'(l\phi_0 - l\alpha - 2b)\} \\ + \overset{l}{\Sigma} Q\{(R^2 + 2k'^2)(\phi_0 + \alpha) + \frac{1}{2}b(2a - l) \\ - 3k'(l - a) + 2bk'\alpha\}\end{array}\right]$$
$$- \frac{m}{A}\left[\begin{array}{l}H_1(k'l + 2R^2\phi_0) + \overset{l}{\Sigma} Pa(l - a) \\ + \overset{l}{\Sigma} Q\{-R^2(\phi_0 + \alpha) - lk' - \frac{1}{2}bl + a(k' + b)\}\end{array}\right], \quad c(103)$$

from which

$$M_1 + M_2 = \frac{A}{l - 2k\phi_0}(\Delta l - et^\circ l)$$

$$- \frac{1}{l - 2k'\phi_0}\left[\begin{array}{l} - H_1(4k'^2\phi_0 - 3k'l + 2R^2\phi_0) \\ + \overset{l}{\Sigma} P\{a(l-a-2k'\alpha) - k'(l\phi_0 - l\alpha - 2b)\} \\ + \overset{l}{\Sigma} Q\{(R^2 + 2k'^2)(\phi_0 + \alpha) + \frac{1}{2}b(2a - l) \\ - 3k(l-a) + 2bk'\alpha\} \end{array}\right]$$

$$+ \frac{m}{l - 2k'\phi_0}\left[\begin{array}{l} H_1(k'l + 2R^2\phi_0) \\ + \overset{l}{\Sigma} Pa(l-a) \\ + \overset{l}{\Sigma} Q\{-R^2(\phi_0 + \alpha) - k'l - \frac{1}{2}bl + a(k'+b)\} \end{array}\right]. \quad c(104)$$

From $c(99)$,

$$\Delta c = l\Delta\phi_0 + \frac{1}{A}\left[\begin{array}{l} (M_1 + M_2)l\phi_0 - (M_1 - M_2)(k' - d) \\ - H_1 l(l - 2k'\phi_0) \\ - \frac{1}{2}\overset{l}{\Sigma} P\{b(l + 2a) + (l - 2a)(\alpha l + d) \\ \qquad - \phi_0 l^2 - 2\alpha R^2\} \\ + \overset{l}{\Sigma} Q\{(l-a)^2 - (k' + b)[l(\alpha + \phi_0) + 2b]\} \\ + \overset{l}{\Sigma} Q\{bl\phi_0 + b(k' - d)\} \end{array}\right]$$

$$- \frac{m}{A}\left[\begin{array}{l} (M_1 - M_2)(k' - d) \\ + \frac{1}{2}\overset{l}{\Sigma} P\{(l - 2a)(b + d) - 2\alpha R^2\} \\ + \overset{l}{\Sigma} Q\{a(l-a) - b(k' - d)\} \end{array}\right], \quad c(105)$$

where

$$d = \frac{2R^2\phi_0}{l}. \quad . \quad . \quad . \quad . \quad . \quad . \quad . \quad . \quad . \quad c(105a)$$

SYMMETRICAL CIRCULAR ARCHES WITH A HINGE AT EACH SUPPORT.

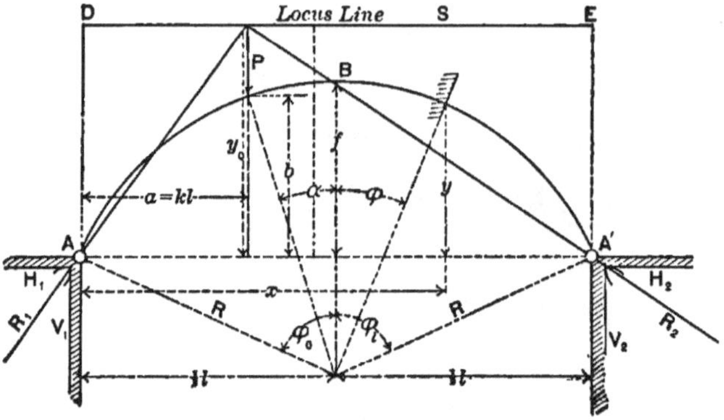

FIG. 31.

In Fig. 31 let ABA' represent a symmetrical circular arch having a hinge at A and A': then there can be no bending-moments at these points; hence M_1 and M_2 are zero, and the resultants R_1 and R_2 will pass through the hinges.

As in the case of parabolic arches, we will consider each class of loading, etc., separately.

(*a*) *Vertical Loads, with the Effect of N_x neglected.*

Assuming that l remains constant, $\Delta l = 0$; and by remembering that M_1 and M_2 are zero, and also that all terms containing Q and m do not appear, we have, from $c(87)$,

$$H_1 = \frac{\overset{l}{\Sigma} P\{a(l-a-2k'\alpha) - k'(l\phi_0 - l\alpha - 2b)\}}{4k'^2\phi_0 + 2R^2\phi_0 - 3k'l}; \quad c(106)$$

or, since $a = R(\sin\phi_0 - \sin\alpha)$,

$b = R(\cos\alpha - \cos\phi_0)$, $k' = R\cos\phi_0$, and $l = 2R\sin\phi_0$.

c(106) becomes

$$H_1 = \tfrac{1}{2}\overset{l}{\Sigma}P\left\{\frac{(\sin^2\phi_0 - \sin^2\alpha) - 2\alpha\cos\phi_0(\sin\phi_0 - \sin\alpha) - 2\cos\phi_0[(\phi_0 - \alpha)\sin\phi_0 + \cos\phi_0 - \cos\alpha]}{2\phi_0\cos^2\phi_0 - 3\sin\phi_0\cos\phi_0 + \phi_0}\right\} \quad c(107)$$

or

$$H_1 = \overset{l}{\Sigma}P\left\{\frac{\tfrac{1}{2}(\sin^2\phi_0 - \sin^2\alpha) + \cos\phi_0(\cos\alpha + \alpha\sin\alpha - \cos\phi_0 - \phi_0\sin\phi_0)}{2\phi_0\cos^2\phi_0 - 3\sin\phi_0\cos\phi_0 + \phi_0}\right\} \quad c(108)$$

or

$$H_1 = \overset{l}{\Sigma}P\frac{A}{B}. \quad . \quad . \quad . \quad . \quad . \quad . \quad c(109)$$

The values of $\frac{A}{B}$ are given in Table XVII

Since all loads are vertical, H_1 and H_2 will be equal in magnitude.

From (47),

$$V_1 = \overset{l}{\Sigma}P\frac{l-a}{l} \quad . \quad . \quad . \quad . \quad . \quad . \quad c(110)$$

or

$$V_1 = \overset{l}{\Sigma}P(1-k), \quad . \quad . \quad . \quad . \quad . \quad . \quad c(111)$$

where $k = a/l$.

c(110) can also be written

$$V_1 = \overset{l}{\Sigma}P\frac{\sin\phi_0 + \sin\alpha}{2\sin\phi_0}. \quad . \quad . \quad . \quad . \quad c(112)$$

Having determined the values of H_1 and V_1, the stresses in the arch can be found by graphics or by means of equations (39) to (48).

From (50), for a single load,

$$y_0 = \frac{V_1}{H_1}a \quad . \quad . \quad . \quad . \quad . \quad . \quad . \quad . \quad c(113)$$

or

$$y_0 = \frac{B}{A} k(1 - k)l. \quad . \quad . \quad . \quad . \quad . \quad c(114)$$

By means of $c(114)$ the curve DE in Fig. 31 can be located and the stresses in the arch found graphically.

The values of A/B can be found from Table XVII, and of $k(1 - k)$ from Table V.

Change of Shape due to the Action of Vertical Loads (N_x neglected).

The values of $\Delta\phi$, Δx, and Δy can be determined from $c(77)$, $c(87)$, and $c(99)$ by remembering that all terms containing Q and m disappear, that M_1 and M_2 are zero, that $g = \frac{1}{2}l$, and that the values of H_1 and V_1 are to be found from $c(109)$ and $c(111)$.

(*b*) *Vertical Loads, Effect of the Axial Stress included.**

From $c(87)$,

$$H_1 = \frac{\left\{ \begin{array}{c} \overset{l}{\Sigma} P\{a(l - a - 2k'a) - k'(l\phi_0 - l\alpha - 2b)\} \\ - m\,\{\overset{l}{\Sigma} Pa(l - a)\} \end{array} \right\}}{4k'^2\phi_0 + 2R^2\phi_0 - 3k'l + m(2R^2\phi_0 + k'l)} \quad c(115)$$

or

$$H_1 = \overset{l}{\Sigma} P \frac{2A - m(\sin^2 \phi_0 - \sin^2 \alpha)}{2B + 2m(\phi_0 + \sin \phi_0 \cos \phi_0)} \quad . \quad . \quad . \quad . \quad c(116)$$

or

$$H_1 = \mathfrak{H} \frac{1 - \frac{m}{2A}(\sin^2 \phi_0 - \sin^2 \alpha)}{+ \frac{m}{B}(\phi_0 + \sin \phi_0 \cos \phi_0)}, \quad . \quad . \quad . \quad . \quad . \quad c(117)$$

See Appendix C.

in which $\mathfrak{H}$ is to be found from $c(109)$ and

$$A = \tfrac{1}{2}(\sin^2 \phi_0 - \sin^2 \alpha) \\ + \cos \phi_0(\cos \alpha + \alpha \sin \alpha - \cos \phi_0 - \phi_0 \sin \phi_0),$$

$$B = 2\phi_0 \cos^2 \phi_0 - 3 \sin \phi_0 \cos \phi_0 + \phi_0.$$

Since the value of B is constant for any particular arch, the values of A can be very easily found from Table XVII by multiplying the tabular quantities by B.

The denominator of $c(117)$ is constant for any particular arch, and hence the value of H_1 can be found with but little labor.

The value of V_1 can be found from $c(111)$.

From (50).

$$y_0 = \frac{V_1}{H_1} a, \quad . \quad . \quad . \quad . \quad . \quad . \quad . \qquad c(118)$$

where V_1 and H_1 are to be found from $c(111)$ and $c(117)$ respectively.

For practical purposes it will be sufficient to compute but a few values of y_0, and then draw the curve DE, Fig. 31, by means of a curved ruler.

The change in shape of the arch can be found by means of $c(77)$, $c(87)$, and $c(99)$.

(*c*) *Horizontal Loads* (N_x *neglected*).

From $c(87)$,

$$H_1 = \mp \overset{l}{\Sigma} Q \left\{ \frac{(R^2 + 2k'^2)(\phi_0 + \alpha) + \frac{1}{2}b(2a - l) - 3k'(l - a) + 2bk'\alpha}{2R^2(\phi_0 - 3 \sin \phi_0 \cos \phi_0 + 2\phi_0 \cos^2 \phi_0)} \right\} \qquad c(119)$$

or

$$H_1 = \tfrac{1}{2}\Sigma Q \left\{ 1 + \frac{\alpha - \sin \alpha \cos \alpha - 2 \cos \phi_0(\sin \alpha - \alpha \cos \alpha)}{\phi_0 - 3 \sin \phi_0 \cos \phi_0 + 2\phi_0 \cos^2 \phi_0} \right\}. \qquad c(120)$$

The values of $\phi_0 - 3 \sin \phi_0 \cos \phi_0 + 2\phi_0 \cos^2 \phi_0$ are given in Table XVIII; of $\alpha - \sin \alpha \cos \alpha$ and $\sin \alpha - \alpha \cos \alpha$, in Table XIX.

From (47),

$$V_1 = \sum^l Q\frac{b}{l} = \sum^l Q\frac{\cos \alpha - \cos \phi_0}{2 \sin \phi_0}, \quad . \quad . \quad c(121)$$

$$V_2 = - V_1. \quad . \quad . \quad . \quad . \quad . \quad . \quad . \quad . \quad . \quad . \quad c(122)$$

Having the values of V_1 and H_1, the stresses can be found graphically or by equations (39) to (48).

As in case of the parabolic arch, we can locate the locus of the points of intersection of R_1 and R_2 by means of the formula

$$x_0 = \frac{H_1}{V_1}b.$$

Substituting the values of H_1 and V_1 from $c(120)$ and $c(121)$ for a single load, we have

$$x_0 = R \sin \phi_0 \left\{ 1 + \frac{\begin{array}{c}\alpha - \sin \alpha \cos \alpha \\ - 2 \cos \phi_0(\sin \alpha - \alpha \cos \alpha)\end{array}}{\phi_0 - 3 \sin \phi_0 \cos \phi_0 + 2\, \phi_0 \cos^2 \phi_0} \right\}, \quad c(123)$$

which is easily evaluated by means of Tables XVIII and XIX.

The change in shape can be determined from $c(77)$, $c(87)$, and $c(99)$.

(*d*) *Horizontal Loads, including Effect of Axial Stress.*

From $c(103)$,

$$H_1 = \sum^l Q \frac{\left\{\begin{array}{c}(R^2 + 2k'^2)(\phi_0 + \alpha) + \frac{1}{2}b(2a - l) - 3k'(l - a) \\ + 2bk'\alpha \\ + m\{R^2(\phi_0 + \alpha) + lk' + \frac{1}{2}bl - a(b + k')\}\end{array}\right\}}{4k'^2 \phi_0 + 2R^2\phi_0 - 3k'l + m(2R^2\phi_0 + k'l)} \quad c(124)$$

or

$$H_1 = \sum^l Q \frac{\left\{\begin{array}{c}\phi_0 - 3 \sin \phi_0 \cos \phi_0 + 2\phi_0 \cos^2 \phi_0 + \alpha \\ - \sin \alpha \cos \alpha - 2 \cos \phi_0(\sin \alpha - \alpha \cos \alpha) \\ + m\{\phi_0 + \sin \phi_0 \cos \phi_0 + \alpha + \sin \alpha \cos \alpha\}\end{array}\right\}}{\begin{array}{c}2(\phi_0 - 3 \sin \phi_0 \cos \phi_0 + 2\phi_0 \cos^2 \phi_0) \\ + 2m(\phi_0 + \sin \phi_0 \cos \phi_0)\end{array}}, \quad c(125)$$

which can be quickly evaluated by means of Tables XVIII and XIX.

$$V_1 = \sum^{l} Q \frac{\cos\alpha - \cos\phi_0}{2\sin\phi_0} \quad . \quad . \quad . \quad . \quad c(126)$$

and

$$x_0 = \frac{H_1}{V_1} b, \quad . \quad . \quad . \quad . \quad . \quad . \quad . \quad . \quad . \quad . \quad c(127)$$

in which H_1 and V_1 are to be found from c(125) and c(126).

The change in shape can be determined from c(77), c(87), and c(99).

(*e*) *Temperature.*

From c(87) or c(103),

$$H_1 = \frac{et^\circ A}{R} \frac{\sin\phi_0}{\phi_0 + 2\phi_0\cos^2\phi_0 - 3\sin\phi_0\cos\phi_0 + m(\phi_0 + \sin\phi_0\cos\phi_0)}, \quad c(128)$$

or, when the effect of the axial stress is neglected,

$$H_1 = \frac{et^\circ A}{R} \frac{\sin\phi_0}{B}. \quad . \quad . \quad . \quad . \quad . \quad . \quad . \quad . \quad . \quad . \quad . \quad . \quad . \quad c(129)$$

c(128) and c(129) are quickly evaluated by the aid of Tables XVIII and XIX.

(*f*) *Change in Length of Span.*

From c(87),

$$H_1 = \frac{-A}{2R^2(B + m(\phi_0 + \sin\phi_0\cos\phi_0))} \Delta l, \quad . \quad c(130)$$

or, if the effect of the axial stress is neglected,

$$H_1 = \frac{-A}{2R^2 B} \Delta l. \quad . \quad . \quad . \quad . \quad . \quad . \quad c(131)$$

These equations are readily evaluated by the aid of Tables XVIII and XIX.

(*g*) *Sinking of a Support.*

In case one of the supports changes its elevation after the arch is in place a slight change in the stresses may result from the change in span, but this usually will be too small to be of any practical importance.

SYMMETRICAL CIRCULAR ARCH WITHOUT HINGES.

(*a*) *Vertical Loads* (*N_x neglected*).

Equating *c*(102) and *c*(104) and solving for H_1, we have

$$H_1 = \frac{\sum^{l} P\{2bl - l(l-2a)(\phi_0 - \alpha) - 2a^2\phi_0\}}{2l(k'\phi_0 + d\phi - l)}, \quad . \quad . \quad . \quad c(132)$$

which reduces to

$$H_1 = \tfrac{1}{2}\sum^{l} P\left\{\frac{2\sin\phi_0[\cos\alpha + \alpha\sin\alpha] - \sin\phi_0[2\cos\phi_0 + \phi_0\sin\phi_0] - \phi_0\sin^2\alpha}{\phi_0^2 + \phi_0\sin\phi_0\cos\phi_0 - 2\sin^2\phi_0}\right\}, c(133)$$

which is easily evaluated by the aid of Tables XX, XXI, and XXII.

Substituting *c*(102) in *c*(105), and then solving for M_1, we have

$$M_1 = \frac{H_1 R}{\phi_0}(\sin\phi_0 - \phi_0\cos\phi_0)$$

$$+ \frac{\sum^{l} PR}{2\phi_0(\sin\phi_0\cos\phi_0 - \phi_0)}\Big\{\sin\alpha\phi_0(\cos\alpha\sin\phi_0 - \cos\phi_0\sin\phi_0 - \phi_0)$$

$$+ \alpha\phi_0\sin\phi_0 + (\sin\phi_0\cos\phi_0 - \phi_0)[\cos\alpha + \alpha\sin\alpha - \cos\phi_0 - \phi_0\sin\phi_0]\Big\}. \; c(134)$$

By the aid of Tables XIX, XXIII, and XXIV *c*(134) can be quickly evaluated.

The value of M_2 can be found from *c*(134) by assuming the load applied at a point on the arch, so that (a) in $c(134)$ will become $(l - a)$.

From (47),

$$V_1 = \frac{1}{l}\{M_2 - M_1 + \sum^{l} Pl(1 - k)\}, \quad . \quad . \quad . \quad c(135)$$

where $k = \frac{a}{l}$.

Having determined the values of H_1, M_1, and V_1, the stresses can be found graphically, or by equations (39) to (41).

The ordinates fixing the locations of the resultants R_1 and R_2 for any particular load can be found from the following equations.

From (50), (51), and (52),

$$y_0 = \frac{M_1 + V_1 a}{H_1}, \quad . \quad . \quad . \quad . \quad . \quad . \quad . \quad (50)$$

$$y_1 = \frac{M_1}{H_1}, \quad . \quad . \quad . \quad . \quad . \quad . \quad . \quad . \quad . \quad . \quad (51)$$

and

$$y_2 = \frac{M_2}{H_2}. \quad . \quad . \quad . \quad . \quad . \quad . \quad . \quad . \quad . \quad . \quad (52)$$

From (51) and (52) we obtain, by substituting the values of M_1 and M_2, and remembering that $H_1 = H_2$ in magnitude,

$$y_2 - y_1 = -\frac{P}{\sin\phi_0 \cos\phi_0 - \phi_0}\left\{\begin{matrix} -\sin\alpha(\cos\phi_0 \sin\phi_0 + \phi_0) \\ +\sin\phi_0(\cos\alpha \sin\alpha + \alpha \end{matrix}\right\}\frac{R}{H_1}, \quad c(136)$$

and

$$y_2 + y_1 = \frac{2R}{\phi_0}(\sin\phi_0 - \phi_0 \cos\phi_0)$$
$$+ \frac{P}{\phi_0}\{\cos\alpha + \alpha \sin\alpha - \cos\phi_0 - \phi_0 \sin\phi_0\}\frac{R}{H_1}. \quad c(137)$$

From $c(136)$ and $c(137)$,

$$y_1 = \frac{R}{\phi_0}(\sin\phi_0 - \phi_0\cos\phi_0)$$

$$+ \frac{\left[\begin{array}{l}(\phi_0^2 + \phi_0\sin\phi_0\cos\phi_0 \\ \quad - 2\sin^2\phi_0)\left\{[\sin\phi_0\cos\phi_0 - \phi_0]\left[\begin{array}{c}\cos\alpha + \alpha\sin\alpha \\ -\cos\phi_0 - \phi_0\sin\phi_0\end{array}\right]\right. \\ \left. + \phi_0\sin\phi_0(\cos\alpha\sin\alpha + \alpha) - \sin\alpha(\phi_0\cos\phi_0\sin\phi_0 + \phi_0^2)\right\}\end{array}\right]R}{\phi_0(\sin\phi_0\cos\phi_0 - \phi_0)\{2\sin\phi_0[\cos\alpha + \alpha\sin\alpha] - \sin\phi_0[2\cos\phi_0 + \phi_0\sin\phi_0] - \phi_0\sin^2\alpha\}}. \quad c(138)$$

By the aid of Tables XX, XXII, XXIII, and XXIV $c(138)$ can be readily evaluated.

Evidently y_2 can be obtained from (138) by making (a) equal $(l - a)$.

From (50),

$$y_0 = y_1 + \frac{\sin\phi_0 - \sin\alpha}{2\sin\phi_0}(y_2 - y_1) + \frac{P}{H_1}\left(\frac{\sin\alpha(\sin\phi_0 - \sin\alpha)}{2\sin\phi_0}\right)R. \quad c(139)$$

The change in shape can be found from $c(77)$, $c(87)$, and $c(99)$.

(*b*) *Horizontal Loads* (N_x *neglected*).

From $c(102)$ and $c(104)$,

$$-H_1 = \overset{l}{\Sigma} Q\left\{\frac{2l(l - a - \alpha(b + k')) + \phi_0(bl - 2a(b + k')) - 2\phi_0 R^2(\phi_0 + \alpha)}{2l(k'\phi_0 + d\phi_0 - l)}\right\}. \quad c(140)$$

or

$$H_1 = +\frac{\overset{l}{\Sigma} Q}{2}\left\{1 + \frac{\phi_0(\sin\alpha\cos\alpha - \alpha) + 2\sin\phi_0(\sin\alpha - \alpha\cos\alpha)}{2\sin^2\phi_0 - \phi_0\sin\phi_0\cos\phi_0 - \phi_0^2}\right\}, \quad c(141)$$

which can be easily evaluated by means of Tables XIX and XX.

Substituting $c(102)$ in $c(105)$, and eliminating M_2 between $c(102)$ and $c(105)$, we have

$$M_1 = H_1\left(\frac{l}{2\phi_0} - k'\right)$$

$$+ \frac{\overset{l}{\Sigma} Q}{2(k' - d)}\{a^2 - al + 3bk' - bd + 2b^2\}$$

$$- \frac{\overset{l}{\Sigma} Q}{2\phi_0}\{l - a - b\alpha - k'(\alpha + \phi_0)\} \quad . \quad . \quad . \quad c(142)$$

or

$$M_1 = \frac{H_1 R}{\phi_0}\{\sin\phi_0 - \phi_0 \cos\phi_0\}$$

$$+\frac{\overset{l}{\Sigma}QR}{2(\sin\phi_0\cos\phi_0-\phi_0)}\left\{\begin{matrix}(\cos\alpha-\cos\phi_0)(\sin\phi_0\cos\phi_0-\phi_0+2\cos\alpha)\\ -\sin\phi_0(\sin^2\phi_0-\sin^2\alpha)\end{matrix}\right\}$$

$$-\frac{\overset{l}{\Sigma}QR}{2\phi_0}\{\sin\phi_0 - \phi_0\cos\phi_0 + \sin\alpha - \alpha\cos\alpha\}, \quad . \quad . \quad . \quad c(143)$$

which can be evaluated by the aid of Tables XIX and XXIII.

The magnitudes of H_2 and M_2 can be found from $c(141)$ and $c(143)$ by making (a) equal $(l - a)$.

From (47),

$$V_1 = \frac{1}{l}(M_2 - M_1 + \overset{l}{\Sigma}Qb). \quad . \quad . \quad . \quad . \quad c(144)$$

Having the values of H_1, H_2, V_1, V_2, M_1, and M_2, the reactions R_1 and R_2 are completely determined, and the stresses can be found graphically or by equations (39) to (41).

The change in shape can be found from $c(77)$, $c(87)$, and $c(99)$.

(c) *Effect of a Change in Temperature, Length of Span, the Angle ϕ_0, etc.*

From $c(102)$ and $c(104)$,

$$H_1 = \frac{2A\phi_0(let^\circ - \Delta l) + A(l - 2k'\phi_0)(\Delta\phi_l - \Delta\phi_0)}{2l(k'\phi_0 + \phi_0 d - l)} \qquad c(145)$$

or

$$H_1 = \frac{2A\phi_0(let^\circ - \Delta l) + A(l - 2k'\phi_0)(\Delta\phi_l - \Delta\phi_0)}{4R^2(\phi_0^2 + \phi_0\sin\phi_0\cos\phi_0 - 2\sin^2\phi_0)}, \quad c(146)$$

where the value of the parenthesis in the denominator can be obtained from Table XX.

From $c(102)$ and $c(105)$,

$$M_1 = H_1\left(\frac{l}{2\phi_0} - k\right)$$

$$+ \frac{A}{4\phi_0(k' - d)}\{(l\phi_0 - k' + d)(\Delta\phi_l - \Delta\phi_0) - 2\phi_0\Delta c\}. \quad c(147)$$

or

$$M_1 = \frac{H_1R}{\phi_0}(\sin\phi_0 - \phi_0\cos\phi_0) + \frac{A\sin\phi_0}{4R\phi_0(\sin\phi_0\cos\phi_0 - \phi_0)}$$

$$\left\{\frac{R(2\phi_0\sin\phi_0 + \sin\phi_0\cos\phi_0 - \phi_0)}{\sin\phi_0}(\Delta\phi_l - \Delta\phi_0) - 2\phi_0\Delta c\right\}, \quad c(148)$$

which can be evaluated by the aid of Table XIX.

From (47),

$$V_1 = \frac{1}{l}(M_2 - M_1). \quad . \quad . \quad . \quad . \quad . \quad c(149)$$

The stresses can be now found from (39) to (41).

The change in shape can be found from $c(77)$, $c(87)$, and $c(99)$.

(*d*) *Effect of the Axial Stress.*

In order to economize space, the expressions for H_1 and M_1 will be given which are perfectly general, applying to cases of vertical loads, horizontal loads, change in temperature, etc.

From $c(102)$ and $c(104)$,

$$H_1 = \frac{\begin{array}{l} + \sum^{l} P(2bl - l(l - 2a)(\phi_0 - \alpha) - 2a^2\phi_0) \\ - \sum^{l} Q\left\{\begin{array}{c} 2l(l - a - \alpha(b + k)) + \phi_0(lb - 2a(b + k)) \\ - 2\phi_0 R^2(\phi_0 + \alpha) \end{array}\right\} \\ + 2m\phi_0 \sum^{l} Pa(l - a) \\ - 2m\phi_0 \sum^{l} Q\{- R^2(\phi_0 + \alpha) - kl - \frac{1}{2}bl + a(b + k)\} \\ + 2A\phi_0(let^\circ - \Delta l) + A(l - 2k\phi_0)(\Delta\phi_l - \Delta\phi_0) \end{array}}{2(k'\phi_0 + \phi_0 d - l) + ml(k' + d)\phi_0}. \quad c(150)$$

The terms containing m show the effect of the axial stress.

From $c(102)$ and $c(105)$,

$$M_1 = H_1 \left\{ \frac{l}{2\phi_0} - k' \right\} + \frac{\Sigma P}{4\phi_0(1+m)(k'-d)} \{(l-2a)(b-d)\phi_0$$
$$+ 2R^2\alpha\phi_0 + (k'-d)(1+m)(2b - 2a\alpha - l\phi_0 + l\alpha)$$
$$- m[\phi_0(l-2a)(b+d) - 2\alpha\phi_0 R^2]\}$$
$$+ \frac{1}{2(1+m)(k'-d)} \overset{l}{\Sigma} Q \{-a(l-a) - 2b(k'+b) - bl\phi_0\}.$$
$$+ \frac{1}{2(1+m)(k'-d)} \overset{l}{\Sigma} Q \{lb\phi_0 + bk - bd\}$$
$$- \frac{1}{2\phi_0} \overset{l}{\Sigma} Q \{l - a - b\alpha - k'(\alpha + \phi_0)\}$$
$$- \frac{m}{2(1+m)(k'-d)} \overset{l}{\Sigma} Q \{a(l-a) - b(k'-d)\}$$
$$+ \frac{A}{4\phi_0(k-d)} \{(l\phi_0 - k' + d)(\Delta\phi_0 + \Delta\phi_1) - 2\phi_0 \Delta c\}. \quad c(151)$$

The terms containing m show the effect of the axial stress.

By the application of $c(150)$, $c(151)$, and (47) the stresses in any symmetrical circular arch without hinges can be completely determined by the ordinary methods of graphics.

CHAPTER V.

SYMMETRICAL ARCHES HAVING A VARIABLE MOMENT OF INERTIA.

THE treatment of symmetrical arches can be considerably simplified by the methods we are about to introduce. The following equations for H_1 and M_1 can be applied to any symmetrical arch when the axis is a curve which can be expressed by a linear equation. We will first consider the case where the arch has no hinges.

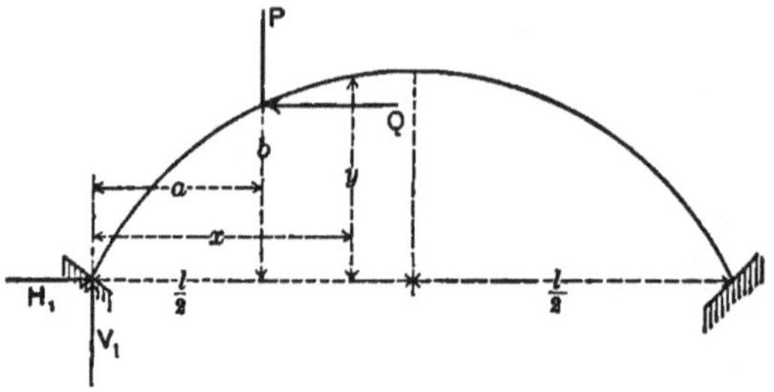

FIG. 32.

SYMMETRICAL ARCH WITHOUT HINGES.

Value of M_1.—From (*d*), (*a*), and (*b*),

$$\Delta\phi = \Delta\phi_0 + \frac{1}{E}\int_0^x \frac{M_x}{\theta_x} ds. \qquad g(59)$$

$$\Delta x = -y\Delta\phi_0 - \frac{1}{E}\int_0^x \frac{M_x}{\theta_x} y ds + et^\circ \int_0^x dx - \frac{1}{E}\int_0^x \frac{N_x}{F_x} dx. \quad g(60)$$

$$\Delta y = x\,\Delta\phi_0 + \frac{1}{E}\int_0^x \frac{M_x}{\theta_x} x ds + ct^\circ \int_0^x dy - \frac{1}{E}\int_0^x \frac{N_x}{F_x} dy. \quad g(61)$$

Assume a single load placed at any point upon the arch; then, since the arch is fixed at the ends and symmetrical, $\Delta\phi^l = \Delta\phi_0$, $\Delta l = 0$, and $\Delta c = 0$. If $x = l$, $g(59)$, $g(60)$, and $g(61)$ become, neglecting temperature for the present,

$$\Delta\phi_l = \Delta\phi_0 + \frac{1}{E}\int_0^l \frac{M_x}{\theta_x} ds = 0, \quad . \quad . \quad . \quad . \quad . \quad g(62)$$

$$\Delta l = -\frac{1}{E}\int_0^l \frac{M_x}{\theta_x} y ds - \frac{1}{E}\int_0^l \frac{N_x}{F_x} dx = 0, \quad . \quad g(63)$$

and

$$\Delta c = \frac{1}{E}\int_0^l \frac{M_x}{\theta_x} x ds - \frac{1}{E}\int_0^l \frac{N_x}{F_x} dy = 0. \quad . \quad . \quad g(64)$$

Now, from (41),

$$M_x = M_1 + V_1 x + K, \quad . \quad . \quad . \quad . \quad . \quad . \quad . \quad . \quad . \quad . \quad . \quad . \quad g(65)$$

where

$$K = -H_1 y - P(x - a) + Q(v - b)\ x > a. \quad . \quad g(66)$$

Substituting $g(65)$ in $g(62)$ and $g(64)$, we have

$$M_1 \int_0^l \frac{ds}{\theta_x} + V_1 \int_0^l \frac{x ds}{\theta_x} + \int_0^l \frac{K ds}{\theta_x} = 0 \quad . \quad . \quad . \quad . \quad . \quad . \quad g(67)$$

and

$$M_1 \int_0^l \frac{x ds}{\theta_x} + V_1 \int_0^l \frac{x^2 ds}{\theta_x} + \int_0^l \frac{K x ds}{\theta_x} - \int_0^l \frac{N_x}{F_x} dy = 0. \quad g(68)$$

From (47),

$$V_1 = \frac{M_2 - M_1}{l} + B, \quad . \quad . \quad . \quad . \quad . \quad . \quad g(69)$$

where

$$B = P(1 - k) + \frac{Qb}{l}. \quad . \quad . \quad . \quad . \quad . \quad g(70)$$

Substituting the value of V_1 in $g(67)$ and $g(68)$, and eliminating M_1, we have

$$M_1 = \frac{\left\{\int_0^l \frac{Kxds}{\theta_x} - \int_0^l \frac{N_x}{F_x}dy\right\}\int_0^l \frac{xds}{\theta_x} - \int_0^l \frac{Kds}{\theta_x}\int_0^l \frac{x^2ds}{\theta_x}}{\int_0^l \frac{ds}{\theta_x}\int_0^l \frac{x^2ds}{\theta_x} - \left(\int_0^l \frac{xds}{\theta_x}\right)^2}, \quad g(71)$$

in which

$$K = -H_1y - P(x-a) + Q(y-b), \quad x > a \quad . \quad g(66)$$

$$N_x = V_x \sin\phi + H_x \cos\phi, \quad . \quad . \quad . \quad . \quad (42)$$

$$V_x = V_1 - P, \qquad x > a. \quad \text{From } (40) \quad . \quad . \quad g(72)$$

$$H_x = H_1 - Q. \qquad x > a. \quad \text{From } (39) \quad . \quad . \quad g(73)$$

Then in $g(71)$ we have two unknown quantities, H_1 and V_x. But V_x occurs in N_x only, which contains the effect of the axial stress; hence for the common method of arch treatment we can neglect the term containing N_x. A method will be given, however, which will enable us to very nearly obtain the actual effect of the axial stress.

Value of H_1.—The value of H_1 can be found as follows:

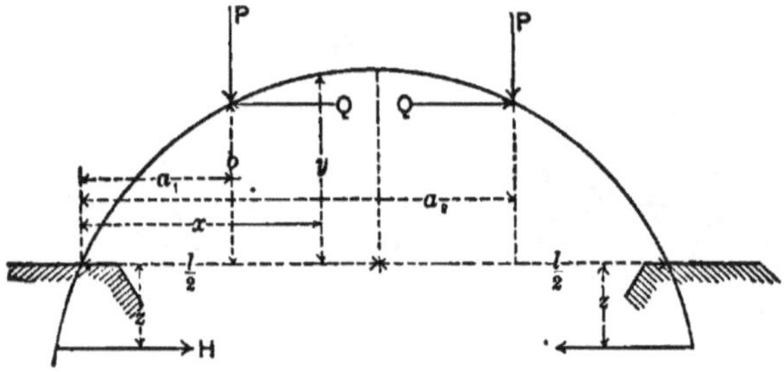

FIG. 33.

Assume the arch free to slide longitudinally upon the supports, and that *two equal and symmetrical* loads are applied; also assume that there are equal and symmetrical moments Hz applied at the supports; then $\Delta\phi_l = \Delta\phi$, the same as if the arch were fixed at the ends, since our loading is symmetrical. From $g(62)$ we have

$$\int_0^l \frac{M_x ds}{\theta_x} = 0. \quad . \quad . \quad . \quad . \quad . \quad . \quad g(74)$$

But from (41),

$$M_x = M_1 + K', \quad . \quad . \quad . \quad . \quad . \quad . \quad g(75)$$

where

$$K' = V_1 x - \sum^x P(x-a) + \sum^x Q(y-b), \quad . \quad . \quad g(76)$$

$H_1 y$ being zero, since the arch is free to slide upon the supports.

Substituting $g(75)$ in $g(74)$,

$$\int_0^l \frac{M_x ds}{\theta_x} = \int^l \frac{M_1 + K'}{\theta_x} ds = M_1 \int_0^l \frac{ds}{\theta_x} + \int_0^l \frac{K'}{\theta_x} ds = 0 \quad g(77)$$

or

$$M_1 = - \frac{\int_0^l \frac{K'}{\theta_x} ds}{\int_0^l \frac{ds}{\theta_x}} \quad . \quad . \quad . \quad . \quad . \quad . \quad g(78)$$

The change in length of the span due to the action of our loading can be found by the aid of $g(63)$. Let $\Delta' l$ be the change in the length of the span; then

$$\Delta' l = - \frac{M_1}{E} \int_0^l \frac{y ds}{\theta_x} - \frac{1}{E} \int_0^l \frac{K' y ds}{\theta_x} - \frac{1}{E} \int_0^l \frac{N_x}{F_x} dx. \quad g(79)$$

Substituting the value of M_1 from $g(78)$,

$$\Delta' l = \frac{1}{E} \frac{\int_0^l \frac{K' ds}{\theta_x}}{\int_0^l \frac{ds}{\theta_x}} \int_0^l \frac{y ds}{\theta_x} - \frac{1}{E} \int_0^l \frac{K' y ds}{\theta_x} - \frac{1}{E} \int_0^l \frac{N_x}{F_x} dx, \quad g(80)$$

where

$$K' = V_1 x - \overset{x}{\Sigma} P(x-a) + \overset{x}{\Sigma} Q(y-b); \quad . \quad . \quad g(76)$$

$$V_1 = \frac{\overset{l}{\Sigma} P(l-a) + \overset{l}{\Sigma} Qb}{l}; \quad . \quad . \quad . \quad . \quad . \quad . \quad . \quad g(81)$$

$$N_x = V_x \sin\phi + H_x \cos\phi; \quad . \quad . \quad . \quad . \quad . \quad . \quad . \quad (42)$$

$$V_x = V_1 - \overset{x}{\Sigma} P, \; x \gtreqless a; \quad . \quad . \quad . \quad . \quad . \quad . \quad . \quad (40)$$

$$H_x = H_1 - \overset{x}{\Sigma} Q = -\overset{x}{\Sigma} Q = 0. \quad . \quad . \quad . \quad . \quad . \quad (39)$$

All of which are known quantities; hence the value of $\Delta' l$ can be accurately determined from $g(80)$ for any *symmetrical loading.*

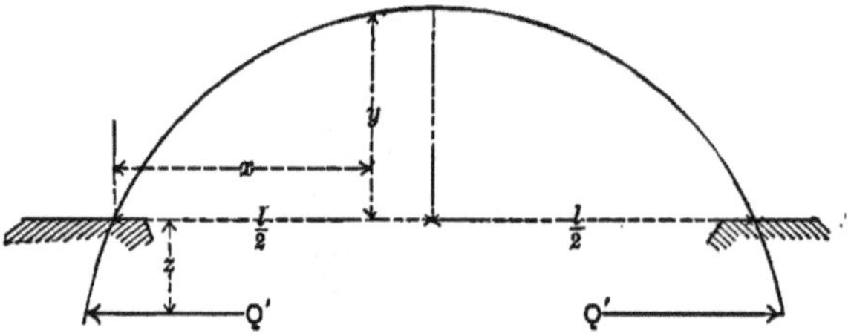

FIG. 34.

Now suppose the arch unloaded and free to slide as before, and let two equal and symmetrical moments $Q'z$ be applied at the supports; then $\Delta\phi_l = \Delta\phi_0$, and we have from $g(62)$

$$\int_0^l \frac{M_x ds}{\theta_x} = 0.$$

But $$M_x = Q'(z+y);$$

hence

$$\int_0^l \frac{M_x ds}{\theta_x} = Q'z \int_0^l \frac{ds}{\theta_x} + Q' \int_0^l \frac{yds}{\theta_x} = 0 \quad . \quad g(82)$$

or

$$z = -\frac{\int_0^l \frac{yds}{\theta_x}}{\int_0^l \frac{ds}{\theta_x}}. \quad . \quad . \quad . \quad . \quad . \quad g(83)$$

The corresponding change in the length of the span is given by $g(63)$, or

$$\Delta''l = -\frac{1}{E}\int_0^l \frac{M_x}{\theta_x} yds - \frac{1}{E}\int_0^l \frac{N_x}{F_x} dx, \quad . \quad . \quad . \quad . \quad . \quad . \quad g(84)$$

where $M_x = Q'(z+y)$ and $N_x = +H_x \cos\phi = +Q' \cos\phi$; hence

$$\Delta''l = -\frac{Q'z}{E}\int_0^l \frac{yds}{\theta_x} - \frac{Q'}{E}\int_0^l \frac{y^2ds}{\theta_x} - \frac{Q'}{E}\int_0^l \frac{dx}{F_x}\cos\phi. \quad . \quad g(85)$$

Substituting $g(83)$ in $g(85)$,

$$\Delta''l = \frac{Q'}{E}\left\{ -\int_0^l \frac{y^2ds}{\theta_x} - \int_0^l \frac{dx}{F_x}\cos\phi + \frac{\left(\int_0^l \frac{yds}{\theta_x}\right)^2}{\int_0^l \frac{ds}{\theta_x}} \right\}. \quad g(86)$$

Let $\mathfrak{H}_1$ be the horizontal thrust at the support necessary to cause a change in the length of the span of $\Delta'l$; then we have

$$\Delta''l : \Delta'l :: Q' : \mathfrak{H}_1 = Q'\frac{\Delta'l}{\Delta''l}.$$

Therefore

$$\mathfrak{H}_1 = \frac{\int_0^l \frac{K'yds}{\theta_x} + \int_0^l \frac{N_x dx}{F_x} - \frac{\int_0^l \frac{K'ds}{\theta_x}}{\int_0^l \frac{ds}{\theta_x}}\int_0^l \frac{yds}{\theta_x}}{+\int_0^l \frac{y^2ds}{\theta_x} + \int_0^l \frac{dx}{F_x}\cos\phi - \frac{\left(\int_0^l \frac{yds}{\theta_x}\right)^2}{\int_0^l \frac{ds}{\theta_x}}}, \quad . \quad g(87)$$

where

$$\left.\begin{aligned} K' &= V_1x - \sum^x P(x-a) + \sum^x Q(y-b); \\ V_1 &= \frac{\sum^l P(l-a) + \sum^l Qb}{l}; \\ N_x &= V_x \sin\phi + H_x \cos\phi; \\ V_x &= V_1 - \sum^x P, \quad x \overset{=}{>} a; \\ H_x &= H_1 - \sum^x Q, \quad x \overset{=}{>} a. \end{aligned}\right\} \text{For two equal and symmetrical loads.}$$

(*a*) *Vertical Loads only.*

If the loads are vertical,

$$\left.\begin{aligned} K' &= V_1x - \sum^x P(x-a) \\ &= Px - \sum^x P(x-a); \\ V_1 &= \sum^l P(1-k) = P; \\ V_x &= V_1 - \sum^x P = P - \sum^x P; \\ N_x &= V_1 \sin\phi - \sum^x P \sin\phi \\ &= P\sin\phi - \sum^x P\sin\phi. \end{aligned}\right\} \text{For two equal and symmetrical vertical loads, } x \overset{=}{>} a.$$

Since our loads are equal and symmetrically placed, and K' is the moment at any point x, considering the arch as an unconfined girder, the value of K' due to one load will have a corresponding equal value due to the other load. Then, since there are symmetrical values of $\frac{yds}{\theta_x}$, the value of $\int_0^l K'y\frac{ds}{\theta_x}$ for one load must be equal to that for the other load.

Therefore for a *single vertical* load we have

$$K' = P(1-k)x - [P(x-a) \text{ when } x > a]$$

and

$$N_x = P(1-k)\sin\phi - [P\sin\phi \text{ when } x > a].$$

For $x = 0$ to $x = a$,

$$K' = P(1-k)x \quad\text{and}\quad N_x = P(1-k)\sin\phi. \quad . \quad g(88)$$

For $x = a$ to $x = l$,

$$K' = Pk(l-x) \quad\text{and}\quad N_x = -Pk\sin\phi; \quad . \quad . \quad g(89)$$

and we have

$$H_1 = \frac{P\left\{\begin{array}{l} (1-k)\int_0^a \frac{xyds}{\theta_x} + k\int_a^l \frac{(l-x)yds}{\theta_x} \\ +(1-k)\int_0^a \frac{dx}{F_x}\sin\phi - k\int_a^l \frac{dx}{F_x}\sin\phi \\ -\frac{(1-k)\int_0^a \frac{xds}{\theta_x} + k\int_a^l \frac{(l-x)ds}{\theta_x}}{\int_0^l \frac{ds}{\theta_x}}\int_0^l \frac{yds}{\theta_x} \end{array}\right\}}{+\int_0^l \frac{y^2ds}{\theta_x} + \int_0^l \frac{dx}{F_x}\cos\phi - \frac{\left(\int_0^l \frac{yds}{\theta_x}\right)^2}{\int_0^l \frac{ds}{\theta_x}} = D}. \quad g(90)$$

From $g(90)$, the horizontal thrust due to any vertical load can be found when the relation between x and y is known. The equation applies equally well to the parabolic, circular, or elliptic arch.

(b) *Horizontal Loads only.*

Here we have

$$\left.\begin{array}{l} K' = -\overset{x}{\Sigma}Q(y-b), \\ V_1 = 0, \\ \text{and} \\ N_x = -\overset{x}{\Sigma}Q\cos\phi. \end{array}\right\} \text{For two equal and symmetrical horizontal loads, } x > a.$$

For $x = 0$ to $x = a_1$,

$$K' = 0 \quad \text{and} \quad N_x = 0; \quad \ldots \ldots \ldots \quad g(91)$$

for $x = a_1$ to $x = a_2$,

$$K' = -Q(y-b) \quad \text{and} \quad N_x = -Q\cos\phi; \quad . \quad g(92)$$

for $x = a_2$ to $x = l$,

$$K' = 0 \quad \text{and} \quad N_x = 0. \quad \ldots \ldots \ldots \quad g(93)$$

Let H_1 = the thrust due to the load on the left;
H_2 = the thrust due to the right load.

Then

$$H_1 - H_2 = Q;$$

but $$H_1 + H_2 = \mathfrak{H}_1;$$

hence $$2H_1 = \mathfrak{H}_1 + Q$$

or

$$H_1 = \tfrac{1}{2}\mathfrak{H}_1 + \tfrac{1}{2}Q. \quad \ldots \ldots \quad g(94)$$

Therefore

$$H_1 = \tfrac{1}{2}\left\{ Q + \frac{\displaystyle\int_0^l \frac{K'yds}{\theta_x} + \int_0^l \frac{N_x}{\theta_x}dx - \frac{\displaystyle\int_0^l \frac{K'ds}{\theta_x}}{\displaystyle\int_0^l \frac{ds}{\theta_x}} \int_0^l \frac{yds}{\theta_x}}{D} \right\}$$

or

$$H_1 = \frac{Q}{2}\left\{ 1 - \frac{\displaystyle\int_{a_2}^{a_1} \frac{(y-b)}{\theta_x}yds - \int_{a_2}^{a_1} \frac{dx}{\theta_x}\cos\phi - \frac{\displaystyle\int_{a_2}^{a_1} \frac{(y-b)ds}{\theta_x}}{\displaystyle\int_0^l \frac{ds}{\theta_x}} \int_0^l \frac{yds}{\theta_x}}{\displaystyle + \int_0^l \frac{y^2ds}{\theta_x} + \int_0^l \frac{dx}{\theta_x}\cos\phi - \frac{\left(\displaystyle\int_0^l \frac{yds}{\theta_x}\right)^2}{\displaystyle\int_0^l \frac{ds}{\theta_x}}} \right\}. \quad g(95)$$

$$H_2 = Q - H_1. \quad \ldots \ldots \quad g(96)$$

$g(95)$ is general, and can be applied to parabolic, circular, and elliptic arches with equal facility.

(*c*) *Moments, Vertical Loads only.*

In $g(71)$, for a single vertical load,

$$K_1 = -H_1y - P(x-a) \qquad x \overline{\gtrless} a,$$

and

$$N_x = V_1 \sin\phi - P\sin\phi + H_1\cos\phi, \qquad x \overline{\gtrless} a,$$

where H_1 can be found from $g(90)$.

There remains then only the term $V_1 \sin\phi$, which is as yet unknown. In case there are equal and symmetrical loads V_1 becomes known, as it is equal to one half the total loading.

If, however, the loading is not symmetrical, the values of M_1 and M_2 can be computed with the term $V_1 \sin\phi$ neglected, and the corresponding value of V_1 found from (47), and then a second calculation made and this value introduced. Generally the value of the expression containing V_1 is very small, and is omitted entirely by nearly all American authors.

Neglecting the term containing V_1, for $x = 0$ to $x = a$,

$$K = -H_1y \quad . \quad . \quad . \quad . \quad . \quad . \quad . \quad . \quad . \quad . \quad . \quad . \quad g(97)$$

and

$$N_x = H_1\cos\phi \text{ (approximately)}; \quad . \quad . \quad . \quad . \quad . \quad g(98)$$

for $x = a$ to $x = l$,

$$K = -H_1y - P(x-a) \quad . \quad . \quad . \quad . \quad . \quad . \quad . \quad . \quad g(99)$$

and

$$N_x = H_1\cos\phi - P\sin\phi \text{ (approximately)}. \quad . \quad g(100)$$

Therefore $g(71)$ becomes

$$M_1 = \frac{\left\{\begin{array}{l} -H_1\int_0^l \frac{xyds}{\theta_x}\int_0^l \frac{xds}{\theta_x} - \int_a^l P(x-a)\frac{xds}{\theta_x}\int_0^l \frac{xds}{\theta_x} \\ +H_1\int_0^l \frac{\cos\phi dy}{\theta_x}\int_0^l \frac{xds}{\theta_x} - \int_a^l P\sin\phi\frac{dy}{\theta_x}\int_0^l \frac{xds}{\theta_x} \\ +H_1\int_0^l \frac{yds}{\theta_x}\int_0^l \frac{x^2ds}{\theta_x} + \int_a^l P(x-a)\frac{ds}{\theta_x}\int_0^l \frac{x^2ds}{\theta_x} \end{array}\right\}}{\int_0^l \frac{ds}{\theta_x}\int_0^l \frac{x^2ds}{\theta_x} - \left(\int_0^l \frac{xds}{\theta_x}\right)^2}, \quad g(101)$$

where the value of H_1 is to be found from $g(90)$.

The value of M_2 can be found from $g(101)$ by replacing a by $(l-a)$.

(d) *Moments, Horizontal Loads only.*

In the case of horizontal loads only,

$$K = -H_1 y + \sum^x Q(y-b), \qquad x \overline{\gtrless} a.$$

$$N_x = V_1 \sin\phi + H_1 \cos\phi - \sum^x Q \cos\phi. \qquad x \overline{\gtrless} a.$$

Then for $x = 0$ to $x = a$,

$$K = -H_1 y \quad . \quad . \quad . \quad . \quad . \quad . \quad . \quad . \quad . \quad . \quad g(102)$$

and

$$N_x = H_1 \cos\phi \text{ (approximately)}; \quad . \quad . \quad . \quad . \quad . \quad g(103)$$

for $x = a$ to $x = l$,

$$K = -H_1 y + Q(y-b) \quad . \quad . \quad . \quad . \quad . \quad . \quad . \quad g(104)$$

and

$$N_x = H_1 \cos\phi - Q\cos\phi \text{ (approximately)}. \quad . \quad . \quad g(105)$$

Therefore $g(71)$ becomes

$$M_1 = \frac{\left\{\begin{array}{l} -H_1 \int_0^l \frac{xyds}{\theta_x} \int_0^l \frac{xds}{\theta_x} + \int_a^l Q(y-b)\frac{xds}{\theta_x} \int_0^l \frac{xds}{\theta_x} \\ -H_1 \int_0^l \frac{dy}{\theta_x}\cos\phi \int_0^l \frac{xds}{\theta_x} + \int_a^l Q\cos\phi\frac{dy}{\theta_x} \int_0^l \frac{xds}{\theta_x} \\ +H_1 \int_0^l \frac{yds}{\theta_x} \int_0^l \frac{x^2ds}{\theta_x} - \int_a^l Q(y-b)\frac{ds}{\theta_x} \int_0^l \frac{x^2ds}{\theta_x} \end{array}\right\}}{\int_0^l \frac{ds}{\theta_x} \int_0^l \frac{x^2ds}{\theta_x} - \left(\int_0^l \frac{xds}{\theta_x}\right)^2}, \quad g(106)$$

where H_1 is to be found from $g(95)$.

The value of M_2 can be found from $g(106)$ by replacing a by $(l-a)$.

(*e*) *Effect of a Change in Temperature.*

Assuming that the span does not change in length,

$$et^\circ \int_0^l dx = et^\circ l = 0; \quad . \quad . \quad . \quad . \quad . \quad g(107)$$

or, if the arch is free to slide upon the supports,

$$et^\circ l = \Delta' l. \quad . \quad . \quad . \quad . \quad . \quad . \quad . \quad g(108)$$

Let H_t be the horizontal thrust necessary to cause a change in the length of the span of $\Delta' l$; then, referring to $g(86)$,

$$\Delta'' l : \Delta' l :: Q' : H_t = \frac{\Delta' l}{\Delta'' l} Q' \quad . \quad . \quad . \quad g(109)$$

or

$$H_t = \frac{Eet^\circ l}{\int_0^l \frac{y^2 ds}{\theta_x} + \int_0^l \frac{dx}{F_x} \cos\phi - \frac{\left(\int_0^l \frac{yds}{\theta_x}\right)^2}{\int_0^l \frac{ds}{\theta_x}}} \quad . \quad . \quad g(110)$$

From $g(83)$,

$$z = -\frac{\int_0^l \frac{yds}{\theta_x}}{\int_0^l \frac{ds}{\theta_x}} \quad . \quad . \quad . \quad . \quad . \quad . \quad . \quad . \quad . \quad . \quad . \quad . \quad g(111)$$

But $M_1 = H_t z$;

hence

$$M_1 = -Eet^\circ l \frac{\int_0^l \frac{yds}{\theta_x}}{\left[\int_0^l \frac{y^2 ds}{\theta_x} + \int_0^l \frac{dx}{F_x}\cos\phi - \frac{\left(\int_0^l \frac{yds}{\theta_x}\right)^2}{\int_0^l \frac{ds}{\theta_x}}\right]\int_0^l \frac{ds}{\theta_x}} . \quad g(112)$$

SYMMETRICAL ARCH WITH A HINGE AT EACH SUPPORT.

In this case we have no moments at the points of support.

Assume that the arch is free to slide upon the supports; then, from $g(63)$,

$$\Delta' l = -\frac{1}{E}\int_0^l \frac{M_x}{\theta_x} yds - \frac{1}{E}\int_0^l \frac{N_x}{F_x} dx. \quad . \quad . \quad g(113)$$

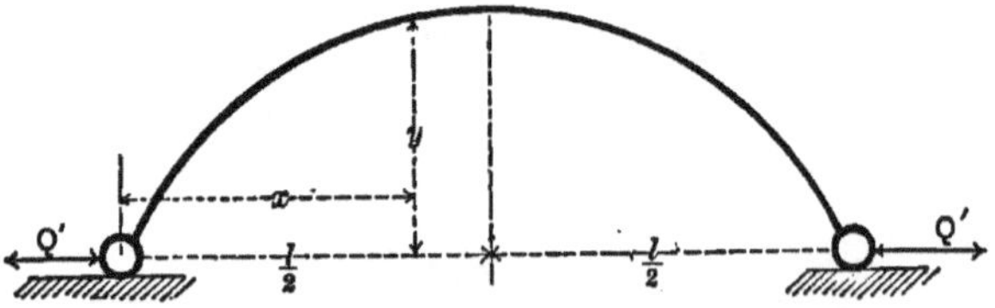

FIG. 35.

Let Q' be any horizontal load at the hinges; then

$$M_x = Q'y \quad . \quad . \quad . \quad . \quad . \quad . \quad . \quad . \quad . \quad . \quad . \quad g(114)$$

and

$$N_x = H_x \cos\phi = Q' \cos\phi. \quad . \quad . \quad . \quad . \quad . \quad . \quad g(115)$$

Then $g(113)$ becomes

$$\Delta' l = -\frac{Q'}{E}\int_0^l \frac{y^2 ds}{\theta_x} - \frac{Q'}{E}\int_0^l \frac{dx}{F_x}\cos\phi. \quad . \quad . \quad g(116)$$

Now suppose the horizontal loads Q' removed and two equal and symmetrical vertical loads applied to the arch; then

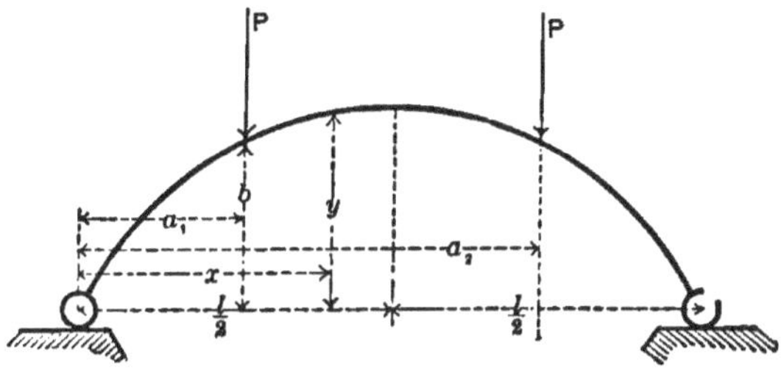

FIG. 36.

$$\Delta''l = -\frac{1}{E}\int_0^l \frac{M_x}{\theta} yds - \frac{1}{E}\int_0^l \frac{N_x}{F_x} dx. \quad . \quad . \quad g(117)$$

$$M_x = V_1 x - \overset{x}{\Sigma} P(x - a)$$

and $\quad V_1 = \overset{l}{\Sigma} P(1 - k);$

hence

$$M_x = \overset{l}{\Sigma} P(1 - k)x - \overset{x}{\Sigma} P(x - a). \quad . \quad . \quad . \quad g(118)$$

$$N_x = V_x \sin\phi + H_x \cos\phi = V_x \sin\phi$$

But $V_x = V_1 - \overset{x}{\Sigma} P$; therefore

$$N_x = \overset{l}{\Sigma} P(1 - k)\sin\phi - \overset{x}{\Sigma} P \sin\phi. \quad . \quad . \quad g(119)$$

Then for $x = 0$ to $x = a_1$

$$M_x = \overset{l}{\Sigma} P(1 - k)x \quad . \quad . \quad . \quad . \quad . \quad . \quad . \quad . \quad g(120)$$

and

$$N_x = \overset{l}{\Sigma} P(1 - k)\sin\phi; \quad . \quad . \quad . \quad . \quad . \quad . \quad . \quad . \quad g(121)$$

for $x = a_1$ to $x = a_2$,

$$M_x = \overset{l}{\Sigma} P(1 - k)x - P(x - a_1) \quad . \quad . \quad . \quad . \quad g(122)$$

and

$$N_x = \overset{l}{\Sigma} P(1 - k) \sin \phi - P \sin \phi ; \quad . \quad . \quad . \quad g(123)$$

for $x = a_2$ to $x = l$,

$$M_x = \overset{l}{\Sigma} P(1 - k)x - \overset{x}{\Sigma} P(x - a) \quad . \quad . \quad . \quad . \quad g(124)$$

and

$$N_x = \overset{l}{\Sigma} P(1 - k) \sin \phi - \overset{x}{\Sigma} P \sin \phi . \quad . \quad . \quad . \quad g(125)$$

Evidently the change in the length of the span due to the left load will equal that due to the right load; hence we have, for $x = 0$ to $x = a$,

$$M_x = P(1 - k)x \quad . \quad . \quad . \quad . \quad . \quad . \quad . \quad . \quad . \quad g(126)$$

and

$$N_x = P(1 - k) \sin \phi ; \quad . \quad . \quad . \quad . \quad . \quad . \quad . \quad g(127)$$

for $x = a$ to $x = l$,

$$M_x = P(1 - k)x - P(x - a) \quad . \quad . \quad . \quad . \quad g(128)$$

and

$$N_x = P(1 - k \sin \phi - P \sin \phi) . \quad . \quad . \quad g(129)$$

Therefore $g(117)$ becomes

$$\frac{\Delta'' l}{2} = -\frac{P}{E}\int_0^l \frac{(1 - k)xyds}{\theta_x} - \frac{P}{E}(1 - k)\int_0^l \frac{\sin \phi dx}{F_x}$$

$$+ \frac{P}{E}\int_a^l \frac{(x - a)yds}{\theta_x} + \frac{P}{E}\int_a^l \frac{\sin \phi dx}{F_x} . \quad . \quad . \quad g(130)$$

Let $\mathfrak{H}_1$ represent the horizontal thrust necessary to cause a change in the length of the span of $\Delta''l$; then

$$\Delta'l : \Delta''l :: Q' : \mathfrak{H}_1 = \frac{\Delta''l}{\Delta'l}Q';$$

and we have, since for *vertical loads* the horizontal thrust is constant and $H_1 = \frac{1}{2}\mathfrak{H}_1$,

$$H_1 = \frac{\left\{ \begin{array}{l} P(1-k)\left\{\int_0^l \frac{xyds}{\theta_x} + \int_0^l \frac{dx}{F_x}\sin\phi\right\} \\ \qquad - P\left\{\int_a^l \frac{(x-a)yds}{\theta_x} + \int_a^l \frac{dx}{F_x}\sin\phi\right\} \end{array} \right\}}{\int_0^l \frac{y^2ds}{\theta_x} + \int_0^l \frac{dx}{F_x}\cos\phi}. \quad g(131)$$

From $g(131)$ the horizontal thrust due to any vertical load can be found with comparatively little labor.

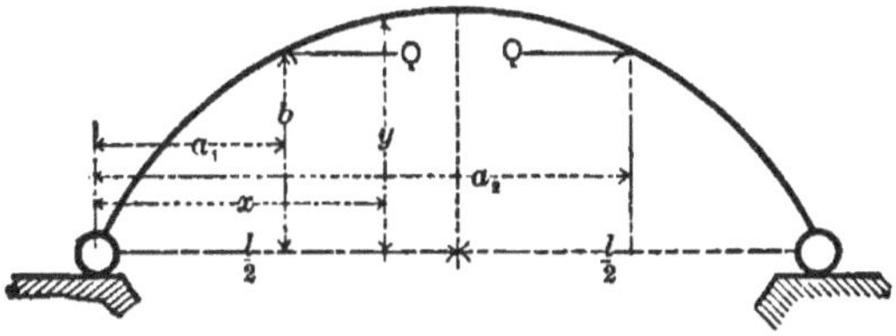

FIG. 37.

For two equal and symmetrical *horizontal loads* we have from (41), assuming the arch free to slide,

$$M_x = V_1x + \overset{x}{\Sigma}Q(y-b);$$

but $\quad V_1 = 0,$

hence

$$M_x = \overset{x}{\Sigma}Q(y-b). \quad \ldots\ldots\ldots\ldots \quad g(132)$$

$$N_x = V_x \sin\phi + H_x \cos\phi = H_x \cos\phi;$$

but

$$H_x = H_1 - \overset{x}{\Sigma} Q,$$

hence

$$N_x = -\overset{x}{\Sigma} Q \cos\phi. \quad . \quad . \quad . \quad . \quad . \quad . \quad . \quad . \quad g(133)$$

Then for $x = 0$ to $x = a_1$, and for $x = a_2$ to $x = l$,

$$M_x = 0 \quad . \quad . \quad . \quad . \quad . \quad . \quad . \quad . \quad . \quad . \quad . \quad . \quad g(134)$$

and

$$N_x = 0; \quad . \quad . \quad . \quad . \quad . \quad . \quad . \quad . \quad . \quad . \quad . \quad . \quad g(135)$$

for $x = a_1$ to $x = a_2$,

$$M_x = +Q(y - b) \quad . \quad . \quad . \quad . \quad . \quad . \quad . \quad . \quad g(136)$$

and

$$N_x = -Q\cos\phi. \quad . \quad . \quad . \quad . \quad . \quad . \quad . \quad . \quad g(137)$$

Hence the corresponding change in the length of the span is

$$\Delta''l = -\frac{Q}{E}\int_{a_1}^{a_2}\frac{(y-b)yds}{\theta_x} + \frac{Q}{E}\int_{a_1}^{a_2}\frac{dx}{F_x}\cos\phi. \quad g(138)$$

Then since

$$\Delta'l : \Delta''l :: Q' : \mathfrak{H}_1 = \frac{\Delta''l}{\Delta'l}Q',$$

$$\mathfrak{H}_1 = \frac{Q\int_{a_1}^{a_2}\frac{(y-b)yds}{\theta_x} - Q\int_{a_1}^{a_2}\frac{dx}{F_x}\cos\phi}{\int_0^l\frac{y^2ds}{\theta_x} + \int_0^l\frac{dx}{F_x}\cos\phi}. \quad . \quad g(139)$$

For a single load, $H_1 = \frac{1}{2}\mathfrak{H}_1 + \frac{1}{2}Q$. Hence

$$H_1 = Q\left\{\frac{1}{2} + \frac{1}{2}\frac{\int_{a_1}^{a_2}\frac{(y-b)yds}{\theta_x} - \int_{a_1}^{a_2}\frac{dx}{F_x}\cos\phi}{\int_0^l\frac{y^2ds}{\theta_x} + \int_0^l\frac{dx}{F_x}\cos\phi}\right\}, \quad g(140)$$

an equation quite simple in its application.

* If the moment of inertia is assumed to vary according to the laws assumed by most writers upon the theory of arches, their equations can be very easily obtained from our general forms.

For example, let the horizontal thrust H_1 for a single vertical load placed upon a parabolic arch having no hinges be required. Assuming $\theta\cos\phi = A =$ a constant, and that the terms containing the effect of the axial stress are neglected, and remembering that $ds\cos\phi = dx$, we have, from $g(90)$,

$$\frac{H_1}{P} = \frac{\left\{\begin{array}{l}(1-k)\int_0^a yxdx + k\int_a^l y(l-x)dx \\ -\dfrac{+(1-k)\int_0^a xdx + k\int_a^l(l-x)dx}{\int_0^l dx}\int_0^l ydx\end{array}\right\}}{\int_0^l y^2dx - \dfrac{\left(\int_0^l ydx\right)^2}{\int_0^l dx}}.$$

where, using the nomenclature employed in Chapter III,

$$(1-k)\int_0^a yxdx = \frac{1-k}{3}fl^2(4k^3 - 3k^4),$$

$$k\int_a^l(l-x)ydx = \tfrac{1}{3}fl^2k(1 - 6k^2 + 8k^3 - 3k^4),$$

* For several examples illustrating the application of these general formulas to special cases, see Appendix E.

$$(1 - k)\int_0^{a} x dx = \tfrac{1}{2}l^2k^2(1 - k),$$

$$k\int_a^l (l - x)dx = \tfrac{1}{2}l^2k(1 - 2k + k^2)$$

$$\int_0^l dx = l,$$

$$\int_0^l y dx = \tfrac{2}{3}fl,$$

$$\int_0^l y^2 dx = \tfrac{8}{15}f^2l,$$

and we have

$$H_1 = \frac{\frac{1}{3}fl^2(k - 2k^3 + k^4) - \frac{1}{3}fl^2(k - k^3)}{\frac{12}{135}f^2l}P,$$

or

$$H_1 = \tfrac{15}{4}Pk^2(1 - k)^2, \quad . \quad . \quad . \quad . \quad . \quad . \quad . \quad . \quad . \quad . \quad p(143)$$

which is the same as obtained by the method employed in Chapter III. (See equation $p(143)$, page 71.)

In a similar manner any of the equations usually employed can be quickly deduced from our general formulas, which have the advantage of being general to the extent that they can be employed for any arch when the relation between x and y can be represented by a linear equation.

SUMMATION FORMULAS.

In many cases it is preferable to replace the sign of integration by that of summation. This is particularly true in arches where the moments of inertia do not change according to some law which permits of readily reducing the above equations to fit the particular case. As examples of such structures may be mentioned the Douro Arch and the Washington Bridge.

The summation formulas are as follows:

(A) ARCH WITHOUT HINGES.

Vertical Load only.

$$M_1 = \frac{\begin{aligned}&+\left\{-H_1\sum_0^l\frac{xy\Delta s}{\theta_x}-P\sum_a^l(x-a)\frac{x\Delta s}{\theta_x}\right\}\sum_0^l\frac{x\Delta s}{\theta_x}\\&+\left\{+H_1\sum_0^l\frac{\cos\phi\,\Delta y}{F_x}-P\sum_a^l\sin\phi\,\frac{\Delta y}{F_x}\right\}\sum_0^l\frac{x\Delta s}{\theta_x}\\&+\left\{+H_1\sum_0^l\frac{y\Delta s}{\theta_x}+P\sum_a^l(x-a)\frac{\Delta s}{\theta_x}\right\}\sum_0^l\frac{x^2\Delta s}{\theta_x}\end{aligned}}{\sum_0^l\frac{\Delta s}{\theta_x}\sum_0^l\frac{x^2\Delta s}{\theta_x}-\left(\sum_0^l\frac{x\Delta s}{\theta_x}\right)^2}. \quad g(141)$$

$$\frac{H_1}{P} = \frac{\begin{aligned}&(1-k)\sum_0^a\frac{xy\Delta s}{\theta_x}+k\sum_a^l\frac{(l-x)y\Delta s}{\theta_x}+(1-k)\sum_0^a\frac{\sin\phi\,\Delta x}{F_x}\\&-k\sum_a^l\sin\phi\frac{\Delta x}{F_x}-\frac{(1-k)\sum_0^a\frac{x\Delta s}{\theta_x}+k\sum_a^l\frac{(l-x)\Delta s}{\theta_x}}{\sum_0^l\frac{\Delta s}{\theta_x}}\sum_0^l\frac{y\Delta s}{\theta_x}\end{aligned}}{\sum_0^l\frac{y^2\Delta s}{\theta_x}+\sum_0^l\frac{\Delta x}{F_x}\cos\phi-\frac{\left(\sum_0^l\frac{y\Delta s}{\theta_x}\right)^2}{\sum_0^l\frac{\Delta s}{\theta_x}}}. \quad g(142)$$

Horizontal Loaa only.

$$M_1 = \frac{\begin{aligned}&+\left\{-H_1\sum_0^l\frac{xy\Delta s}{\theta_x}-Q\sum_a^l(y-b)\frac{x\Delta s}{\theta_x}\right\}\sum_0^l\frac{x\Delta s}{\theta_x}\\&+\left\{-H_1\sum_0^l\frac{\Delta y}{F_x}\cos\phi+Q\sum_a^l\frac{\Delta y}{F_x}\cos\phi\right\}\sum_0^l\frac{x\Delta s}{\theta_x}\\&+\left\{+H_1\sum_0^l\frac{y\Delta s}{\theta_x}-Q\sum_a^l(y-b)\frac{\Delta s}{\theta_x}\right\}\sum_0^l\frac{x^2\Delta s}{\theta_x}\end{aligned}}{\sum_0^l\frac{\Delta s}{\theta_x}\sum_0^l\frac{x^2\Delta s}{\theta_x}-\left(\sum_0^l\frac{x\Delta s}{\theta_x}\right)^2}. \quad g(143)$$

$$H_1=\frac{Q}{2}\left\{1-\frac{\sum\limits_{a_2}^{a_1}\frac{(y-b)y\Delta s}{\theta_x}-\sum\limits_{a_2}^{a_1}\frac{\Delta s}{\theta_x}-\frac{\sum\limits_{a_2}^{a_1}\frac{(y-b)\Delta s}{\theta_x}}{\sum\limits_0^l\frac{\Delta s}{\theta_x}}\sum\limits_0^l\frac{y\Delta s}{\theta_x}}{\sum\limits_0^l\frac{y^2\Delta s}{\theta_x}+\sum\limits_0^l\frac{\Delta x}{F_x}\cos\phi-\frac{\left(\sum\limits_0^l\frac{y\Delta s}{\theta_x}\right)^2}{\sum\limits_0^l\frac{\Delta s}{\theta_x}}}\right\}. \quad g(144)$$

Temperature.

$$H_1=\frac{Et^\circ l}{\sum\limits_0^l\frac{y^2\Delta s}{\theta_x}+\sum\limits_0^l\frac{\Delta x}{\theta_x}\cos\phi-\frac{\left(\sum\limits_0^l\frac{y\Delta s}{\theta_x}\right)^2}{\sum\limits_0^l\frac{\Delta s}{\theta_x}}}. \quad . \quad g(145)$$

(*b*) ARCH WITH TWO HINGES (ONE AT EACH SUPPORT).

Vertical Load only.

$$H_1=\frac{P(1-k)\left\{\sum\limits_0^l\frac{xy\Delta s}{\theta_x}+\sum\limits_0^l\frac{\Delta x}{F_x}\sin\phi\right\}-P\left\{\sum\limits_a^l\frac{(x-a)y\Delta s}{\theta_x}+\sum\limits_a^l\frac{\Delta x}{F_x}\sin\phi\right\}}{\sum\limits_0^l\frac{y^2\Delta s}{\theta_x}+\sum\limits_0^l\frac{\Delta x}{F_x}\cos\phi}. \quad . \quad g(146)$$

Horizontal Load only.

$$H_1=Q\left\{\frac{1}{2}+\frac{1}{2}\,\frac{\sum\limits_{a_1}^{a_2}\frac{(y-b)y\Delta s}{\theta_x}-\sum\limits_{a_1}^{a_2}\frac{\Delta x}{F_x}\cos\phi}{\sum\limits_0^l\frac{y^2\Delta s}{\theta_x}+\sum\limits_0^l\frac{\Delta x}{F_x}\cos\phi}\right\}. \quad g(147)$$

Temperature.

$$H_1=\frac{Et^\circ l}{\sum\limits_0^l\frac{y^2\Delta s}{\theta_x}+\sum\limits_0^l\frac{\Delta x}{F_x}\cos\phi}. \quad . \quad . \quad . \quad g(148)$$

ARCH WITH ONE HINGE AT THE CROWN.

This type of arch is seldom, if ever, employed by American engineers. French and German engineers sometimes consider masonry arches having lead * or iron hinges at the skew-backs and the crown as one-hinge arches for *moving loads.*

For this case we will neglect the effect of the axial stress as being of little importance in cases where this form of arch would be employed.

Vertical Loads.

Value of H_1.—Let two equal and symmetrically placed loads be applied to the arch; then $\Delta\phi_0 = \Delta\phi_l$.

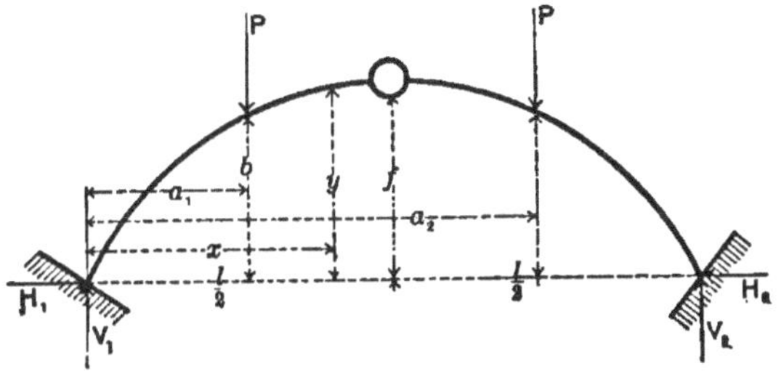

FIG. 38

From $g(63)$ we have

$$\Delta l = \int_0^l \frac{M_x yds}{\theta_x} = 0, \quad . \quad . \quad . \quad . \quad g(149)$$

where

$$M_x = M_1 + V_1 x - \mathfrak{H}_1 y - \sum^x P(x - a). \quad . \quad g(150)$$

When $x = \frac{l}{2}$, $M_x = 0$, since there can be no bending-moment at the hinge; hence

$$M_1 = -V_1\frac{l}{2} + \mathfrak{H}_1 f + P\left(\frac{l}{2} - a_1\right). \quad . \quad . \quad g(151)$$

* See pages 229 and 230.

But since our loads are equal and symmetrically placed, $V_1 = P$, and

$$M_1 = \mathfrak{H}_1 f - Pa_1. \qquad \qquad g(152)$$

Substituting this value in $g(150)$ and then the value of M_x in $g(149)$, we have

$$-\mathfrak{H}_1 \int_0^l \frac{y^2 ds}{\theta_x} + \mathfrak{H}_1 f \int_0^l \frac{yds}{\theta_x} - Pa_1 \int_0^l \frac{yds}{\theta_x} + P \int_0^l \frac{xyds}{\theta_x}$$

$$- \int_0^l \sum^x P(x-a) \frac{yds}{\theta_a} = 0 \qquad \qquad g(153)$$

or

$$\mathfrak{H}_1 = \frac{P \int_0^l (x - a_1) \frac{yds}{\theta_x} - \int_0^l \sum^x P(x-a) \frac{yds}{\theta_x}}{\int_0^l \frac{y^2 ds}{\theta_x} - f \int_0^l \frac{yds}{\theta_x}}$$

or

$$H = \frac{P}{2} \frac{\int_0^l \frac{xyds}{\theta_x} - a_1 \int_0^l \frac{yds}{\theta_x} - \int_{a_1} (x - a_1) \frac{yds}{\theta_x} - \int_{a_2}^l (x - a_2) \frac{yds}{\theta_x}}{\int_0^l \frac{y^2 ds}{\theta_x} - f \int_0^l \frac{yds}{\theta_x}}, \qquad g(154)$$

where $a_2 = l - a_1$.

Value of V_1 and V_2 for Vertical Loads.

From $g(61)$ we have

$$\Delta y = \frac{1}{E} \int_0^x M_x \frac{xds}{\theta_x}, \qquad \qquad g(155)$$

which becomes, for $x = \frac{l}{2}$,

$$\Delta f = \frac{1}{E}\int_0^{l/2} M_x \frac{xds}{\theta_x}. \quad . \quad . \quad . \quad . \quad . \quad g(156)$$

Let two equal and symmetrically placed loads be applied to the arch; then $V_1 = P$, and for our two loads we have, from (41),

$$M_x = M_1 + Px - \mathfrak{H}_1 y - \sum^x P(x - a). \quad . \quad . \quad . \quad g(157)$$

But

$$M_1 = \mathfrak{H}_1 f - Pa_1. \quad . \quad . \quad . \quad . \quad . \quad . \quad . \quad . \quad . \quad . \quad g(158)$$

Hence

$$M_x = \mathfrak{H}_1 f - \mathfrak{H}_1 y - Pa_1 + Px - \sum^x P(x - a). \quad g(159)$$

Then if $\Delta_2 f$ be the vertical displacement of the crown due to the action of these two loads,

$$E\Delta_2 f = \mathfrak{H}_1 f \int_0^{l/2} \frac{xds}{\theta_x} - \mathfrak{H}_1 \int_0^{l/2} \frac{xyds}{\theta_x}$$

$$- Pa_1 \int_0^{l/2} \frac{xds}{\theta_x} + P\int_0^{l/2} \frac{x^2ds}{\theta_x} - P\int_0^{l/2} (x - a)\frac{xds}{\theta_x}. \quad g(160)$$

For a single vertical load,

$$M_1 = - V_1\frac{l}{2} + H f + \sum^{l/2} P\left(\frac{l}{2} - a\right). \quad . \quad . \quad . \quad . \quad . \quad . \quad g(161)$$

Therefore,

$$M_x = - V_1\frac{l}{2} + V_1 x + H_1 f - H_1 y + \sum^{l/2} P\left(\frac{l}{2} - a\right) - \sum^x P(x - a). \quad g(162)$$

If $\Delta_1 f$ be the vertical displacement of the crown due to a single load, we have

$$E\Delta_1 f = H_1 f\int_0^{l/2}\frac{xds}{\theta_x} - H_1\int_0^{l/2}\frac{yxds}{\theta_x} + P\left(\frac{l}{2} - a\right)\int_0^{l/2}\frac{xds}{\theta_x}$$
$$- P\int_0^{l/2}(x - a)\frac{xds}{\theta_x} - V_1\frac{l}{2}\int_0^{l/2}\frac{xds}{\theta_x} + V_1\int_0^{l/2}\frac{x^2ds}{\theta_x}. \quad g(163)$$

Since the vertical deflection of the crown due to one of two equal and symmetrical loads must be one half that due to both loads, $\Delta_2 f = 2\Delta_1 f$. Equating these two values and solving for V_1, we obtain

$$V_1 = \frac{1}{2}\,\frac{P(l - a)\displaystyle\int_0^{l/2}\frac{xds}{\theta_x} - P\int_a^{l/2}(x - a)\frac{xds}{\theta_x} - P\int_0^{l/2}\frac{x^2ds}{\theta_x}}{\displaystyle\frac{l}{2}\int_0^{l/2}\frac{xds}{\theta_x} - \int_0^{l/2}\frac{x^2ds}{\theta_x}}. \quad g(164)$$

This equation is to be employed for all loads on the left of the crown.

$$V_2 = P - V_1$$

These equations enable us to find the values of V_1 and V_2 for *all loads.*

Values of M_1 and M_2 for Vertical Loads.

From (41), making $x = \frac{l}{2}$ and solving for M_1, we have, for a *single load,*

$$M_1 = - V_1\frac{l}{2} + H_1 f + P\left(\frac{l}{2} - a\right), \quad . \quad . \quad g(165)$$

in which the values of V_1 and H_1 are given by $g(164)$ and $g(154)$.

This equation gives the values of M_1 for any load on the *left of the crown.*

From (49),

$$M_2 = M_1 + V_1 l - P(l - a). \quad . \quad . \quad . \quad . \quad g(167)$$

HORIZONTAL LOADS.

Value of H_1 for a Single Horizontal Load.

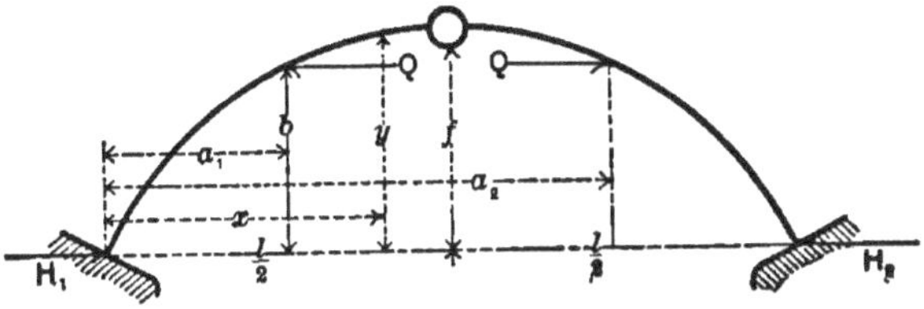

FIG. 39.

Let two equal and symmetrically placed horizontal loads act upon the arch; then $V_1 = 0$ and (41) becomes

$$M_x = M_1 - \mathfrak{H}_1 y + \overset{x}{\Sigma} Q(y - b). \quad . \quad . \quad . \quad . \quad . \quad g(168)$$

If $x = \frac{l}{2}$, then $M_x = 0$; hence

$$M_1 = \mathfrak{H}_1 f - Q(f - b) \quad . \quad . \quad . \quad . \quad . \quad . \quad . \quad . \quad g(169)$$

and

$$M_x = \mathfrak{H}_1 f - \mathfrak{H}_1 y - Q(f - b) + \overset{x}{\Sigma} Q(y - b). \quad . \quad g(170)$$

From $g(63)$,

$$\int_0^l M_x \frac{yds}{\theta_x} = 0. \quad . \quad . \quad . \quad . \quad . \quad . \quad g(171)$$

Substituting the value of M_x and solving for $\mathfrak{H}_1$, we have,

$$\mathfrak{H}_1 = \frac{(f - b)\int_0^l \frac{yds}{\theta_x} - \int_{a_1}^{a_2} (y - b)\frac{yds}{\theta_x}}{f\int_0^l \frac{yds}{\theta_x} - \int_0^l \frac{y^2 ds}{\theta_x}} Q, \quad . \quad g(172)$$

$$H_1 = \tfrac{1}{2}(\mathfrak{H}_1 + Q). \quad . \quad . \quad . \quad . \quad . \quad . \quad . \quad . \quad . \quad . \quad g(173)$$

Value of V_1 for a Single Horizontal Load.

This case will be treated in a manner similar to that employed for vertical loads. From $g(156)$,

$$\Delta f = \frac{1}{E}\int_0^{\frac{l}{2}} M_x \frac{xds}{\theta_x} \quad \ldots \ldots \quad g(174)$$

From (41), for two equal and symmetrical loads,

$$M_x = M_1 - \mathfrak{H}_1 y + \sum^x Q(y - b). \quad \ldots \ldots \quad g(175)$$

But

$$M_1 = \mathfrak{H}_1 f - Q(f - b); \quad \ldots \ldots \ldots \quad g(176)$$

hence

$$M_x = \mathfrak{H}_1 f - \mathfrak{H}_1 y - Q(f - b) + \sum^x Q(y - b). \quad g(177)$$

The vertical displacement due to two equal and symmetrical loads is

$$\Delta_1 f = \frac{1}{E}\left\{ \mathfrak{H}_1 f \int_0^{\frac{l}{2}} \frac{xds}{\theta_x} - \mathfrak{H}_1 \int_0^{\frac{l}{2}} \frac{xyds}{\theta_x} \right.$$

$$\left. Q(f - b)\int_0^{\frac{l}{2}} \frac{xds}{\theta_x} + Q\int_0^{\frac{l}{2}} (y - b)\frac{xds}{\theta_x} \right\}. \quad g(178)$$

For a single load,

$$M_x = H_1 f - H_1 y - Q(f - b) + Q(y - b) - V_1\frac{l}{2} + V_1 x. \quad g(179)$$

But
$$H_1 = \tfrac{1}{2}(\mathfrak{H}_1 + Q); \quad . \quad . \quad . \quad . \quad . \quad . \quad g(180)$$
hence

$$M_x = \tfrac{1}{2}\mathfrak{H}_1 f + \tfrac{1}{2}Qf - \tfrac{1}{2}\mathfrak{H}_1 y - \tfrac{1}{2}Qy - Q(f-b) + Q(y-b) - V_1\frac{l}{2} + V_1 x \quad . \quad g(181)$$

and

$$2\varDelta_1 f = \varDelta_2 f = \Big\{ \mathfrak{H}_1 f \int_0^{\frac{l}{2}} \frac{xds}{\theta_x} - \mathfrak{H}_1 \int_0^{\frac{l}{2}} \frac{xyds}{\theta_x} + Qf \int_0^{\frac{l}{2}} \frac{xds}{\theta_x} - Q\int_0^{\frac{l}{2}} \frac{xyds}{\theta_x} - 2Q(f-b)\int_0^{\frac{l}{2}} \frac{xyds}{\theta_x} + 2Q\int_0^{\frac{l}{2}} (y-b)\frac{xds}{\theta_x} - 2\Big[V_1\frac{l}{2} - V_1 x\Big]\Big\} \cdot g(182)$$

Equating the two values of $\varDelta_2 f$ and solving for V_1, we obtain

$$V_1 = \frac{1}{2}\,\frac{Q(f-b)\int_0^{\frac{l}{2}} \frac{xds}{\theta_x} - Q\int_a^{\frac{l}{2}} (y-b)\frac{xyds}{\theta_x} - Qf\int_0^{\frac{l}{2}} \frac{xds}{\theta_x} + Q\int_0^{\frac{l}{2}} \frac{xyds}{\theta_x}}{\int_0^{\frac{l}{2}} \frac{x^2ds}{\theta_x} - \frac{l}{2}\int_0^{\frac{l}{2}} \frac{xds}{\theta_x}}, \quad g(183)$$

which reduces to

$$V_1 = \frac{Q}{2}\,\frac{\int_0^a (y-b)\frac{xds}{\theta_x}}{\int_0^{\frac{l}{2}} \frac{x^2ds}{\theta_x} - \frac{l}{2}\int_0^{\frac{l}{2}} \frac{xds}{\theta_x}}, \quad . \quad . \quad . \quad . \quad . \quad . \quad . \quad . \quad g(184)$$

which holds good for *all loads on the left of the crown.*

Values of M_1 and M_2 for a Single Horizontal Load.

From (41),

$$M_1 = -V_1x + H_1y - Q(y - b), \quad . \quad . \quad g(185)$$

and from (49),

$$M_2 = M_1 + V_1l - Qb. \quad . \quad . \quad . \quad . \quad . \quad . \quad g(186)$$

Temperature.

Assuming that $\Delta l = 0$, and that the hinge at the crown remains midway between the supports, we have, from $g(60)$,

$$-\frac{1}{E}\int_0^l M_x\frac{yds}{\theta_x} + et^\circ\int_0^l dx = 0. \quad . \quad . \quad g(187)$$

From (41),

$$M_x^1 = M_1 - H_1y; \quad . \quad . \quad . \quad . \quad . \quad g(188)$$

but for $x = \frac{l}{2}$, $M_x = 0$, and hence

$$M_1 = H_1f. \quad . \quad . \quad . \quad . \quad . \quad . \quad . \quad . \quad g(189)$$

Therefore

$$M_x = H_1f - H_1y. \quad . \quad . \quad . \quad . \quad g(190)$$

Substituting this value of M_x in $g(187)$ and solving for H_1,

$$H_1 = \frac{Eet^\circ l}{f\int_0^l \frac{yds}{\theta_x} - \int_0^l \frac{y^2ds}{\theta_x}}. \quad . \quad . \quad . \quad . \quad g(191)$$

The above equations are perfectly general, and in their integral form can be applied to any symmetrical arch which has a regular curve for an axis. In the summation form the equations apply in the case of any symmetrical arch.

If the axis is parabolic in form and $E\theta \cos\phi = A =$ a constant, our equations become quite simple.

SYMMETRICAL PARABOLIC ARCH WITH A HINGE AT THE CROWN AND $E\theta$ COS ϕ = A CONSTANT.

(*a*) *Single Vertical Load.*

$$H = \frac{1}{2}\,\frac{P\int_0^l (x - a_1)ydx - \int_a^l \overset{x}{\Sigma} P(x-a)ydx}{\int_0^l y^2dx - f\int_0^l ydx}, \qquad g(192)$$

where

$$\int_a^l \overset{x}{\Sigma} P(x-a)ydx = \int_{a_1}^l P(x-a_1)ydx + \int_{a_2}^l P(x-a_2)ydx.$$

Substituting the value of y and integrating,

$$H_1 = \frac{5}{2}\frac{l}{f}P(2-k)k^3 \text{ for } k \overset{=}{>} \tfrac{1}{2}; \quad . \quad . \quad . \quad g(193)$$

$$H_1 = \frac{5}{2}\frac{l}{f}P(1-k^2)(1-k)^2 \text{ for } k \overset{=}{<} \tfrac{1}{2}. \quad g(194)$$

From $g(164)$,

$$V = \frac{P}{\frac{l}{2}}\cdot\frac{(l-a)\int_0^{\frac{l}{2}} xdx - \int_a^{\frac{l}{2}} (x-a)xdx - \int_0^{\frac{l}{2}} x^2dx}{\frac{l}{2}\int_0^{\frac{l}{2}} xdx - \int_0^{\frac{l}{2}} x^2dx} \qquad g(195)$$

or

$$\left.\begin{aligned} V_1 &= P(1-4k^3), &\text{for } k \overset{=}{<} \tfrac{1}{2}\\ V_1 &= P[1-4(1-k)^3], &\text{for } k \overset{=}{>} \tfrac{1}{2}\end{aligned}\right\}, \quad . \quad . \quad . \quad . \quad g(196)$$

$$M_1 = \frac{Pl}{2}(-2k + 14k^3 - 5k^4), \quad \text{for } k \overset{=}{<} \tfrac{1}{2}, \quad . \quad . \quad . \quad . \quad g(197)$$

and

$$M = \frac{Pl}{2}(6k^3 - 5k^4), \qquad \text{for } k \overset{=}{<} \tfrac{1}{2}. \quad . \quad . \quad . \quad . \quad g(198)$$

(b) Single Horizontal Load.

From $g(172)$,

$$H_1 = \frac{Q}{2} \frac{(f-b)\int_0^l ydx - \int_{a_1}^{a_2}(y-b)ydx}{f\int_0^l ydx - \int_0^l y^2dx} \quad . \quad g(199)$$

or

$$H_1 = Q(1 - 20k^3 + 40k^4 - 16k^5). \quad . \quad . \quad . \quad . \quad g(200)$$

From $g(184)$,

$$V_1 = \frac{Q}{2} \frac{\int_0^a (y-b)xdx}{\int_0^{\frac{l}{2}} x^2dx - \frac{l}{2}\int_0^{\frac{l}{2}} xdx} \quad . \quad . \quad . \quad . \quad . \quad g(201)$$

or

$$V_1 = 4\frac{f}{l}(4k^3 - 6k^4)Q, \quad . \quad . \quad . \quad . \quad . \quad . \quad . \quad . \quad . \quad g(202)$$

$$M_1 = -V_1x + H_1y - Q(y-b), \quad . \quad . \quad . \quad . \quad g(203)$$

and

$$M_2 = M_1 + V_1l - Qb. \quad . \quad . \quad . \quad . \quad . \quad . \quad . \quad . \quad g(204)$$

(c) Temperature.

From $g(191)$,

$$H_1 = \frac{Aet^\circ l}{f\int_0^l ydx - \int_0^l y^2dx} \quad . \quad . \quad . \quad . \quad g(205)$$

or

$$H_1 = \frac{15Aet^\circ}{2f^2}. \quad . \quad . \quad . \quad . \quad . \quad . \quad . \quad . \quad . \quad g(206)$$

Given the values of H_1 and V_1 for any vertical load on the left of the crown, to determine M_1, M_2, V_2, and H_2 for this load, and also for an equal and symmetrically placed load on the right of the crown.

Our formulas have been deduced for loads on the left of the crown, but they are sufficient for the complete determination of all the outer forces for any load. In fact we need only the values of $H.$ and V_1 if graphics be employed.

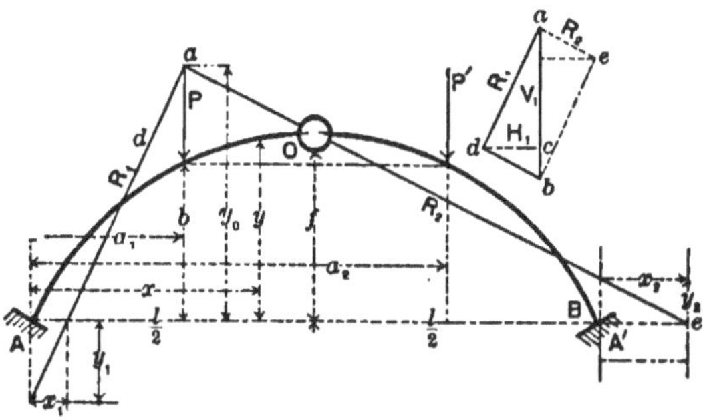

FIG. 40.

In Fig. 40, let P be any load on the left of the crown. Make $ab = P$, $ac = V_1$, and $dc = H_1$; then $ad = R_1$ and $ae = R_2$.

Through O draw aOe parallel to ae. From a, where this line cuts P, draw ad parallel to ad. Then we have the true equilibrium polygon for the load P, from which all the outer forces can be readily obtained.

Since the arch is symmetrical, evidently the values of H_1, V_1, M_1, etc., for the equal and symmetrically placed load P' are equal to the values of H_2, V_2, M_2, etc., for the load P.

The fields of loading which cause like stresses can be found in a manner similar to that given on page 25 for arches having two hinges.

THE THREE-HINGED ARCH.

The three-hinged arch as usually constructed is symmetrical, and has a hinge at each support and one at the crown. The introduction of the third hinge materially simplifies the determination of H_1 and H_2; in fact the problem is practically one of graphic statics in its simplest form.

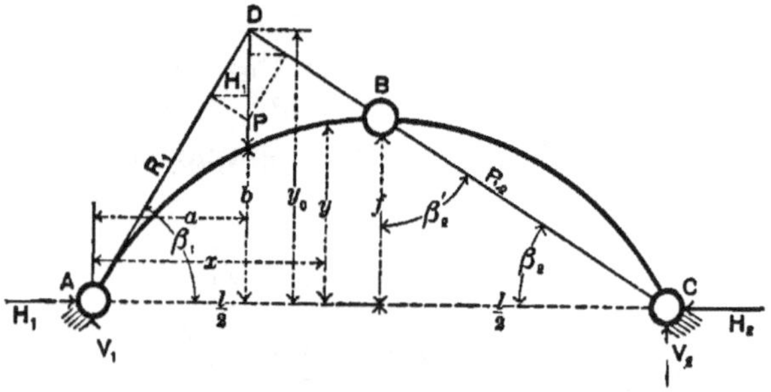

FIG. 41.

In Fig. 41 let ABC be any arch having three hinges, and let A, B, and C be the location of the hinges; then, evidently, there can be no bending-moments at A, B, or C, and the reaction R_1 will pass through A and R_2 through C. For a single *vertical load* on the left of B the reaction R_2 must also pass through B, since there can be no moment at this point. The determination of H_1, V_1, H_2, and V_2 now becomes quite simple, as follows:

Draw BC through B and C until it cuts P in D, and then draw DA through A. By simple resolution of the forces meeting in D the values of H_1, V_1, etc., are readily found.

The same results can be obtained by applying the following formulas:

From Fig. 41,

$$\tan \beta_2 = 2\frac{f}{l}; \quad \ldots \ldots \quad g(207)$$

$$y_0 = (l - a) \tan \beta_2 = 2(1 - k)f; \quad \ldots \quad g(208)$$

$$\tan\beta_1 = \frac{y_0}{kl} = 2\frac{1-k}{k}\frac{f}{l}; \quad . \quad . \quad . \quad . \quad . \quad . \quad . \quad g(209)$$

$$V_1 = P(1-k). \quad . \quad . \quad . \quad . \quad . \quad . \quad . \quad . \quad g(210)$$

From (50), by transposition,

$$H_1 = \frac{V_1}{y_0}kl \quad . \quad . \quad . \quad . \quad . \quad . \quad . \quad . \quad g(211)$$

or

$$H_1 = P\frac{(1-k)}{2(-k)f}kl = \frac{P}{2}\frac{kl}{f}. \quad . \quad . \quad . \quad g(212)$$

The stresses in the various members of the arch can now be found by the ordinary methods.

The determination of the fields of loading which produce the maximum stresses has been fully explained on page 25 *et seq.*

The treatment of horizontal loads differs but little from that outlined above.

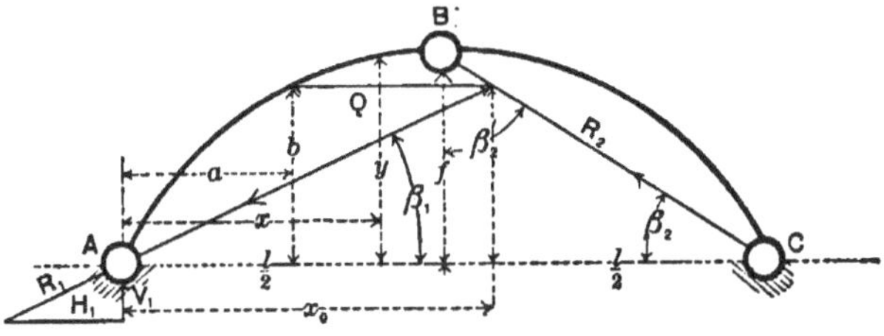

FIG. 42.

Fig. 42 clearly shows the method for locating R_1 and R_2.

$V_1 = Q\frac{b}{l}$ and acts upward or downward as Q acts towards the ~~right~~ left or the ~~left~~ right respectively.

$$\tan\beta_2' = \frac{l}{2f} \quad . \quad . \quad . \quad . \quad . \quad . \quad . \quad . \quad g(213)$$

$$x_0 = l - b\tan\beta_2' = \frac{l}{2f}(2f-b). . \quad . \quad g(214)$$

From Fig. 42,

$$H_1 = \frac{V_1}{b}x_0 = Q\frac{1}{2f}(2f - b). \quad . \quad . \quad . \quad g(215)$$

As the Q loads are almost without exception due to the action of wind and are treated as static loads, the best way to obtain the stresses in the various members of the rib is to determine the resultant values of H_1 and V_1 and then treat the problem graphically.

CHAPTER VI.

COMPARISON OF FOUR TYPES OF ARCHES.

WE will take four types of the parabolic arch having $E\theta \cos \phi =$ a constant and show graphically the relations between the values of the outer forces for the different types.

Let Type 1° = arch with no hinges;
" 2° = " " one hinge;
" 3° = " " two hinges;
" 4° = " " three hinges.

(*a*) VERTICAL LOADS.

Comparison of H_1.

The formulas* are:

$$1°.\quad \frac{H_1}{P}\frac{f}{l} = \frac{15}{4}k^2(1-k)^2;$$

$$2°.\quad \frac{H_1}{P}\frac{f}{l} = \frac{5}{8}k^2(8-4k);$$

$$3°.\quad \frac{H_1}{P}\frac{f}{l} = \frac{5}{8}k(1-2k^2+k^3).$$

$$4°.\quad \frac{H_1}{P}\frac{f}{l} = \frac{1}{2}k.$$

* See pages 29, 139, 20, and 143, respectively.

These values are represented graphically in Fig. 43, from which we see that the 2° type differs quite considerably from the others, particularly for loads near the crown and those near the springing.

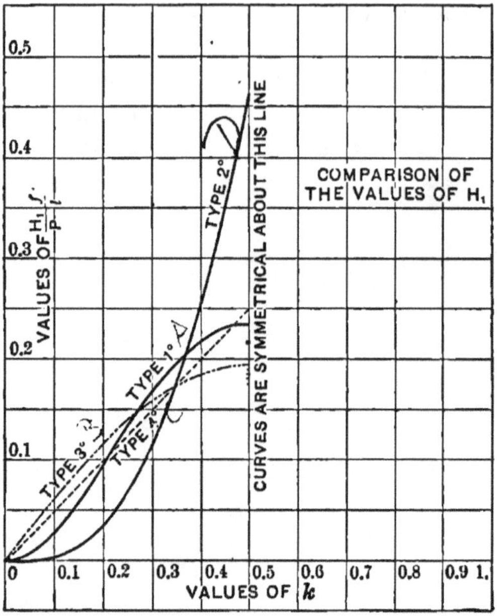

FIG. 43

Comparison of V_1.

*Formulas.**—Type 1°. $\frac{V_1}{P} = (1 - k)^2 (1 + 2k)$;

2° $\frac{V_1}{P} = (1 - 4k^3)$

3° $\frac{V_1}{P} = (1 - k)$;

4°. $\frac{V_1}{P} = (1 - k)$.

* See pages 30, 139, 21, and 143, respectively.

These values are represented graphically in Fig. 44.

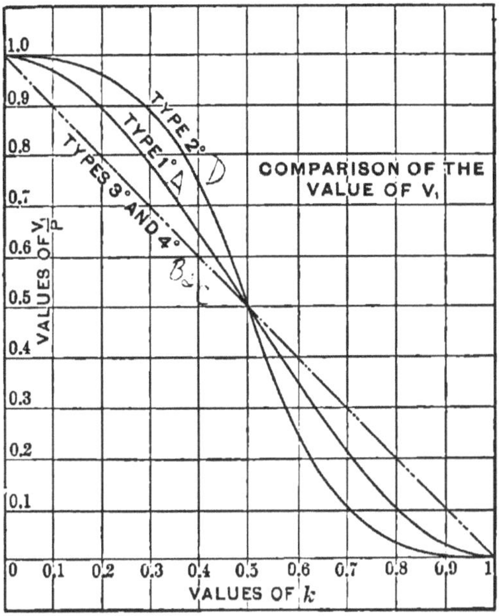

FIG. 44.

The equation for intermediate vertical shear is

$$V_x = V_1 - \overset{x}{\Sigma} P \quad \text{or} \quad V_x = KP.$$

We give below the values of V_x for the arch without hinges and that with a hinge at each support. The span is divided into twenty equal divisions, and the load P assumed to occupy *each point of division*. The values of V_x are given for *each division*.

These tables and those given later for maximum bending-moments are principally useful in preliminary computations unless θ is assumed to vary as the secant of ϕ, when of course they very materially decrease the labor of calculation.

SYMMETRICAL PARABOLIC ARCH WITHOUT HINGES.*

$V_x = K_0 P$ values of K_0.

Point of Division.	1	2	3	4	5	6	7	8	9	10	Division Number.
P on 1	+.961−	−.036+	−.033+	−.029+	−.026+	−.022+	−.019+	−.016+	−.012+	−.009+	19 *P* on
" 2	+.857−	+.869−	−.119+	−.107+	−.095+	−.083+	−.071+	−.058+	−.046+	−.034+	18 "
" 3	+.708−	+.732−	+.756−	−.219+	−.195+	−.170+	−.146+	−.122+	−.097+	−.073+	17 "
" 4	+.531−	+.570−	+.608−	+.645−	−.315+	−.277+	−.238+	−.200+	−.162+	−.123+	16 "
" 5	+.343−	+.396−	+.448−	+.501−	+.554−	−.394+	−.341+	−.288+	−.235+	−.183+	15 "
" 6	+.156−	+.222−	+.288−	+.354−	+.420−	+.486−	−.448+	−.381+	−.315+	−.249+	14 "
" 7	−.019+	+.058−	+.130−	+.214−	+.291−	+.369−	+.447−	−.476+	−.398+	−.321+	13 "
" 8	−.173+	−.086+	0	+.086−	+.173−	+.259−	+.346−	+.432−	−.482+	−.395+	12 "
" 9	−.298+	−.206+	−.115+	−.023+	+.069−	+.161−	+.253−	+.345−	+.437−	−.471+	11 "
" 10	−.390+	−.297+	−.203+	−.109+	−.016+	+.078−	+.172−	+.266−	+.359−	+.453−	10 "
" 11	−.448+	−.356+	−.264+	−.172+	−.081+	+.012−	+.103−	+.195−	+.287−	+.379−	9 "
" 12	−.469+	−.382+	−.296+	−.210+	−.123+	−.037+	+.050−	+.136−	+.222−	+.309−	8 "
" 13	−.456+	−.378+	−.301+	−.223+	−.145+	−.068+	+.010−	+.088−	+.165−	+.243−	7 "
" 14	−.413+	−.346+	−.280+	−.214+	−.148+	−.082+	−.016+	+.051−	+.117−	+.183−	6 "
" 15	−.345+	−.292+	−.239+	−.186+	−.134+	−.081+	−.028+	+.025−	+.077−	+.120−	5 "
" 16	−.261+	−.222+	−.184+	−.146+	−.107+	−.069+	−.030+	+.008−	+.046−	+.085−	4 "
" 17	−.171+	−.147+	−.122+	−.098+	−.073+	−.049+	−.025+	0	+.024−	+.049−	3 "
" 18	−.087+	−.075+	−.063+	−.051+	−.039+	−.027+	−.015+	−.002+	+.010−	+.022−	2 "
" 19	−.025+	−.022+	−.018+	−.015+	−.011+	−.008+	−.005+	−.001+	+.002−	+.006−	1 "
Division Number.	20	19	18	17	16	15	14	13	12	11	Point of Division.

* First published by Prof. Greene in *Engineering News*, vol. IV.

SYMMETRICAL PARABOLIC ARCH WITH TWO HINGES.*

$V_x = K'_2 P$ values of K_2.

Point of Division.	1	2	3	4	5	6	7	8	9	10	Division Number.
P on 1	+.832−	−.155+	−.143+	−.131+	−.118+	−.106+	−.093+	−.081+	−.069+	−.056+	19 *P* on
" 2	+.666−	+.691−	−.284+	−.260+	−.235+	−.211+	−.186+	−.161+	−.137+	−.112+	18 "
" 3	+.508−	+.544−	+.580−	−.384+	−.348+	−.312+	−.276+	−.240+	−.204+	−.168+	17 "
" 4	+.359−	+.406−	+.452−	+.498−	−.455+	−.409+	−.362+	−.316+	−.270+	−.223+	16 "
" 5	+.222−	+.277−	+.333−	+.389−	+.444−	−.500+	−.445+	−.389+	−.333+	−.278+	15 "
" 6	+.096−	+.159−	+.223−	+.287−	+.350−	+.414−	−.523+	−.459+	−.395+	−.332+	14 "
" 7	−.013+	+.057−	+.126−	+.196−	+.266−	+.336−	+.406−	−.524+	−.455+	−.385+	13 "
" 8	−.107+	−.032+	+.042−	+.116−	+.191−	+.265−	+.340−	+.414−	−.512+	−.437+	12 "
" 9	−.183+	−.106+	−.029+	+.048−	+.125−	+.203−	+.280−	+.357−	+.434−	−.489+	11 "
" 10	−.243+	−.165+	−.086+	−.008+	+.070−	+.148−	+.226−	+.304−	+.383−	+.461−	10 "
" 11	−.283+	−.206+	−.129+	−.052+	+.025−	+.103−	+.180−	+.257−	+.334−	+.411−	9 "
" 12	−.307+	−.232+	−.158+	−.084+	−.009+	+.065−	+.140−	+.214−	+.288−	+.363−	8 "
" 13	−.313+	−.243+	−.173+	−.104+	−.034+	+.036−	+.106−	+.175−	+.245−	+.315−	7 "
" 14	−.304+	−.241+	−.177+	−.113+	−.050+	+.014−	+.077−	+.141−	+.205−	+.268−	6 "
" 15	−.278+	−.223+	−.167+	−.111+	−.056+	0	+.055−	+.111−	+.167−	+.222−	5 "
" 16	−.241+	−.194+	−.148+	−.102+	−.055+	−.009+	+.038−	+.084−	+.130−	+.177−	4 "
" 17	−.192+	−.156+	−.120+	−.084+	−.048+	−.012+	+.024−	+.060−	+.096−	+.132−	3 "
" 18	−.134+	−.109+	−.084+	−.060+	−.035+	−.011+	+.014−	+.038−	+.063−	+.088−	2 "
" 19	−.068+	−.055+	−.043+	−.031+	−.018+	−.006+	+.007−	+.019−	+.031−	+.044−	1 "
Division Number.	20	19	18	17	16	15	14	13	12	11	Point of Division.

* First published by Prof. Greene in *Engineering News*, vol. IV.

Comparison of the Maximum Values of M_x.

In each of the four types of arches which we are considering, if the values of M_1, V_1, and H_1 be substituted in (41), we find that

$$M_x = \frac{l}{2}P(K)(Z),$$

where K depends upon $k = \frac{a}{l}$ and Z upon $z = \frac{x}{l}$, showing that for parabolic arches the value of M_x varies *with the span alone* for given values of k and z. Then we may write

$$M_x = \frac{Pl}{2}J$$

If the values of J be computed for each load for every value of x and tabulated, the maximum values of M_x are readily found by taking the sum of the values of J having like signs.

We give below the values of J for types 1° and 3° which are most common in practice.

It will be noticed that the positive and negative moments are approximately equal, although the arch is divided into but twenty equal divisions. For a uniform horizontal load covering the entire structure the positive and negative moments would be equal, since the equilibrium polygon would be a parabola coinciding with the axis of the rib.

In Fig. 45 * is shown relatively the *maximum* values of M_x for the four types.

It appears from this diagram that type 1° has moments which vary more nearly according to the variation of the section of the rib than either of the others.

The second type has very large moments near the springing, which rapidly decrease until about the quarter-point, and then

---

* This diagram is from a note by M. Souleyre: "Note sur l'emploi de quatre types d'arcs dans les Ponts, Viaducts et Fermes Métalliques de grande portée." Annales des Ponts et Chaussées, mai, 1896.

after increasing slightly, decrease rapidly, becoming zero at the crown.

A crescent-shaped rib corresponds more nearly with the variation of the maximum moments in the third type.

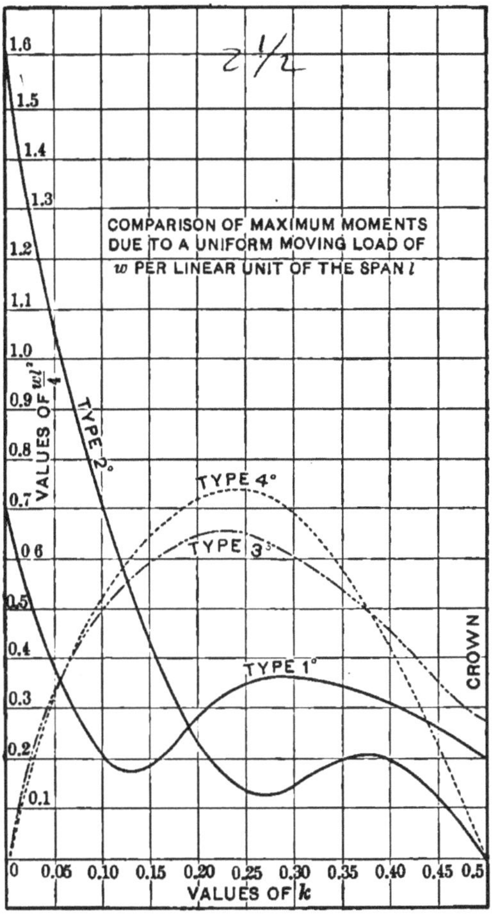

FIG. 45.

As the formulas of the fourth type do not depend upon the values of θ, the rib can be designed to correspond with the variation in the moments.

Thus far we have considered only the live or moving load effects.

Symmetrical Parabolic Arch without Hinges.*

$$M_x = \frac{Pl}{2} J_0 \text{ values of } J_0.$$

Point of Division.	0	1	2	3	4	5	6	7	8	9	Crown 10	
P on 1	−.0789	+.0171	+.0135	+.0102	+.0073	+.0047	+.0025	+.0005	−.0010	−.0023	−.0031	19
" 2	−.1214	−.0358	+.0510	+.0391	+.0284	+.0189	+.0106	+.0036	−.0022	−.0069	−.0103	18
" 3	−.1354	−.0647	+.0085	+.0841	+.0622	+.0427	+.0256	+.0110	−.0011	−.0109	−.0182	17
" 4	−.1280	−.0749	−.0179	+.0429	+.1075	+.0760	+.0484	+.0245	+.0045	−.0116	−.0239	16
" 5	−.1054	−.0712	−.0316	+.0132	+.0633	+.1186	+.0793	+.0452	+.0164	−.0071	−.0254	15
" 6	−.0734	−.0579	−.0357	−.0069	+.0284	+.0705	+.1191	+.0744	+.0363	+.0047	−.0202	14
" 7	−.0369	−.0388	−.0330	−.0194	+.0019	+.0311	+.0679	+.1126	+.0650	+.0252	−.0069	13
" 8	0	−.0173	−.0259	−.0259	−.0173	0	+.0259	+.0605	+.1037	+.0555	+.0160	12
" 9	+.0340	+.0041	−.0165	−.0280	−.0303	−.0234	−.0074	+.0179	+.0524	+.0965	+.0494	11
" 10	+.0625	+.0234	−.0062	−.0266	−.0375	−.0390	−.0313	−.0141	+.0125	+.0484	+.0937	10
" 11	+.0836	+.0388	+.0032	−.0232	−.0404	−.0485	−.0473	−.0370	−.0174	+.0114	+.0494	9
" 12	+.0959	+.0490	+.0108	−.0188	−.0397	−.0521	−.0557	−.0508	−.0372	−.0149	+.0160	8
" 13	+.0994	+.0538	+.0160	−.0141	−.0363	−.0509	−.0576	−.0566	−.0478	−.0313	−.0069	7
" 14	+.0946	+.0534	+.0187	−.0093	−.0307	−.0455	−.0537	−.0552	−.0501	−.0385	−.0202	6
" 15	+.0820	+.0475	+.0183	−.0056	−.0242	−.0376	−.0457	−.0485	−.0461	−.0384	−.0254	5
" 16	+.0640	+.0379	+.0157	−.0027	−.0173	−.0280	−.0348	−.0379	−.0371	−.0324	−.0239	4
" 17	+.0430	+.0259	+.0113	−.0009	−.0107	−.0180	−.0229	−.0254	−.0254	−.0230	−.0182	3
" 18	+.0225	+.0137	+.0062	−.0001	−.0052	−.0091	−.0117	−.0132	−.0134	−.0124	−.0103	2
" 19	+.0065	+.0040	+.0019	+.0001	−.0014	−.0025	−.0033	−.0038	−.0039	−.0037	−.0031	1

First published by Prof. Greene in *Engineering News*, vol. IV.

Symmetrical Parabolic Arch with Two Hinges.*

$$M_x = \frac{Pl}{2} J_2 \text{ values of } J_2.$$

Point of Division.	1	2	3	4	5	6	7	8	9	Crown 10	
P on 1	+.0832	+.0676	+.0523	+.0402	+.0289	+.0178	+.0084	+.0003	−.0066	−.0122	19
" 2	+.0667	+.1359	+.1075	+.0815	+.0580	+.0370	+.0184	+.0023	−.0114	−.0226	18
" 3	+.0509	+.1051	+.1634	+.1250	+.0902	+.0590	+.0315	+.0075	−.0129	−.0297	17
" 4	+.0359	+.0765	+.1217	+.1715	+.1260	+.0851	+.0489	+.0173	−.0097	−.0320	16
" 5	+.0221	+.0498	+.0831	+.1219	+.1663	+.1162	+.0717	+.0328	−.0006	−.0283	15
" 6	+.0097	+.0257	+.0480	+.0767	+.1118	+.1532	+.1016	+.0551	+.0155	−.0176	14
" 7	−.0013	+.0043	+.0170	+.0366	+.0632	+.0968	+.1374	+.0849	+.0394	+.0010	13
" 8	−.0107	−.0139	−.0097	+.0019	+.0210	+.0475	+.0815	+.1229	+.0717	+.0280	12
" 9	−.0183	−.0289	−.0318	−.0270	−.0145	+.0058	+.0338	+.0695	+.1129	+.0641	11
" 10	−.0242	−.0406	−.0492	−.0500	−.0430	−.0281	−.0055	+.0250	+.0633	+.1094	10
" 11	−.0283	−.048[illegible]	−.0618	−.0670	−.0644	−.0542	−.0362	−.0105	+.0229	+.0641	9
" 12	−.0307	−.0539	−.0697	−.0780	−.0790	−.0725	−.0585	−.0371	−.0083	+.2080	8
" 13	−.0313	−.0556	−.0730	−.0834	−.0868	−.0832	−.0726	−.0551	−.0305	+.0010	7
" 14	−.0304	−.0545	−.0720	−.0833	−.0882	−.0868	−.0791	−.0649	−.0445	−.0176	6
" 15	−.0279	−.0502	−.0670	−.0781	−.0837	−.0838	−.0783	−.0672	−.0505	−.0283	5
" 16	−.0241	−.0435	−.0583	−.0685	−.0740	−.0749	−.0711	−.0627	−.0497	−.0320	4
" 17	−.0191	−.0347	−.0467	−.0550	−.0598	−.0609	−.0585	−.0525	−.0429	−.0297	3
" 18	−.0133	−.0241	−.0325	−.0385	−.0420	−.0430	−.0416	−.0365	−.0314	−.0226	2
" 19	−.0068	−.0124	−.0167	−.0198	−.0216	−.0222	−.0216	−.0197	−.0166	−.0122	1

* First published by Prof. Greene in *Engineering News*, vol. IV.

The dead load is very nearly a uniform horizontally distributed load, and hence the moments due to this load are practically zero in the four types.

If in the 1° and 2° types the dead-load stresses are computed as if the ribs were hinged at the springing and then built with hinges at the springing, when the falseworks are removed the rib will settle into position and the dead-load stresses will be practically those computed.

From Fig. 45 we see that the live-load flange-stresses are a minimum for the 1° type, or the arch without hinges. A rib constructed with pins at the springing is very easily made into a rib with fixed ends by arranging the details so that the flanges may be rigidly connected with the piers or abutments *after the falseworks are removed.*

This method is followed by French and German engineers in many cases, especially for masonry arches and metal arches with solid webs.

The arch without hinges, or type 1°, appears to be the most economical of the four for the dead and live loads.

There remains to be considered the effect of temperature.

Comparison of Temperature Effects.

Type 1°. $H_1 = \frac{45}{4f^2}Aet°. \quad M_1 = H_1\frac{2}{3}f.$

2°. $H_1 = \frac{15}{2f^2}Aet°. \quad M_1 = H_1 f.$

3°. $H_1 = \frac{15}{8f^2}Aet°. \quad M_1 = 0.$

4°. $H_1 = 0. \quad M_1 = 0.$

Hence for

$$\text{Type } 1°. \quad M_x = H_1(\tfrac{2}{3}f - y) = \frac{15}{8f^2}Aet°(4f - 6y).$$

$$\text{Type } 2^\circ. \quad M_x = H_1(f - y) = \frac{15}{8f^2}Aet^\circ(4f - 4y).$$

$$3^\circ. \quad M_x = H_1 y = \frac{15}{8f^2}Aet^\circ(y).$$

From which we see that the effect of temperature is greatest in the 1° type and least in the 4° type; also that the effect in the 2° type is greater than that in the 3° type.

For structures carrying moving loads the second and fourth types are not desirable on account of vertical vibration of the

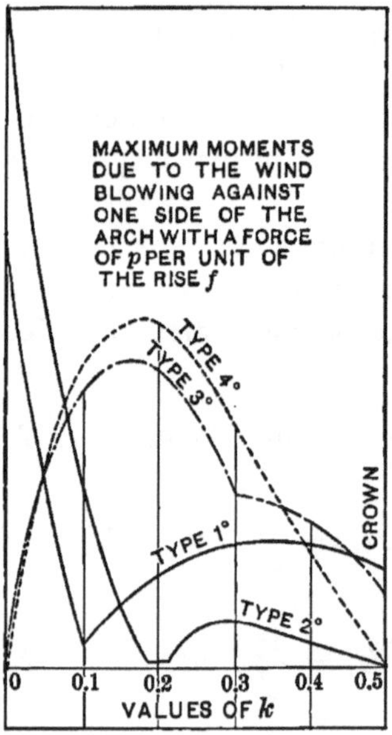

Fig. 46.

structure, leaving the first and third types to be selected from. Complete calculations show that for large structures the *first type is more economical as well as more rigid*, and practice proves this type well adapted to the work required for a railway bridge.

(*b*) HORIZONTAL LOADS.

Horizontal loads, being usually due to wind, may be considered as a dead load, covering the arch on one side from the crown to the springing.

For a uniform load p per unit of height of the arch,* Fig. 46 shows the relative values of max M_x for the four types.

Here we see that the first type more nearly agrees with the variation of θ in the variation of the moments, so that the conclusion drawn above remains unchanged for structures carrying a moving load. In case there is no moving load, as in roof-trusses, the fourth type appears to be most economical. This type is almost always employed by American engineers for large roof-trusses.

Comparison of Types 1°, 3°, *and* 4° *designed for a Single-track Railway Bridge having a Span of* 416 *Feet.*†

To more clearly show the relation between the three types 1°, 3°, and 4°, a comparison of the maximum stresses in the individual members of a trussed parabolic arch rib are shown in Figs. 47, 48, 49, and 50.

The diagrams show the maximum stresses due to dead load, live load, wind, and changes in temperature.

Figs. 47 and 48 clearly indicate the superiority of the arch without hinges for economy in the flanges.

Figs. 49 and 50 show that there is little choice between the types as far as the web is concerned, there being a remarkably close agreement between the stresses for the three types.

The principal data employed are as follows (see Fig. 51):

Span	416′ 0″
Rise	67′ 0″

* See note under Fig. 45.

† The computations for this comparison were made by Messrs. Crockwell, Wiggins, and Shaneberger in connection with their theses for graduation from the Rose Polytechnic Institute.

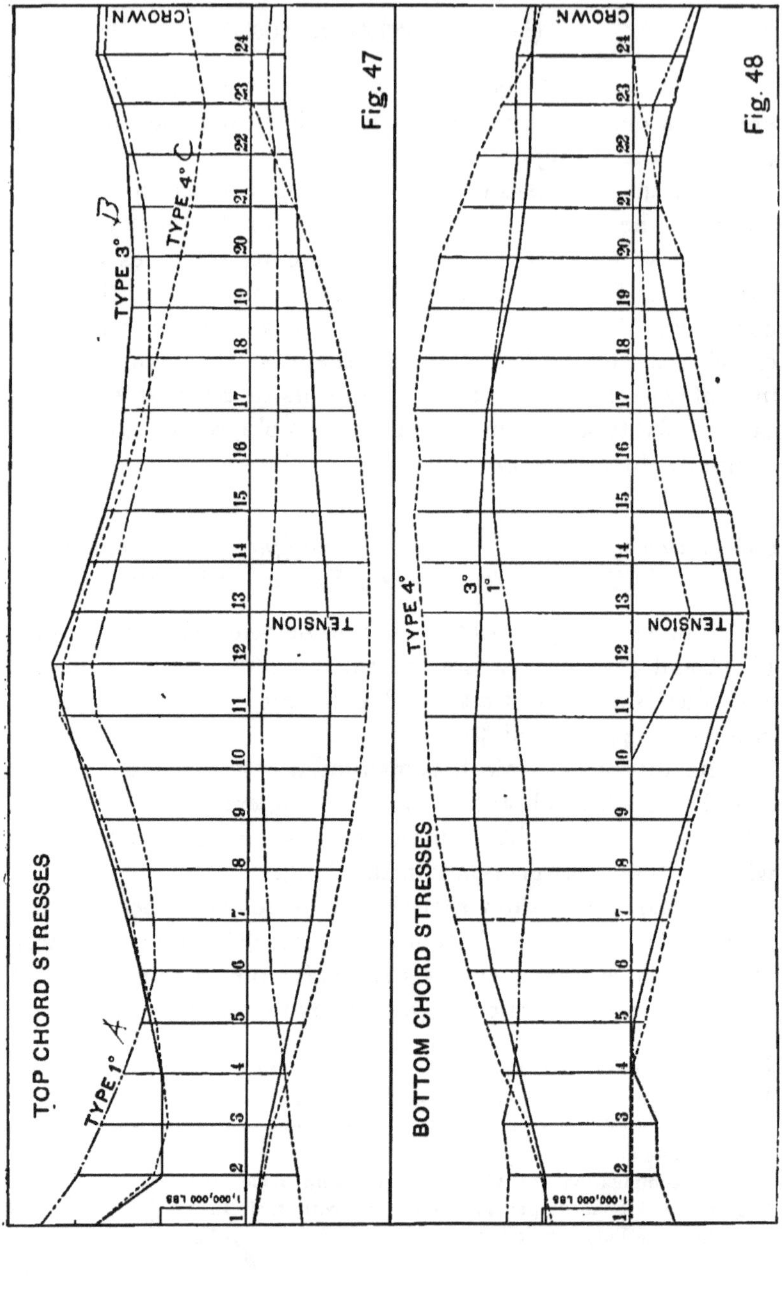
TOP CHORD STRESSES
TYPE 1° A
TYPE 3° B
TYPE 4° C
TENSION
CROWN
1,000,000 LBS
1 2 3 4 5 6 7 8 9 10 11 12 13 14 15 16 17 18 19 20 21 22 23 24
BOTTOM CHORD STRESSES
TYPE 4°
3°
1°
TENSION
CROWN
1,000,000 LBS
1 2 3 4 5 6 7 8 9 10 11 12 13 14 15 16 17 18 19 20 21 22 23 24

Fig. 47

Fig. 48

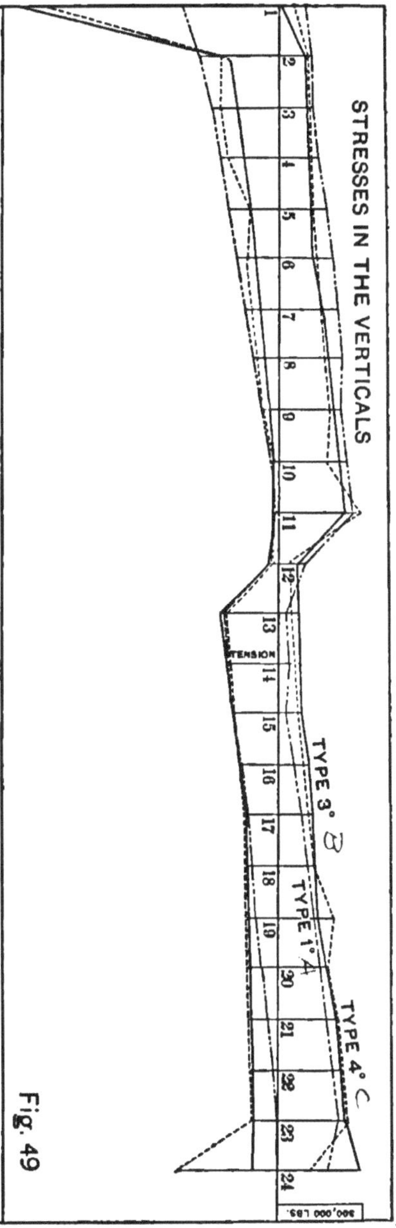
STRESSES IN THE VERTICALS
TENSION
TYPE 3°
TYPE 1°
TYPE 4°
300,000 LBS.

Fig. 49

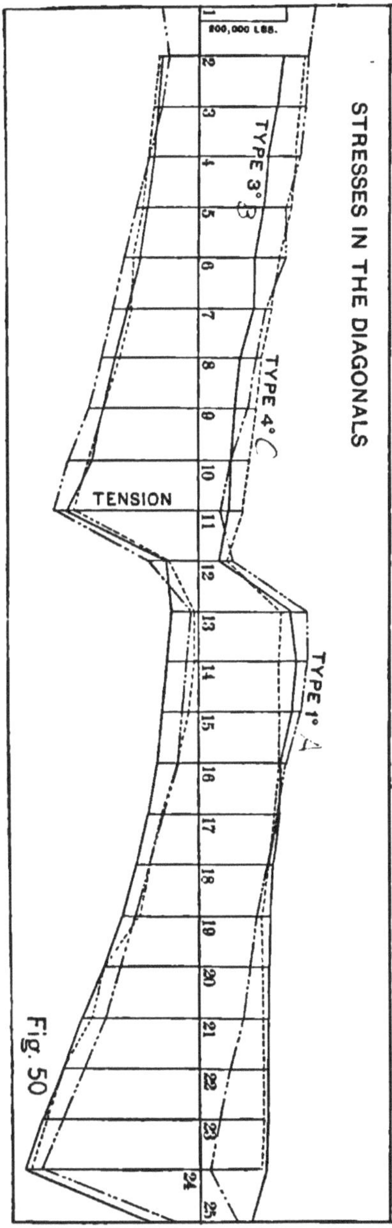
STRESSES IN THE DIAGONALS
200,000 LBS.
TYPE 3°
TYPE 4°
TENSION
TYPE 1°

Fig. 50

Batter of arch planes............	1 in 3
Depth of rib at crown.........................	6′ 0″
" " " " skewbacks	10′ 0″
Moving load per lineal foot of span	4000 lbs.
Dead " " " " " superstructure.......	1500 "
" " " " " " arch................	1000 "
Wind " " " " " span (live)...........	300 "
" " " " " " " (dead)..........	600 "
Range of temperature...........................	± 80° F.

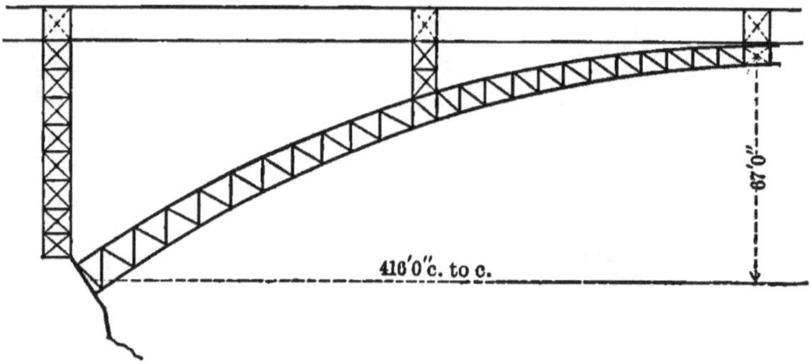

FIG. 51.

Relative Weights of Steel in One Arch Rib, including Gusset-plates, Rivets, etc.

Type 1°, without hinges....................	1.00
" 3°, two "	1.21
" 4°, three "	1.30

CHAPTER VII.

APPLICATIONS.

IN the preceding pages we have deduced formulas for determining the various reactions and moments which result from the application of vertical and horizontal forces to the linear elastic arch, that is, we have assumed that the forces were applied upon the central line or neutral axis of the arch rib. In practice this evidently is not always the case, especially where a superstructure is supported by arch ribs having considerable depth.

The weight of the arch rib alone may without serious error be assumed as applied to the centre line or neutral axis.

Vertical Loads.—Vertical forces due to the superstructure and moving loads may be assumed to act where they intersect the neutral axis in *flat ribs* and in trussed ribs where one system of the web bracing is *vertical.** The same assumption may be made for plate-girder ribs, as they are either very shallow, as in bridges of short spans, or the forces due to the superstructure are applied to the rib quite close together.

For the condition where a vertical force does not intersect the neutral axis of the rib, as in the case of a large semicircular rib near the supports, the following method may be employed. In Fig. 52 let P be a vertical force applied at B which does not intersect the neutral axis. At the centre of the strut BD place the two equal and opposite forces P; then we have for the equivalent of the force P applied at B the force P applied at C and the couple Pd. The reactions can now be found by applying the formula for a vertical load and that for a couple. In passing we may say that this method is general and can be ap-

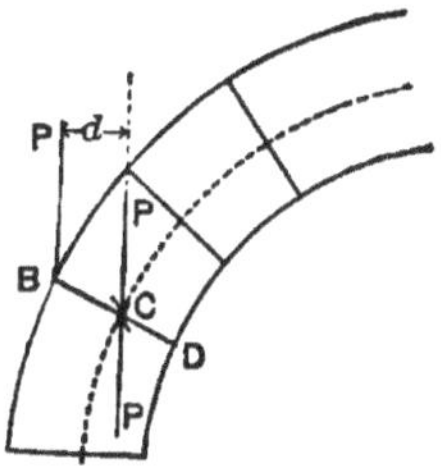

FIG. 52.

* See Fig. 51.

plied for any load whether its direction intersects the neutral axis or not.

Horizontal Loads.—Horizontal forces in the plane of the arch-rib seldom occur in practice, excepting in the case where the arch is employed for supporting a large roof. In this case the horizontal force is the horizontal component of the wind load.

If the stresses due to the wind are small in comparison with those caused by the total dead weight of the structure, the wind forces may be assumed to act upon the neutral axis where the normal components intersect it in the determination of reactions, etc. If greater accuracy is desired, then the force may be replaced by an equal force and a couple, as explained above for vertical forces.

Wind Loads.—We have just explained how to consider wind loads in the plane of the arch. There remains to be discussed the action of the wind against the arch and superstructure perpendicular to their plane.

The superstructure is usually composed of a roadway supported by columns or towers according to the magnitude and design of the structure.

The action of the wind against the roadway creates a horizontal reaction at the top of each post or tower. This reaction is transmitted to the arch-rib in the form of an equal horizontal force at the foot of the column or tower, and a couple which is equivalent to a vertical force acting upward on the wind side of the structure and an equal vertical force acting downward on the opposite side as illustrated in Fig. 53. The vertical forces are treated as explained above.

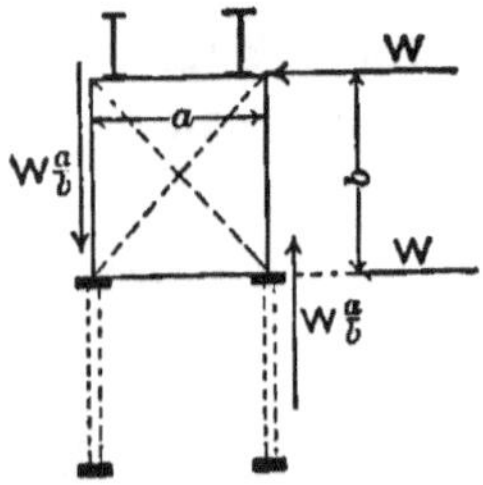

FIG. 53.

The horizontal force with that due to the direct action of the wind against the rib must be considered differently. The actual action of these forces is very complex unless we make the assumption that the arch-ribs act as the chords of a cantilevered beam having a length equal to one half the length of the axis of the arch. Under this assumption the lateral systems may be developed and the stresses in the different members found by ordinary methods. Although this method is not correct, yet its simplicity and probable safety commend its use.

Maximum Stresses.—We have explained in Chapter II the methods for selecting those forces which cause the maximum shears and moments at any point. Another method may be employed which has many features in its favor. The values of the reactions, etc., may be found for each load, and then the stresses in each member of the rib due to each individual load. These stresses being tabulated, the maximum positive and negative stresses are readily determined by simple addition. This method is long, but has the advantage of being free from errors, and if each load is taken as unity, the stresses obtained for the individual loads will be coefficients which can be applied to any load. The latter feature is of considerable importance, as very often the magnitudes of the loads are changed before the final computation is made. In *very large* structures where the moving load is *small* in comparison with the dead load it is customary to make but two computations for the moving load: one for the moving load covering the entire structure, and a second for the load covering one half of the span.

Character of Reactions.—In the hinged or fixed arch the *vertical reactions* (V_1 and V_2) always act *upward* when the vertical forces which are applied to the arch act *downward*, and the *horizontal reactions* (H_1 and H_2) act from the supports *towards* the centre of the span. In case the vertical forces act *upward*, V_1 and V_2 act *downward* and H_1 and H_2 act *away* from the centre of the span.

In the case of horizontal loads, if the load acts from the ~~left~~ towards the ~~right~~ left, V_1 acts ~~upward~~ downward and V_2 acts ~~*downward*~~. Both of the horizontal reactions act from the ~~*left*~~ right towards the ~~right.~~ left.

Co-ordinates $y_0, x_0, y_1, x_1, y_2, x_2$.

Vertical Loads.—The ordinate y_0 is always measured *upward* from the long chord of the arch.

The ordinate y_1 is always measured *upward* at the *left support* for loads on the *right* of the crown. For loads adjacent to the *left* support y_1 is measured *downward*. y_1 is *zero* for a load near a point which is *four tenths* the span from the *left* support, this distance varying with different arches.

The abscissa x_1 is measured to the *left* of the *left* support when y_1 is measured *upward*, and to the *right* when y_1 is measured *downward*.

x_1 and y_1 are zero when the arch has a hinge at the left support.

The directions of x_2 and y_2 are easily determined from what has been said concerning x_1 and y_1.

Horizontal Loads.—x_0 is always measured towards the *right* from the *left* support for loads on the *left* of the crown.

y_1 and y_2 are always measured upward at the left and right support respectively.

x_1 is always measured to the *left* of the *left* support, and x_2 to the *right* of the *right* support.

Bending Moments at the Supports.—For arches with hinges at the supports M_1 and M_2 are zero.

When y_1 is zero M_1 is also zero. When the extremity of y_1 lies between the flanges of the arch-rib both flanges have the same kind of stress; for vertical loads acting downward this stress is compression.

When the extremity of y_1 lies above the rib the upper flange is in compression and the lower in tension for *vertical loads acting downward*, or M_1 is *positive*. When y_1 is measured downward M_1 is *negative* and the upper flange is in tension and the lower in compression unless the extremity of y_1 falls between the flanges, when both are in compression *for vertical loads acting downward.*

To illustrate the application of our formulas we will now solve various examples in detail.

1°. Given a parabolic arch, with a hinge at each support, having a span of 100 and a rise of 25, determine H_1 for a load P placed at a distance 25 from the left support.

Here $\quad l = 100, \quad f = 25, \quad$ and $\quad k = 0.25.$

From (64a) we have

$$H_1 = \frac{5}{8} \cdot \frac{100}{25}(0.2227)P = 0.5568P.$$

The vertical reaction V_1 is found from (65), or

$$V_1 = (1 - 0.25)P = 0.75P.$$

From (66*a*) we have

$$y_0 = 25(1.3474) = 33.68.$$

In a like manner the values of H_1, V_1, and y_0 can be found for any other vertical load.

The method employed above was the common method neglecting the effect of the axial stress. Although this is of little consequence in this case (see Appendix C), we will, however, give the solution which includes the axial stress.

For this we need the values of

$$m = (\text{the radius of gyration})^2, \quad p = \text{parameter} = \frac{l^2}{8f}, \quad \text{and } \phi_0.$$

Let m be assumed $= 4$ (*an average value*).

$$p = 50 \quad \text{and} \quad \phi_0 = 0.7854.$$

Then, from (74),

$$H_1 = \frac{15}{8 \times 100(25)^2 + 30 \times 4 \times 50 \times 0.7854}\left\{\frac{8 \times 100(25)^2}{15}\mathfrak{H}_1 - \frac{4(100)^2}{2(50 + 50)}Pk(1 - k)\right\},$$

or

$$H_1 = 0.0000297\{33333\mathfrak{H}_1 - 200Pk(1 - k)\},$$

which is general for this particular arch.

Substituting the values of $\mathfrak{H}_1$ and k, we have

$$H_1 = 0.000297\{18559.8 - 37.5\} = 0.550P.$$

From the approximate equation (75),

$$H_1 = 0.5568(0.9885) = 0.550P,$$

the difference in results being in the fourth decimal place.

The value of V_1 remains unaffected by the axial stress.

From (76),

$$y_0 = \frac{V_1}{H_1}a = \frac{0.75}{0.55}25 = 34.09.$$

By the common method $y_0 = 33.68$, which is but 0.41 less than obtained above.

In a similar manner any other vertical load may be treated.

2°. Let a horizontal load Q be applied in place of the vertical load P. Then, by the common method from (77) or (77*a*),

$$H_1 = 0.5742Q.$$

Note that the values of H_1 are given by Table III when $Q =$ unity.

From (78*a*),

$$V_1 = 4 \times 0.25 \times 0.1875Q = 0.1875Q.$$

From (79),

$$x_0 = 0.5742l = 57.42.$$

If the axial stress is included in our calculations we have to apply (83), which contains the factor $\frac{1}{B}$. But

$$\frac{1}{B} = \frac{15}{8lf^2 + 30mp\phi_0} = 0.0000297;$$

$$\frac{4lf^2}{15B} = 0.0000297\left(\frac{33333}{2}\right) = 0.4949;$$

$$\frac{mp(\alpha + \phi_0)}{B} = \frac{4 \times 50(0.463 + 0.785)}{B} = 0.0074.$$

Hence

$$H_1 = 0.4949\{2(0.5742)\}Q + 0.0074Q = 0.5757Q,$$

which is but a very small amount larger than the result found by the common method.

$V_1 = 0.1875Q$, as before.

From (85),

$$x_0 = \frac{H_1}{V_1}b = \frac{0.5757}{0.1875}4k(1 - k)f = 57.57.$$

3°. In place of the loads P and Q, suppose the arch-rib constructed of metal having a modulus of elasticity $E =$ 28,000,000, and let the temperature rise 50°. What will be the value of H_1 if the coefficient of expansion of the metal is 0.0000055?

From (86),

$$H_1 = \frac{15 \,.\, E\theta \cos \phi}{8(25)^2}(0.0000055)50.$$

If θ is taken at the crown, $\cos \phi = 1$.

Let $\theta = 4$; then

$$H_1 = 92.4.$$

From (87), which includes the axial stress,

$$H_1\frac{168000000000}{500000 + 4712}(0.0000055)50 = 91.53.$$

Since a rise in temperature tends to lengthen the arch rib,

the span will tend to increase, hence H_1 must act from left towards the right.

4°. Let the arch be assumed parabolic in shape and fixed at the ends. Let a load P be applied at the quarter-point and determine the reactions, etc.

The following data will be used:

$$l = 100, \qquad f = 25, \qquad \phi_0 = 0.7854, \qquad \alpha = 0.463.$$

For the value of H_1 we have, from (91) or (91a),

$$H_1 = \tfrac{15}{4} \cdot \tfrac{100}{25}(0.0351)P = 0.5265P.$$

From (92) or (92a) we have

$$M_1 = \tfrac{100}{2}(-0.1054)P = -5.27P.$$

From (93) or (93a),

$$V_1 = 0.8437P.$$

From (92) or (92a),

$$M_2 = \tfrac{100}{2}(+0.0820)P = +4.10P.$$

From (93) or (93a), letting $k = 1 - k = 0.75$,

$$V_2 = 0.1562P.$$

From (94),

$y_0 = \frac{6}{5}25 = 30,$ measured up.

From (95),

$y_1 = -0.4(25) = -10,$ measured down.

From (96),

$y_2 = +0.3111(25) = +7.777,$ measured up.

From (97) or (97*a*),

$$x_1 = -\tfrac{100}{10}(-0.625) = +6.25,\ \text{measured to the right.}$$

From (98) or (98*a*),

$$x_2 = -\tfrac{100}{10}(2.625) = -26.25,\ \text{measured to the right.}$$

A good check upon the above work is to lay off the ordi nates and see if the two reaction lines meet on the load line P as indicated in the figure below.

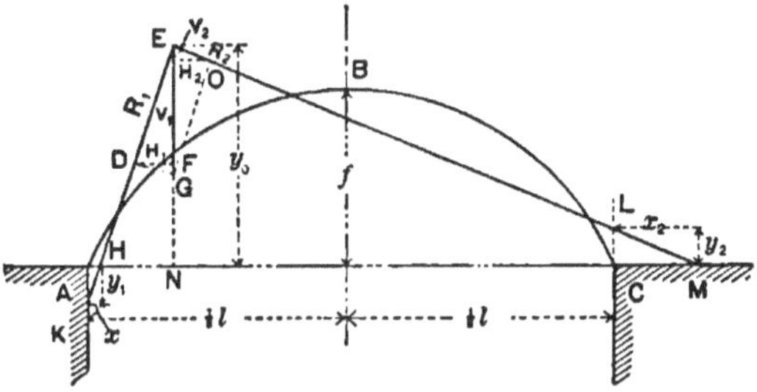

Thus far the formulas of the common method have been employed. We will now consider the effect of the axial stress.

For this case we apply (101) to obtain the value of H_1, letting $m = 4$ and $p = 50$. From (102),

$$C = \frac{15 \times 100 \times 25}{4 \times 100 \times 625 + 90 \times 4 \times 50 \times 0.7854} = 0.1419;$$

$$\frac{3lm}{2f(p+2f)} = \frac{3 \times 100 \times 4}{50(100)} = 0.24.$$

Then

$$H_1 = 0.1419\{100(0.0351) - 0.24(0.1875)\}P$$

or

$$H_1 = 0.491P.$$

By the approximate formula (103),

$$H_1 = 0.935(0.5265)P = 0.492P.$$

For the bending-moment M_2 we employ (107), in which there are several coefficients which are constant for this arch. We will first compute these:

$$D = 1 + \frac{3 \times 4}{(100)^2} - \frac{6 \times 4 \times 50 \times 0.7854}{(100)^3} = 0.99025.$$

$$\frac{8D}{5}f - f + \frac{6mp\phi_0}{fl}D = 39.61 - 25.0 + 0.37 = 14.98.$$

$$\frac{3mlD}{2f(p + 2f)} = 0.2377.$$

$$\frac{3mp}{l^2} = 0.6.$$

$$\phi_0 = 0.7854. \qquad \alpha = 0.463.$$

Then, from (107)

$$M_2(0.4805) = H_1\{14.98\} - 99.025Pk(1 - 2k^2 + k^3)$$
$$+ 50P(2k - 3k^2 + k^3)$$
$$+ 0.2377P(1 - k)k - 0.6P\{0.7854(2k - 1) + 0.463\}.$$

In the case we are considering $H_1 = 0.491P$ and $k = 0.25$. Making the proper substitutions, we have

$$M_2(0.4805) = + 7.355P - 22.053P$$
$$+ 16.405P$$
$$+ 0.044P - 0.043P$$

or

$$M_2 = \frac{+ 23.804 - 22.096}{0.4805}P = + 3.555P.$$

To determine the value of M, we substitute $1 - k$ for k in the above formula, or

$$M_1(0.4805) = +\ 7.355P \quad 22.053P$$
$$+\ 11.715P$$
$$+\ 0.044P - 0.043P$$

or

$$M_1 = P\frac{-\ 2.939}{0.4805} = -\ 6.116P.$$

As a check we will apply (112) to this case; then

$$M_1 = -\ \ 3.555P + 19.640P$$
$$-\ 22.270P +\ \ 0.185P$$
$$+\ \ 0.047P$$

or

$$M_1 = -\ 5.953P.$$

$(6.116 - 5.953)P = 0.16P$, or an error of about 2%, caused by neglecting decimals.

From (113),

$$V_1 = \tfrac{1}{100}\{3.555P - (-\ 6.116P) + 75P\}$$

or

$$V_1 = 0.8467P,$$

which differs but $0.003P$ from the value obtained by the common method.

From (51),

$$y_1 = \frac{M_1}{H_1} \quad \text{and} \quad y_2 = \frac{M_2}{H_2},$$

or

$$y_1' = \frac{-\ 6.116}{0.491} = -\ 12.46, \text{ measured downward,}$$

and

$$y_2 = \frac{+3.555}{0.491} = +7.24, \text{ measured upward.}$$

From (54),

$$x_1 = \frac{M_1}{V_1} \quad \text{and} \quad x_2 = \frac{M_2}{V_2} = \frac{M_2}{1 - V_1}.$$

Therefore

$$x_1 = \frac{6.116}{0.8467} = 7.2, \text{ measured to the right,}$$

and

$$x_2 = \frac{3.555}{1 - 0.8467} = 23.2, \text{ measured to the right.}$$

From (50),

$$y_0 = \frac{M_1 + V_1 a}{H_1} = \frac{-6.116 + 21.18}{0.491} = 30.6.$$

To show the effect of the axial stress upon each of the quantities we will tabulate our results:

COMPARISON OF RESULTS.

Function.	Common Method.	Exact Method.	Difference.	Percentage of Common Method.
H_1	$0.5265P$	$0.491P$	$0.035P$	6.6
H_2	$0.5265P$	$0.491P$	$0.035P$	6.6
V_1	$0.8437P$	$0.8467P$	$0.003P$	0.3
V_2	$0.1563P$	$0.1533P$	$0.003P$	2.0
M	$5.27P$	$6.116P$	$0.84P$	16.0
M_2	$4.10P$	$3.555P$	$0.55P$	15.5
y_0	30.0	30.6	0.6	2.0
y_1	10.0	12.46	2.46	24.6
y_2	7.77	7.24	0.53	7.0
x_1	6.25	7.2	0.95	15.2
x_2	26.25	23.2	3.05	11.6

5°. Let the vertical load be replaced by a horizontal load acting from left to right.

For the value of H_1 by the common method we have, from (115) or (116),

$$H_1 = 0.6329Q \text{ acting towards the left.}$$

From (117) or (118),

$$M_1 = 25(0.2109)Q = 5.272Q.$$

Now since Q acts towards the right, M_1 will be *negative.*

From (119) or (120),

$$M_2 = -25(0.1171)Q = -2.927Q.$$

Since Q acts towards the right and our formula was deduced for Q acting towards the *left*, M_2 will be *positive.*

From (121) or (121*a*),

$$V_1 = 12\tfrac{25}{100}(0.0351)Q = 0.1053Q, \text{ acting downward ;}$$

$$V_2 = 0.1053Q, \text{ acting upward.}$$

From (123) or (123*a*),

$$y_1 = 25(0.3332) = 8.33, \text{ measured upward.}$$

From (124) or (124*a*),

$$y_2 = 25(0.3189) = 7.97, \text{ measured upward.}$$

From (125) or (125*a*),

$$x_1 = 100(0.5) = 50, \text{ measured to the left.}$$

From (126) or (126*a*),

$$x_2 = 100(0.2778) = 27.78, \text{ measured to the right.}$$

From (127) or (127*a*),

$$x_0 = \frac{100}{2}1.25 = 62.5, \text{ measured to the right.}$$

Note that for horizontal loads x_1 and x_2 are always measured outward and y_1 and y_2 upward, without regard to the direction of Q.

We will now consider the effect of the axial stress.

The value of H_1 is given by (128) or (130), which contain several constants, which we will compute first.

$$C = 0.1419, \qquad (\alpha + \phi_0) = 1.248, \qquad \frac{3mp}{lf} = 0.24.$$

Then, from (130),

$$H_1 = 0.1419\{6\tfrac{3}{8}Q\Delta_{12} + 0.24(1.248)Q\}$$

or

$$H_1 = 0.1419\{4.219 + 0.299\}Q = 0.6411Q.$$

The value of M_2 is given by (132).

$$D = 0.99025.$$

$$\frac{8D}{5}f - f + \frac{6mp\phi_0}{fl}D = 14.98.$$

$$\frac{3m}{2f}\left\{\frac{p}{p+2f} + 1 - \frac{2p\phi_0}{l}\right\} = 0.1715.$$

$$\frac{3mp}{lf}(\alpha + \phi_0) = 0.299.$$

Then (132) becomes

$$\begin{aligned} M_2(0.4805) = {} & 14.98H_1 - 0.1715Qk(1-k) \\ & + 25Q\{\Delta_4\} - 39\,61Q\{\Delta_2\} \\ & - 0.298Q \end{aligned}$$

or

$$M_2(0.4805) = +\ 9.604Q - 0.032Q + 12.012Q - 22.744Q - 0.298Q;$$

hence

$$M_2 = \frac{-1.458}{0.4805}Q = -3.03Q.$$

As in the common method, M_2 is positive, since Q acts from left to right.

By making k equal $1 - k$ and substituting H_2 for H the value of M_1 becomes

$$M_1(0.4805) = +\ 5.376Q - 0.032Q + 8.887Q - 16.865Q - 0.298Q.$$

Therefore

$$M_1 = Q\frac{2.932}{0.4805} = -6.122Q.$$

From (137),

$$V_1 = \frac{1}{100}\left\{3.03 + 6.122 - 18.75\right\}Q$$

or

$$V_1 = 0.0960Q.$$

From (51),

$$y_1 = \frac{6.122}{0.6411} = 9.5, \text{ measured upward,}$$

and

$$y_2 = \frac{3.03}{0.3589} = 8.4, \text{ measured upward.}$$

From (54),

$$x_1 = \frac{6.122}{0.096} = 63.8, \text{ measured to the left,}$$

and

$$x_2 = \frac{3.03}{0.096} = 31.6, \text{ measured to the right.}$$

From (58),

$$x_0 = \frac{1196.2 - 606.1}{9.5} = 62.1, \text{ measured to the right.}$$

6°. For temperature stresses let the arch be of metal, having a modulus of elasticity $E = 28{,}000{,}000$ and a coefficient of expansion $e = 0.0000055$. Assume the temperature to rise 50°.

From (140), neglecting the effect of the axial stress we have

$$H_1 = \frac{45E\theta \cos \phi}{4f^2} et°.$$

Let $\theta \cos \phi = 4$; then

$$H_1 = 554.4.$$

From (144),

$$M_2 = M_1 = 554.4\tfrac{50}{3} = 9240.$$

From (145),

$$y_1 = y_2 = y_0 = 16\tfrac{2}{3}.$$

If the effect of the axial stress is included, we have, from (139),

$$H_1 = 0.1419(3 \times E\theta \cos \phi et°)l = 525.$$

From (141),

$$M_2(0.4805) = \frac{4500}{264137}(14.98)Aet° - 0.1188Aet°.$$

Now $Aet^\circ = 28000000 \times 4 \times 50 \times 0.0000055$

or $Aet^\circ = 30800.$

Therefore

$$M_2 = \frac{4201}{0.4805} = 8744.$$

From (51),

$$y_1 = \frac{M_1}{H_1} = \frac{8744}{525} = 16.6.$$

7°. Let the arch be loaded from the left support to the crown with a uniform *horizontally distributed* load; then $k'' = 0$ and $k' = 0.5$.

From (147),

$$H_1 = w\frac{(100)^2}{8(25)}\left(\left(\frac{1}{2}\right)\left[10 - \frac{15}{2} + \frac{6}{4}\right]\right) = 25w.$$

From (148),

$$M_1 = w\frac{(100)^2}{2}\left[-\frac{1}{32}\right] = -156.2w.$$

From (149),

$$M_2 = w\frac{(100)^2}{2}\left[+\frac{1}{32}\right] = +156.2w.$$

From (150),

$$V_1 = w\frac{100}{2}\left[\frac{13}{16}\right] = 40.6w.$$

The location of the point where the true equilibrium polygon starts can be found from (51).

$$y_1 = \frac{1562}{25} = 6.2, \text{ measured downward}\cdot$$

$$y_2 = \frac{1562}{25} = 6.2, \text{ measured upward.}$$

8°. What will be the vertical deflection of the arch at the crown when there are two equal and symmetrical loads placed at the quarter points?

Let $E = 28000000$ and $\theta \cos \phi = 4$;
$p = 50$.

Since the arch is fixed at the ends, $\Delta\phi_0 = 0$; then $p(84)$, page 55, becomes, making $x = l/2$,

$$\delta y = \frac{l^3}{24A}\left\{3M_1 + V_1\frac{l}{2} - H_1\frac{3l^2}{800} - \frac{4}{l^2}\sum^{l/2} P\left(\frac{l}{2} - a\right)^3\right\}.$$

(We have neglected the effect of the axial stress.)

$$\frac{l^3}{24A} = 0.0000037 \text{ about.}$$

$$+3M_1 = -15.81P \qquad\qquad + V_1\frac{l}{2} = +42.185P$$

$$\begin{array}{r} +12.30P \\ \hline -3.51P \end{array} \qquad\qquad \begin{array}{r} +7.815P \\ \hline +50.000P \end{array}$$

$$-H_1\frac{3l^2}{800} = -19.74P$$

$$\begin{array}{r} -19.74P \\ \hline -39.48P \end{array}$$

$$-\frac{4}{l^2}\sum^{l/2} P\left(\frac{l}{2} - a\right)^3 = -6.25P.$$

Therefore

$$\delta y = 0.0000037(0.76)Q = 0.0000028Q.$$

Suppose $Q = 30000$, then $\delta y = 0.084$; and if our span is measured in feet

$$\delta y = 1.008 \text{ inches.}$$

The sign being positive indicates that the crown rises under the action of these two loads placed at the quarter points.

Thus far we have considered only parabolic arches. We will now solve a few similar problems for a circular arch having a span of 100 and a rise of 25. The following data will be employed:

$$l = 100, \qquad f = 25, \qquad R = 62.5,$$

$$k' = 37.5, \quad \phi_0 = 53^\circ\ 7\tfrac{3}{4}', \qquad m = \frac{\theta}{FR^2} = 0.00102,$$

$$k = 25, \qquad \alpha = 23^\circ\ 35'.$$

9°. Determine H_1, V_1, etc., by the common method, assuming the arch to have a hinge at each support. Vertical load P.

From (160*a*),

$$H_1 = P\frac{A}{B} = P\Delta_{17}.$$

In order to use Table XVII we must determine the values of $\frac{2\phi_0}{\pi}$ and $\frac{\alpha}{\phi_0}$.

$$\frac{2\phi_0}{\pi} = 0.590, \qquad \frac{\alpha}{\phi_0} = 0.443.$$

Entering Table XVII with these values, we have by interpolation $\frac{A}{B} = 0.570$. Hence

$$H_1 = 0.570P.$$

From (161),

$$V_1 = 0.750P.$$

From (163),

$$y_0 = \frac{V_1}{H_1}a = \frac{0.750}{0.570}25 = 32.87.$$

From (164),

$$H_1 = 0.57\frac{1 - \frac{0.00102}{0.17670}(0.64 - 0.16)}{1 + \frac{0.00102}{0.155}(1.407)}P$$

or

$$H_1 = 0.57\frac{1 - 0.00277}{1 + 0.00925}P = 0.563P.$$

$V_1 = 0.75\ P$, as before.

From (50),

$$y_0 = \frac{0.75}{0.563}25 = 33.3.$$

From the approximate equation, $H_1 = \mathfrak{H}_1(1 - e)$, as explained in Appendix C, the value of H_1 becomes

$$H_1 = 0.57(0.9885)P = 0.5634P,$$

no change in the figures occurring until the fourth decimal is reached.

10°. Let the vertical local P be replaced by a horizontal load Q acting towards the left. Then we have to employ formula (172), if the axial stress is neglected.

$$\alpha - \sin\alpha\cos\alpha = \beta_{18} \quad = 0.04491;$$

$$\sin\alpha - \alpha\cos\alpha = \Delta\Delta_{18} = 0.02288;$$

$$2\cos\phi_0 = 1.2;$$

$$\phi_0 - 3\sin\phi_0\cos\phi_0 + 2\phi_0\cos^2\phi_0 = \Delta_{18} = 0.155.$$

Then

$$H_1 = \frac{1}{2}Q\left\{1 + \frac{0.04491 - 1.2(0.02288)}{0.155}\right\}$$

or

$$H_1 = 0.5562Q.$$

From (176),

$$V_1 = -V_2 = Q\frac{18.75}{100} = 0.1875Q.$$

From (178),

$$x_0 = \sin\phi_0\{1.1125\}R$$

or

$$x_0 = 0.799\{1.1125\}62.5 = 55.55, \text{ measured to the right.}$$

From (177),

$$x_0 = \frac{H_1}{V_1}b = 55.62.$$

The difference in these values is due to the omission of decimal figures.

If the axial stress is not neglected the operation becomes considerably longer. The formula to be employed is (180) or (181).

All but two of the terms in (180) have been evaluated above.

$$\phi_0 + \sin\phi_0 \cos\phi_0 = \varDelta_{10} = 1.4073,$$
$$\alpha + \sin\alpha \cos\alpha = \varDelta_{19} = 0.7782.$$

Then

$$H_1 = Q\left\{\frac{0.155 + 0.0449 - 0.01745 + 0.00102(1.4073 + 0.7782)}{0.310 + 0.00204(1.4073)}\right\}$$

or

$$H_1 = 0.558Q.$$

As before,

$$V_1 = 0.1875Q.$$

Then, from (183),

$$x_0 = \frac{0.558}{0.1875} 18.75 = 55.8.$$

11°. Let the arch be fixed at the ends, and let a load P be placed at the quarter-point on the left of the crown; then by the common method H_1 will be found from (192).

The following quantities will be required:

$$2 \sin \phi_0 = 1.60,$$
$$\cos \alpha + \alpha \sin \alpha = \Delta_{22} = 1.0811,$$
$$\sin \phi_0 [2 \cos \phi_0 + \phi_0 \sin \phi_0] = \Delta_{21} = 1.5535,$$
$$\phi_0^2 + \phi_0 \sin \phi_0 \cos \phi_0 - 2 \sin^2 \phi_0 = \Delta_{20} = 0.0250.$$

Hence

$$H_1 = \frac{P}{2} \left\{ \frac{1.60(1.0811) - 1.5535 - 0.931(0.16)}{0.025} \right\}$$

or

$$H_1 = 0.546P.$$

The value of M_1 is given by (196), in which the following terms appear:

$$2\phi_0 = 1.862, \quad \sin \alpha = 0.400, \quad \cos \alpha = 0.9164, \quad \sin \phi_0 = 0.800,$$
$$\cos \phi_0 = 0.600.$$

$$\phi_0 + \sin \phi_0 \cos \phi_0 = \Delta_{19} = 1.4073,$$
$$-\phi_0 + \sin \phi_0 \cos \phi_0 = -\beta_{19} = -0.4474,$$
$$\cos \alpha + \alpha \sin \alpha = \Delta_{22} = 1.0811,$$
$$\cos \phi_0 + \phi_0 \sin \phi_0 = \Delta_{22} = 1.3407,$$
$$\sin \phi_0 - \phi_0 \cos \phi_0 = \Delta\Delta_{19} = .2437.$$

Then

$$M_1 = P\frac{0.546(62.5)}{0.931}(0.2437) + P\frac{62.5}{1.862(-0.4474)}$$
$$\{0.4(0.931)[(0.9164)(0.8) - 1.4073] + (0.41)(0.931)(0.8)$$
$$+ (-0.4474)[1.0811 - 1.3407]\},$$

which reduces to

$$M_1 = +8.931P - 75\left\{\begin{matrix} -0.2510 + 0.3054 \\ +0.1161 \end{matrix}\right\}P$$

or

$$M_1 = +8.931P - 12.788P = -3.856P.$$

We see from this problem that the solutions of the formulas for the fixed circular arch are considerably longer than those for the parabolic arch, even with the aid of the Tables.

In practice it will be found that if H_1, M_1, etc., are determined for loads placed so that α will be in even degrees, and then the values of H_1 and M_1 interpolated from a diagram, much time will be saved. This method also is less liable to have errors in individual terms.

CHAPTER VIII.

APPLICATION OF THE GENERAL SUMMATION FORMULAS TO ARCHES HAVING A HINGE AT EACH SUPPORT.

WE have selected an arch over the Douro in Portugal to illustrate the application of the summation formulas, as the form of the rib is such that none of the common formulas can be applied. The rib is crescent-shaped, with $\theta = 4.6$ at the crown and 0.2 near the hinges.

The data are taken from *Mémoires et Compte Rendu des Travaux de la Société des Ingénieurs Civils* (Sept. and Oct. 1878), *Mémoire par T. Seyrig.*

The superstructure is supported symmetrically at points *A*, *B*, *C*, *D*, *E*, etc., located on the rib as shown in Fig. 54.

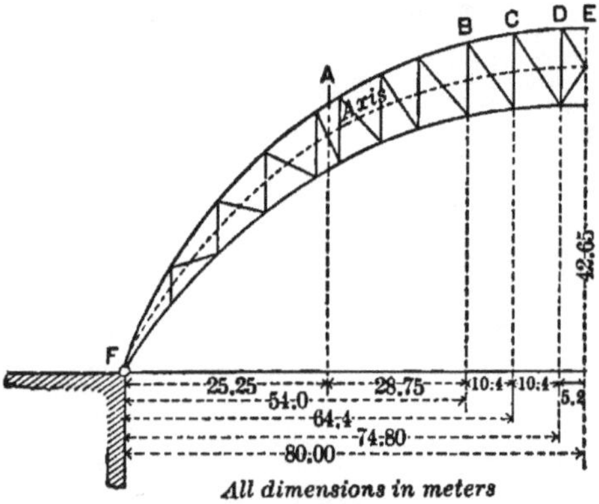

All dimensions in meters

FIG. 54.

The linear arch used in the computations lies midway between the flanges of the arch rib, and is divided into twenty-two sections as indicated in Fig. 55. The coördinates of the points of division are given in the following table of data.

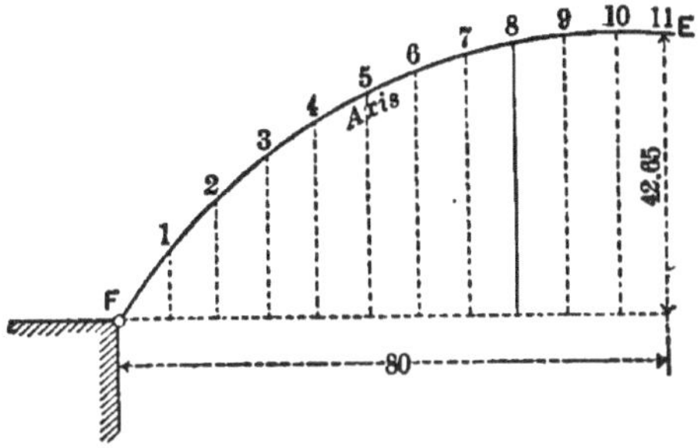

FIG. 55.

Section.	x (1)	y (2)	Δs (3)	θ_x (4)	F_x (5)	$\frac{xy\Delta s}{\theta_x}$ (6)	$\frac{y\Delta s}{\theta_x}$ (7)	$\frac{y^2\Delta s}{\theta_x}$ (8)	$\frac{\Delta s}{F_x}$ (9)
0									
1	2.80	3.00	8.10	0.246	0.293	276.58	98.78	296.34	27.64
2	8.40	9.00	8.15	0.588	0.274	1047.81	124.74	1122.66	29.70
3	14.10	14.55	8.15	1.153	0.264	1450.04	102.84	1496.32	30.87
4	20.40	20.42	8.80	1.848	0.253	1983.49	97.23	1985.43	34.78
5	25.25	24.20	3.80	2.463	0.242	941.58	37.33	903.38	15.70
6	31.00	28.30	9.80	2.863	0.236	3002.97	96.87	2741.42	41.52
7	39.75	32.75	10.05	3.486	0.225	3752.79	94.41	3091.92	44.66
8	49.15	36.85	10.40	3.758	0.222	5012.31	101.98	3757.96	46.84
9	59.20	40.35	10.75	4.220	0.222	6084.57	102.78	4147.17	48.42
10	69.60	42.25	10.50	4.609	0.228	6699.00	96.25	4066.56	46.05
11	80.00	42.65	5.25	4.696	0.228	3814.40	47.68	2033.55	23.02
						34066.54		25642.71	389.20

In the above table columns (1) to (5) inclusive contain the data necessary for the determination of the values of the horizontal thrusts due to a loading of unity at any point being considered. Columns (6) to (9) inclusive have been computed from the data given.

The formula to be applied in this case is (242), page 49, or

$$H_1 = \frac{\overline{\left| + \left\{ \sum_0^{l/2} xy \frac{\Delta s}{\theta_x} - \sum_a^{l/2} xy \frac{\Delta s}{\theta_x} + a \sum_a^{l/2} y \frac{\Delta s}{\theta_x} \right\} P \right.} \\ \underline{\left. + \left\{ \sum_0^{l/2} \frac{\Delta x \sin \phi}{F_x} - \sum_a^{l/2} \frac{\Delta x \sin \phi}{F_x} \right\} P \right|}}{2 \left\{ \sum_0^{l/2} y^2 \frac{\Delta s}{\theta_x} + \sum_0^{l/2} \frac{\Delta x \cos \phi}{F_x} \right\}}.$$

The term containing F_x in the numerator is very small and can be neglected without serious error.

$\Delta x \cos \phi = \Delta s$ approximately. Then we have

$$H_1 = \frac{+\sum_0^{l/2} xy\frac{\Delta s}{\theta_x} - \sum_a^{l/2}(x-a)y\frac{\Delta s}{\theta_x}}{2\left\{\sum_0^{l/2} y^2\frac{\Delta s}{\theta_x} + \sum_0^{l/2}\frac{\Delta s}{F_x}\right\}}P.$$

Since the first term of the numerator and the entire denominator are constant for this case, we may place them at once.

$$\text{Denominator} = 2(25642.71 + 389.20) = 52064;$$

$$\sum_0^{l/2} xy\frac{\Delta s}{\theta_x} = 34066.54.$$

Therefore

$$H_1 = \frac{34066.54 - \sum_a^{l/2}(x-a)y\frac{\Delta s}{\theta_x}}{52064}P.$$

There remains then but one term to compute as the position of the load changes.

Following are the necessary computations for determining the values of H_1 for loads at A, B, C, D, and E respectively:

Load at A.

$a = 25.25, \qquad k = 0.157.$

Point.	$x - a$	$(x-a)y\frac{\Delta s}{\theta_x}$	
5	0	0	Since a must be less than x, where the term $(x - a)$ is employed, we need to compute only the quantities given in the table.
6	5.75	557.00	
7	14.50	1368.94	
8	23.90	2437.32	
9	33.95	3489.38	
10	44.35	4268.68	
11	54.75	2610.48	
		14731.80	

We have, substituting the several values given above,

$$H_1 = P\frac{34066.54 - 14731.80}{52064} = \frac{19334.74}{52064}P$$

or

$$H_1 = 0.3713P.$$

Load at B.

$a = 54.00.$

Point.	$x - a$	$(x-a)y\frac{\Delta s}{\theta_x}$
9	5.2	534.45
10	15.6	1501.50
11	26.0	1239.68
		3275.63

$$H_1 = P\frac{34066.54 - 3275.63}{52064} = \frac{30790.91}{52064}P$$

or

$$H_1 = 0.5914P.$$

Load at C.

$a = 64.4.$

Point.	$x - a$	$(x-a)y\frac{\Delta s}{\theta_x}$
10	5.20	500.50
11	15.60	743.80
		1244.30

$$H_1 = \frac{34066.54 - 1244.30}{52064}P = \frac{32822.24}{52064}P$$

or

$$H_1 = 0.6304P.$$

Load at D.

$a = 74.8.$

Point.	$x - a$	$(x - a)y\frac{\Delta s}{\theta_x}$
10	0	0
11	5.02	247.93
		247.93

$$H_1 = \frac{34066.54 - 247.93}{52064}P = \frac{33818.61}{52064}P$$

or

$$H_1 = 0.6495P.$$

Load at E.

$a = 80.00.$ $x - a = 0$ at E. Hence

$$H_1 = \frac{34066.54}{52064}P$$

or

$$H_1 = 0.6543P.$$

Having now determined the values of H_1 for each load, the stresses in the arch can be found graphically for any given value of P. Since the arch is hinged at the ends, the values of V_1 will be the same as for a straight unconfined beam.

The following table shows the values of H_1 obtained above with those given by Seyrig:

Load at	H_1 (1)	H_1 (Seyrig) (2)	Diff. (3)	Formula (91), page 29 (4)	Diff. (1) and (4) (5)
A	$0.3713P$	$0.370P$	$-0.0013P$	$0.358P$	$0.0133P$
B	$0.5914P$	$0.592P$	$-0.0006P$	$0.653P$	$0.0616P$
C	$0.6304P$	*$0.631P$	$-0.0006P$	$0.712P$	$0.0816P$
D	$0.6495P$	$0.650P$	$-0.0005P$	$0.742P$	$0.0925P$
E	$0.6543P$	....	....	$0.746P$	$0.0917P$

* As given by Seyrig this is 0.637; but as he gives it as the quotient of 2048.36 ÷ 3246.84, which is 0.6309, it is evidently a typographical error.

The differences in column (3) are very small and unimportant.

To show the error in applying the common formula to arches where the moments of inertia do not vary according to the law making $\theta \cos \phi =$ a constant, columns (4) and (5) have been computed, from which it is seen that an error of about sixteen per cent would be made in applying formula (91), page 29, to this particular arch.

The following tables give the loading which was assumed in designing the arch, the unit being 1000 kilograms:

I°. MOVING LOAD COVERING THE ENTIRE ROADWAY.

Load at	Load in 1000^k.	Coefficient of H_1.	H_1.	
A	126.8	0.3713	47.08	Since the arch is symmetrically loaded, the total value of
B	62.4	0.5914	36.90	
C	47.0	0.6304	29.62	
D	40.5	0.6495	26.30	$H_1 = 139.90 \times 2$
			139.90	$= 279.80$

II°. SYMMETRICAL MOVING LOAD COVERING 80 METRES IN THE CENTRE OF THE SPAN.

Load at	Load in 1000^k.	Coefficient of H_1.	H_1.	
A	18.6	0.3713	6.90	Since the arch is symmetrically loaded, the total value of
B	53.8	0.5914	31.81	
C	47.0	0.6304	29.62	
D	40.5	0.6495	26.30	$H_1 = 94.63 \times 2$
			94.63	$= 189.26$

III°. NON-SYMMETRICAL LOADING.

Load at	Load in 1000^k.	Coefficient of H_1.	H_1.	
A	113.0	0.3713	41.95	
B	58.0	0.5914	34.30	
C	48.0	0.6304	30.25	
D	36.6	0.6495	23.78	
D'	3.1	0.6495	2.01	
C'	0	......	.....	
B'	4.6	0.5914	2.72	
A'	13.5	0.3713	4.91	
			139.92 =	Total H_1

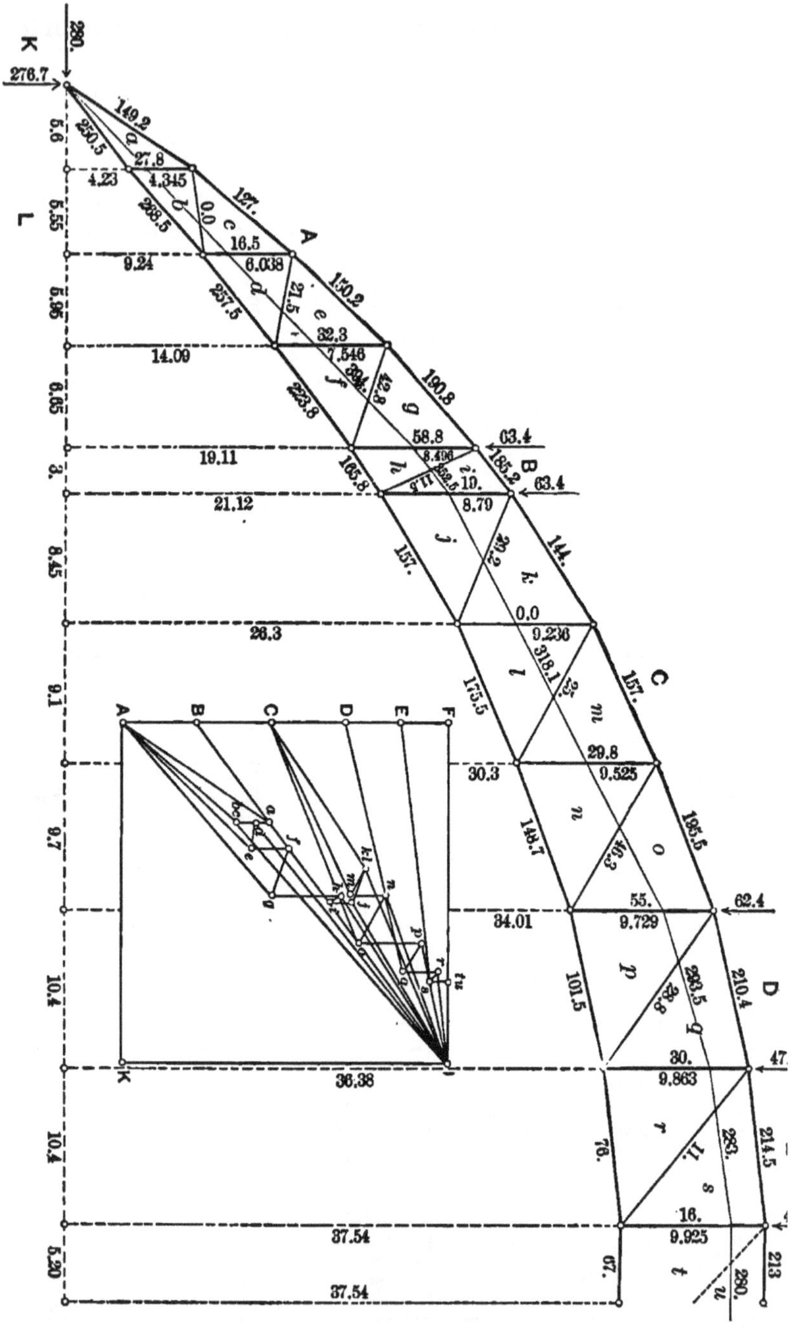
K
L
276.7
280.
149.2
250.5
27.8
4.23
4.345
127.
268.5
A
16.5
9.24
6.038
150.2
257.5
32.3
14.09
7.546
190.8
223.8
58.8
63.4
19.11
8.496
185.2
B
63.4
10.
21.12
8.79
144.
157.
0.0
26.3
9.236
318.1
C
157.
175.5
29.8
9.525
30.3
195.5
148.7
55.
62.4
34.01
9.729
D
210.4
101.5
293.5
30.
47
9.863
36.38
214.5
283.
76.
16.
9.925
37.54
213
67.
280.
37.54
A
B
C
D
E
F
K
5.6
5.55
5.96
6.65
3.
8.45
9.1
9.7
10.4
10.4
5.20

FIG. 56.

The stresses in the various pieces composing the arch can now be found either by computation or by graphics.

The stress diagram for a load over all is shown in Fig. 56.

From this example we see that arches having a variable moment of inertia, not following the law $\theta \cos \phi =$ a constant, can be treated with but very little labor. Most of the computations can be made with the slide-rule, and furthermore the computer need not be familiar with the theory at all, merely deducing certain quantities mechanically, these quantities to be used by the person who is responsible for the designing of the structure.

The effect of the axial stress can be readily seen, since, approximately, it occurs only in the denominator. In our example the axial stress term $= 778.4$. Then the denominator, neglecting the axial stress, amounts to $52{,}064 - 778 = 51{,}286$, and the results obtained by using this denominator correspond to those obtained by the common method if the effect of the variable θ could be considered.

The relative error made in omitting the axial stress term is $\frac{778}{51284} = 0.015$, or 1.5 per cent, an error of no practical importance.

CHAPTER IX.

APPLICATION OF THE GENERAL SUMMATION FORMULAS TO ARCHES WITHOUT HINGES.

In order to illustrate the application of our formulas and to show to what degree of accuracy they lead, we will compute the values of H_1, M_1, etc., for a parabolic arch having moments of inertia *varying according to the law* $\theta \cos \phi =$ *a constant*, by the summation method and by the formulas demonstrated in Chapter III.

DATA.

Span $= l = 190$; Rise $= f = 25$;
Load = a concentration P or Q = unity at points designated.

VERTICAL LOADS. (P = unity).

(*a*) *Determination of H_1 by Summation.*

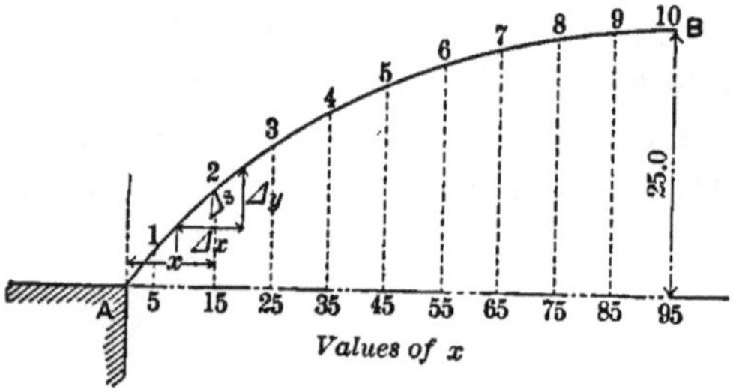

Fig. 57.

Let the semi-arch be divided into ten parts as shown in the figure, and the quantities shown in the tables determined.

$$b = 4k(1 - k)f \quad \text{and} \quad \tan \phi = \frac{8f}{l^2}(\tfrac{1}{2}l - x).$$

The moments of inertia are determined for the section at the *middle points* of Δs or points having the abscissas x. Only the relative values need be determined now, as we propose to neglect the effect of the axial stress.

DATA.

Point.	k	x	y	Δs	θ	$\frac{\Delta s}{\theta_x}$	Approximate Φ
1	0.026	5	2.5	11.2	1.12	10.0	26° 30′
2	.079	15	7.3	10.9	1.09	"	23° 54′
3	.132	25	11.5	10.7	1.07	"	21° 12′
4	.184	35	15.0	10.5	1.05	"	18° 23′
5	.237	45	18.1	10.3	1.03	"	15° 29′
6	.289	55	20.5	10.2	1.02	"	12° 30′
7	.342	65	22.5	10.1	1.01	"	9° 26′
8	.395	75	23.9	10.1	1.01	"	6° 20′
9	.447	85	24.7	10.0	1.00	"	3° 10′
10	.500	95	25.0	5.0(½)	1.00	5.0	0° 0′
						95.0	

SUMMATION TERMS.

Point.	$y\frac{\Delta s}{\theta_x}$	$y^2\frac{\Delta s}{\theta_x}$	$x\frac{\Delta s}{\theta_x}$	$xy\frac{\Delta s}{\theta_x}$
1	25	62.5	50	125
2	73	532.9	150	1095
3	115	1322.5	250	2875
4	150	2250.0	350	5250
5	181	3276.1	450	8145
6	205	4202.5	550	11275
7	225	5062.5	650	14625
8	239	5712.1	750	17925
9	247	6100.9	850	20995
10	125	3125.0	475	11875
	1585	31647.0	4525	94185

From (221), page 46, remembering that the terms containing N_x and F_x are to be omitted, since we propose to neglect the axial stress, we have

$$H_1 = \frac{\sum_0^{\frac{1}{2}l} \frac{K' y \Delta s}{\theta_x} - \frac{\sum_0^{\frac{1}{2}l} \frac{K' \Delta s}{\theta_x}}{\sum_0^{\frac{1}{2}l} \frac{\Delta s}{\theta_x}} \sum_0^{\frac{1}{2}l} \frac{y \Delta s}{\theta_x}}{2\left\{ \sum_0^{\frac{1}{2}l} \frac{y^2 \Delta s}{\theta_x} - \frac{\left(\sum_0^{\frac{1}{2}l} \frac{y \Delta s}{\theta_x}\right)^2}{\sum_0^{\frac{1}{2}l} \frac{\Delta s}{\theta_x}} \right\}},$$

where $K' = V_1 x - \Sigma P(x - \overset{x>a}{a})$; or for the left half of the arch,

$$K' = Px - P(x - \overset{x>a}{a}).$$

But $P =$ unity, and hence

$$K' = x - (x - \overset{x>a}{a}).$$

Then

$$\sum_0^{\frac{1}{2}l} \frac{K' y \Delta s}{\theta_x} = \sum_0^{\frac{1}{2}l} \frac{x y \Delta s}{\theta_x} - \left(\sum_a^{\frac{1}{2}l} \frac{x y \Delta s}{\theta_x} - \sum_a^{\frac{1}{2}l} \frac{a y \Delta s}{\theta_x} \right) \quad . \quad (222)$$

and

$$\sum_0^{\frac{1}{2}l} \frac{K' \Delta s}{\theta_x} = \sum_0^{\frac{1}{2}l} \frac{x \Delta s}{\theta_x} - \left(\sum_a^{\frac{1}{2}l} \frac{x \Delta s}{\theta_x} - \sum_a^{\frac{1}{2}l} \frac{a \Delta s}{\theta_x} \right). \quad . \quad . \quad (223)$$

We will first determine the constants in our expression for H_1.

The denominator becomes

$$2\left(31647 - \frac{(1585)^2}{95}\right) = 13747;$$

$$\frac{\sum_0^{\frac{1}{2}l} \frac{y \Delta s}{\theta}}{\sum_0^{\frac{1}{2}l} \frac{\Delta s}{\theta}} = \frac{1585}{95} = 15.6;$$

and we have

$$H_1 = \frac{\sum_0^{\frac{1}{2}l} \frac{K' y \Delta s}{\theta_x} - 15.6 \sum_0^{\frac{1}{2}l} \frac{K' \Delta s}{\theta_x}}{13747}.$$

The following table contains the quantities to be substituted in this equation

PARTIAL SUMS.

Load at	a	(1) $\sum_a^{\frac{1}{2}l} \frac{xy\Delta ds}{\theta}$	(2) $a\sum_a^{\frac{1}{2}l} \frac{y\Delta s}{\theta}$	(3) $\sum_a^{\frac{1}{2}l} \frac{x\Delta s}{\theta}$	(4) $a\sum_a^{\frac{1}{2}l} \frac{\Delta s}{\theta}$
1	5	94060	7800	4475	425
2	15	92965	22305	4325	1125
3	25	90090	34300	4075	1625
4	35	84840	42770	3725	1925
5	45	76695	46845	3275	2025
6	55	65420	45980	2725	1925
7	65	50795	39715	2075	1625
8	75	32870	27900	1325	1125
9	85	11875	10625	475	425
10	95	0		0	

In this table the summation is actually taken between $\frac{1}{2}l$ and $(a + 1)$, for when $x = a$ the combination of columns 1 and 2 and 3 and 4 respectively equal zero.

For a load at (1), our equation gives us

$$H_1 = \frac{94185 - (94060 - 7800) - \{4525 - (4475 - 425)\} 15.6}{13747},$$

or

$$H_1 = 0.037.$$

In like manner the values of H_1 can be found for all the points from 1 to 10 inclusive. These values are given in the annexed table.

VALUES OF H_1 FOR A LOAD UNITY AT—

Point.	k	H_1	
1	.026	0.037	In case any load other than unity is placed at any point, the corresponding value of H_1 is found by multiplying the load by the corresponding coefficient H_1 in this table.
2	.079	0.207	
3	.132	0.438	
4	.184	0.698	
5	.237	0.964	
6	.289	1.210	
7	.342	1.421	
8	.395	1.581	
9	.447	1.682	
10	.500	1.714	

(b) Determination of H_1 by Integration.

From (91), page 29, we have

$$H_1 = \frac{15}{4n} Pk^2(1 - k)^2,$$

which becomes for our arch with load unity

$$H_1 = 28.6k^2(1 - k)^2 \quad \text{or} \quad H_1 = 28.6\Delta_{11}$$

if the tables are employed.

The values of H_1 are as follows:

VALUES OF H_1 FOR A LOAD UNITY AT—

Point.	k	H_1	
1	.026	.018	In case any load other than unity is placed at any point the value of H_1 is found by multiplying the load by the corresponding value of H_1 in this table.
2	.079	.152	
3	.132	.377	
4	.184	.643	
5	.237	.938	
6	.289	1.201	
7	.342	1.447	
8	.395	1.633	
9	.447	1.744	
10	.500	1.787	

(c) *Comparison of Results.*

VALUES OF H_1 FOR LOAD UNITY AT—

Point.	H_1, by Summation.	H_1, by Integration.	Difference.	Relative Diff. in Per Cent.
1	.037	.018	$-.019$	By Summation. Too large. 106
2	.207	.152	$-.055$	36
3	.438	.377	$-.061$	17
4	.698	.643	$-.055$	8.5
5	.964	.938	$-.026$	2.7
6	1.210	1.201	$-.009$	0.7
7	1.421	1.447	$+.026$	Too small. 1.8
8	1.581	1.633	$+.052$	3.2
9	1.682	1.744	$+.062$	3.5
10	1.714	1.787	$+.073$	4.1

Let a load of one ton per horizontal foot of the arch be assumed, and determine the value of H_1 for a load over all. Then, by *summation*,

$$H_1 = 10(9.095)2 = 181.9 \text{ tons};$$

by *integration* (*for concentrations*),

$$H_1 = 10(9.0475)2 = 180.95 \text{ tons};$$

$$181.9 - 180.95 = 0.95;$$

and relative error equals

$$\frac{0.95}{180.95} = 0.52\%.$$

We will now determine the values of M_1 by both methods.

(d) DETERMINATION OF M_1.

From (225), page 47,

$$M_1 = \frac{\sum_0^l \frac{Kx\Delta s}{\theta_x} \sum_0^l \frac{x\Delta s}{\theta_x} - \sum_0^l \frac{K\Delta s}{\theta_x} \sum_0^l \frac{x^2\Delta s}{\theta_x}}{\sum_0^l \frac{\Delta s}{\theta_x} \sum_0^l \frac{x^2\Delta s}{\theta_x} - \left(\sum_0^l \frac{\Delta s}{\theta_x}\right)^2},$$

in which

$$\sum_0^l \frac{Kx\Delta s}{\theta_x} = -H_1 \sum_0^l \frac{xy\Delta s}{\theta_x} - \left(\sum_0^l \frac{x^2\Delta s}{\theta_x} - a\sum_a^l \frac{x\Delta s}{\theta_x}\right)P. \quad (226)$$

$$\sum_0^l \frac{K\Delta s}{\theta_x} = -H_1 \sum_0^l \frac{y\Delta s}{\theta_x} - \left(\sum_a^l \frac{x\Delta s}{\theta_x} - a\sum_a^l \frac{\Delta s}{\theta_x}\right)P. \quad (227)$$

The following table contains all the constants entering the above equation. Substituting these constants and the values of K, we obtain an equation quite simple in its application.

CONSTANTS.

Point.	x	y	$y\frac{\Delta s}{\theta}$	$y^2\frac{\Delta s}{\theta}$	$x\frac{\Delta s}{\theta}$	$x^2\frac{\Delta s}{\theta}$	$xy\frac{\Delta s}{\theta}$
1	5	2.5	25	62.5	50	250	125
2	15	7.3	73	532.9	150	2250	1095
3	25	11.5	115	1322.5	250	6250	2875
4	35	15.0	150	2250.0	350	12250	5250
5	45	18.1	181	3276.1	450	20250	8145
6	55	20.5	205	4202.5	550	30250	11275
7	65	22.5	225	5062.5	650	42250	14625
8	75	23.9	239	5712.1	750	56250	17925
9	85	24.7	247	6100.9	850	72250	20995
10	95	25.0	250	6250.0	950	90250	23750
9′	105	24.7	247	6100.9	1050	110250	25935
8′	115	23.9	239	5712.1	1150	132250	27485
7′	125	22.5	225	5062.5	1250	156250	28125
6′	135	20.5	205	4202.5	1350	182250	27675
5′	145	18.1	181	3276.1	1450	210250	26245
4′	155	15.0	150	2250.0	1550	240250	23250
3′	165	11.5	115	1322.5	1650	272250	18975
2′	175	7.3	73	532.9	1750	306250	12775
1′	185	2.5	25	62.5	1850	342250	4625
			3170	63294.	18050	2.284750	301150
			1585	31647			

Determination of Constant Factors.

$$\sum_0^l \frac{\Delta s}{\theta_x} \sum_0^l \frac{x^2 \Delta s}{\theta_x} = 190(2284750) = 434,102,500;$$

$$\left(\sum_0^l \frac{x \Delta s}{\theta_x}\right)^2 = (18050)^2 = 325,802,500.$$

Hence

$$\text{Denominator} = 108,300,000;$$

$$\sum_0^l \frac{x \Delta s}{\theta_x} = 18050 \quad \text{and} \quad \sum_0^l \frac{x^2 \Delta s}{\theta_x} = 2,284,750;$$

$$\frac{18,050}{108,300,000} = .00016\tfrac{2}{3};$$

$$\frac{2,284,750}{108,300,000} = 0.211.$$

Therefore

$$M_1 = -.00016\tfrac{2}{3}\left\{H_1\sum_0^l\frac{xy\Delta s}{\theta_x} + \sum_a^l\frac{x^2\Delta s}{\theta_x} - \sum_a^l\frac{ax\Delta s}{\theta_x}\right\}$$
$$+ .0211\left\{H_1\sum_0^l\frac{y\Delta s}{\theta_x} + \sum_a^l\frac{x\Delta s}{\theta_x} - \sum_a^l\frac{a\Delta s}{\theta_x}\right\}.$$

We are now prepared to determine the value of M_1 for a load at any of the points 1 to 10 inclusive.

The substitutions in the above formula are quite simple as illustrated by the detailed deduction of M_1 for a load at point 6 (see table of Partial Sums).

PARTIAL SUMS.

Point.	$\sum_a^l\frac{x\Delta s}{\theta}$	$\sum_a^l\frac{x^2\Delta s}{\theta}$	$\sum_a^l\frac{xy\Delta s}{\theta}$	$\sum_a\frac{\Delta s}{\theta}$	$\sum_a^l\frac{y\Delta s}{\theta}$
1	18000	2284500	301025	180	3145
2	17850	2282250	299930	170	3072
3	17600	2276000	297055	160	2957
4	17250	2263750	291805	150	2807
5	16800	2243500	283660	140	2626
6	16250	2213250	272385	130	2421
7	15600	2171000	257760	120	2196
8	14850	2114750	239835	110	1957
9	14000	2042500	218840	100	1710
10	13050	1952250	195090	90	1460
9′	12000	1842000	169155	80	1213
8′	10850	1709750	141670	70	974
7′	9600	1553500	113545	60	749
6′	8250	1371250	85870	50	544
5′	6800	1161000	59625	40	363
4′	5250	920750	36375	30	213
3′	3600	648500	17400	20	98
2′	1850	342250	4625	10	25
1′	0	0	0	0	0

Load at Point 6.

$a = 55.$

$$M. = \left\{\begin{array}{l} -.00016\tfrac{2}{3}\{1.21(301150) + 2213250 - 55(16250)\} \\ +.0211\{1.21(3170) + 16250 - 55(130)\} \end{array}\right\};$$

$$M_1 = \left\{\begin{array}{ll} -.00016\tfrac{2}{3}(1683891) & = -280.648 \\ +.0211(12935) & = +272.928 \end{array}\right\}.$$

Hence

$$M_1 = 272.928 - 280.648 = -7.720.$$

In like manner the values of M_1 for loads at any other points are determined. We have tabulated below the values obtained by this method.

Load at	Value of M_1.	Load at	Value of M_1.
1	− 4.328	9′	+ 7.022
2	− 9.436	8′	+ 8.459
3	− 11.738	7′	+ 9.105
4	− 11.807	6′	+ 8.920
5	− 10.291	5′	+ 8.009
6	− 7.720	4′	+ 6.487
7	− 4.542	3′	+ 4.550
8	− 1.194	2′	+ 2.443
9	+ 2.050	1′	+ 0.611
10	+ 4.827		

The corresponding values of M_1, as obtained from Table VI, are as follows:

Load at	k	Δ_6	$M_1 = 95\Delta_6$, page 30.	M_1, by Summation.	Diff.
1	.026	−.046	− 4.370	− 4.328	+ .042
2	.079	−.108	− 10.260	− 9.436	+ .824
3	.132	−.133	− 12.635	− 11.738	+ .897
4	.184	−.132	− 12.540	− 11.807	+ .733
5	.237	−.112	− 10.640	− 10.291	+ .349
6	.289	−.081	− 7.695	− 7.720	− .025
7	.342	−.043	− 4.085	− 4.542	− .457
8	.395	−.003	− 0.285	− 1.194	− .909
9	.447	+.032	+ 3.040	+ 2.050	+ .990
10	.500	+.062	+ 5.890	+ 4.827	+ 1.063
9′	.553	+.084	+ 7.980	+ 7.022	+ .858
8′	.605	+.096	+ 9.120	+ 8.459	+ .661
7′	.658	+.099	+ 9.405	+ 9.105	+ .300
6′	.711	+.092	+ 8.740	+ 8.920	− .180
5′	.763	+.078	+ 7.410	+ 8.009	− .599
4′	.816	+.057	+ 5.415	+ 6.487	− 1.072
3′	.868	+.035	+ 3.325	+ 4.550	− 1.225
2′	.921	+.015	+ 1.425	+ 2.443	− 1.018
1′	.974	+.002	+ 0.190	+ 0.611	− .421
			− 62.510	− 61.056	
			+ 61.940	+ 62.483	
			− 0.570	+ 1.427	

If our points had been taken closer together, the positive and negative moments would have been practically equal for a load over all.

The values of V_1 can now be found from the formula

$$V_1 = \frac{M_2 - M_1}{l} + (1 - k). \qquad (47), \text{ page } 16$$

For a load at 6,

$$V_1 = \frac{8.920 + 7.720}{190} + 0.711 = 0.798.$$

The values of y_0 are determined from

$$y_0 = \frac{M_1 + V_1 a}{H_1} \qquad (50), \text{ page } 17.$$

For a load at 6,

$$y_0 = \frac{-7.720 + 0.798(55)}{1.210} = 29.9.$$

For a load at 10,

$$y_0 = \frac{4.827 + 0.500(95)}{1.714} = 30.5,$$

the correct value for all loads being in this particular case $\frac{6}{5}f = \frac{6}{5}(25) = 30$, showing that the above values are sufficiently exact for practical purposes.

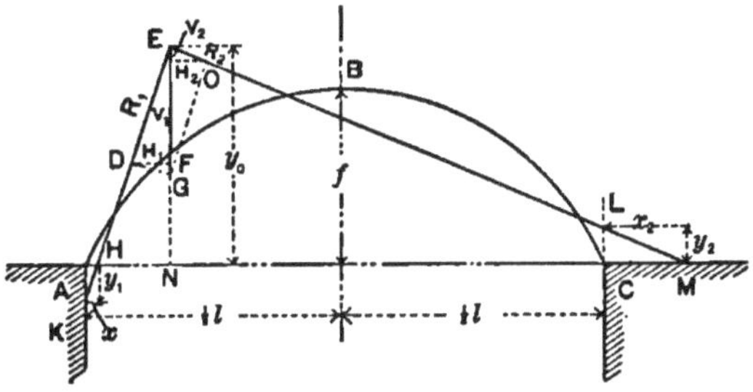

FIG. 58.

Having the values of H_1, V_1, and y_0 determined, the magnitudes and directions of the resultants can be found as follows, and then the stresses determined by the usual methods:

Construct the centre line of the arch to any scale. For any load as 6, make EN equal the corresponding value of y_0. Make $EF = V_1$, $DF = H_1$, and $EG = P$, and complete the parallelogram of forces: then $ED = R_1$ and $DG = EO = R_2$; also, $AH = x_1$, $AK = y_1$, $LC = y_2$, and $CM = x_2$.

HORIZONTAL LOADS. (Q = unity.)

From (232), page 48, introducing the constants already found, and modifying the form, we have

$$\mathfrak{H}_1 = \frac{2\left\{\sum_0^{\frac{1}{2}l}\frac{K'y\Delta s}{\theta_x} - 15.6\sum_0^{\frac{1}{2}l}\frac{K'\Delta s}{\theta_x}\right\}}{13747},$$

where $$K' = (y - b)^{x>a}$$

or

$$\mathfrak{H}_1 = \frac{2\left\{\begin{array}{l} + \left\{\sum_a^{\frac{1}{2}l}\frac{y^2\Delta s}{\theta_x} - b\sum_a^{\frac{1}{2}l}\frac{y\Delta s}{\theta_x}\right\} \\ - 15.6\left\{\sum_a^{\frac{1}{2}l}\frac{y\Delta s}{\theta_x} - b\sum_a^{\frac{1}{2}l}\frac{\Delta s}{\theta_x}\right\}\end{array}\right\}}{13747}$$

$$H_1 = \tfrac{1}{2}(\mathfrak{H}_1 + Q).$$

From the tables computed for vertical loads the partia sums required above are readily found.

PARTIAL SUMS.

Point.	$\sum_a^{\frac{1}{2}l}\frac{y^2\Delta s}{\theta_x}$	$\sum_a^{\frac{1}{2}l}\frac{y\Delta s}{\theta_x}$	$\sum_a^{\frac{1}{2}l}\frac{\Delta s}{\theta_x}$
1	31584.5	1560	85
2	31051.6	1487	75
3	29729.1	1372	65
4	27479.1	1222	55
5	24203.0	1041	45
6	20000.5	836	35
7	14938.0	611	25
8	9225.9	372	15
9	3125.0	125	5
10	0	0	0

To illustrate the application of the formula we will determine the value of H_1 for a load at point 7.

Load at 7,

$$b = 22.5$$

$$\mathfrak{H}_1 = \frac{2\begin{pmatrix} +[14938 - 22.5(611)] \\ -15.6[611 - 22.5(25)] \end{pmatrix}}{13747}.$$

$$\mathfrak{H}_1 = \frac{2[1190.5 - 15.6(756.6)]}{13747} = .064.$$

Then

$$H_1 = \frac{1.064}{2} = 0.532.$$

By Table XII, $H_1 = \Delta_{11} = 0.537$.

The following table contains the values of H_1 as found by two methods for loads at points 1 to 10 inclusive.

Point.	k	H_1. by Summation.	H_1 by Table XII.	Diff.
1	.026	0.984	0.991	.007
2	.079	0.910	0.930	.020
3	.132	0 806	0.836	.030
4	.184	0.715	0.740	.025
5	.237	0.633	0.651	.018
6	.289	0.574	0.584	.010
7	.342	0.532	0.537	.005
8	.395	0.509	0.511	.002
9	.447	0.501	0.502	.001
10	.500	0.500	0.500	.000

For loads on the right of the point 10 we have merely to subtract the value of H_1 for the corresponding load on the left of the crown from unity.

The above values of H_1 are for loads acting from the right towards the left, and hence they are positive and the same in character as for loads acting vertically downward.

For bending-moments M_1 we have from (236), page 48, introducing the constants already found,

$$M_1 = \begin{cases} -0.00016\tfrac{2}{3}\left\{H_1\sum_0^l\frac{xy\Delta s}{\theta_x} - \sum_a^l\frac{xy\Delta s}{\theta_x} + b\sum_a^l\frac{x\Delta s}{\theta_x}\right\}, \\ +0.0211\left\{H_1\sum_0^l\frac{y\Delta s}{\theta_x} - \sum_a^l\frac{y\Delta s}{\theta_x} + b\sum_a^l\frac{\Delta s}{\theta_x}\right\}. \end{cases}$$

The partial sums required above are given on page 197.

As the application of this formula is precisely the same in method as that for vertical loads, we will only illustrate its application in a few cases.

Load at Point 10.

$$b = 25.$$

$$M_1 = \begin{cases} -0.00016\tfrac{2}{3}\left\{0.500(301150) - 195090 + 25(13050)\right\}, \\ +0.0211\left\{0.500(3170) - 1460 + 25(90)\right\}; \end{cases}$$

$$M_1 = \begin{cases} -0.00016\tfrac{2}{3}(281735) = -46.956, \\ +0.0211(2375) = +50113; \end{cases}$$

or

$$M_1 = +3.157.$$

By the use of Table XIII, $M_1 = +0.1250(25) = +3.125$.

Load at Point 1.

$$b = 2.5.$$

$$M_1 = \begin{cases} -0.00016\tfrac{2}{3}\left\{0.984(301150) - 301025 + 2.5(18000)\right\}, \\ +0.0211\left\{0.984(3170) - 3145 + 2.5(180)\right\}; \end{cases}$$

$$M_1 = \begin{cases} -0.00016\tfrac{2}{3}(40306) = -6.7177, \\ +0.0211(423) = +8.9253. \end{cases}$$

Hence

$$M_1 = +2.208.$$

By the use of Table XIII, $M_1 = +2.231$.

The above results indicate a closer agreement in the two methods than was found for vertical loads.

To determine M_2 it is necessary to merely consider a as $l - a$.

The method of procedure is now parallel with that outlined for vertical loads.

Fig. 59 shows graphically the results obtained by the two methods.

The close agreement of the curves shows clearly that the approximate method of summation is quite accurate enough for practical purposes. This method requires considerable

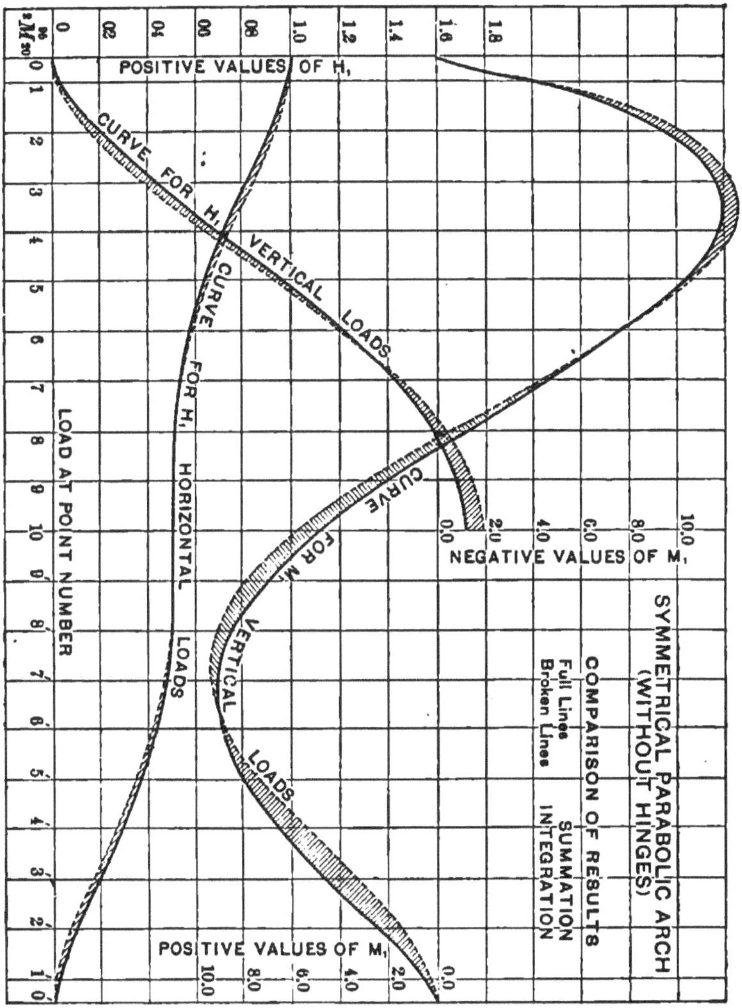

FIG. 59.

more work, but it has the advantage of being approximately correct for any form of arch and any values of θ, the circular or elliptical arch requiring no more labor in calculating the values of H_1, M_1, etc., than the parabolic arch.

CHAPTER X.

THE ST. LOUIS ARCH.*

To further show the accuracy of the results obtained by the use of the summation formulas we will compute the values of H_1 and M_1 for the well-known St. Louis or Eads Bridge, using the data given in the History* of the bridge. The results given by Prof. Woodward were computed with great care from formulas deduced to fit the peculiarities of the arch-rib.

θ has but two values throughout the rib. For a distance equal to one twelfth of the span from each support θ has a constant value, and between these two sections another value which is uniform throughout that section; thus the use of the formulas of Chapter IV is prohibited.

DATA.

Span $= l =$ 519.2328 ft. Rise $= f =$ 47.31 ft.

Radius $= R =$ 736.0 ft. $\phi_0 = 20° \; 39' \; 17''.92$.

Area of each flange for $\frac{1}{12}$ the span at the ends $= F = 67$ sq. in.

Area of each flange in centre section $= F =$ 100.5 sq. in.

Depth centre to centre of flanges = 12 ft.

Dead load = 1 ton per running foot horizontal.

Live " = 0.8 " " " " "

In applying our formulas the linear arch will be assumed to lie midway between the flanges of the rib. We will divide this linear arch into fifty-one divisions, as shown in the first table. The co-ordinates x and y will be computed for the *centre* points

* See "A History of the St. Louis Bridge," by C. M. Woodward (St. Louis, G. J. Jones & Co., 1881).

of these divisions, and the moments of inertia taken at the same points.

Since the areas of the flanges are 67 and 100.5 sq. in., and the distance centre to centre of the flanges 12 ft. throughout, the moments of inertia will be in the ratio of *two* to *three*. As we propose to neglect the influence of the axial stress—as was done by the computers for the structure as built—we need not concern ourselves about the actual values of θ, but use relative values. The following data will be used throughout in the computation of H_1 and M_1:

TABLE OF CO-ORDINATES, ETC.

Point.	x	y	Δy	Δx	Δs	Relative θ.
1	6.6	2.33	4.83	13.3	14.14	3
2	18.3	6.52	3.46	10.0	10.58	3
3	28.3	9.98	3.31	"	10.53	3
4	38.3	13.22	3.16	"	10.48	3
5	48.3	16.31	3.02	"	10.44	2
6	58.3	19.18	2.87	"	10.40	2
7	68.3	21.98	2.65	"	10.34	2
8	78.3	24.55	2.50	"	10.31	2
9	88.3	27.06	2.43	"	10.29	2
10	98.3	29.34	2.21	"	10.24	2
11	108.3	31.55	2.14	"	10.22	2
12	118.3	33.61	1.98	"	10.19	2
13	128.3	35.45	1.77	"	10.15	2
14	138.3	37.21	1.69	"	10.14	2
15	148.3	38.76	1.55	"	10.12	2
16	158.3	40.23	1.40	"	10.10	2
17	168.3	41.56	1.25	"	10.08	2
18	178.3	42.73	1.10	"	10.06	2
19	188.3	43.76	0.96	"	10.04	2
20	198.3	44.72	0.81	"	10.03	2
21	208.3	45.46	0.74	"	10.03	2
22	218.3	46.12	0.51	"	10.01	2
23	228.3	46.63	0.44	"	10.01	2
24	238.3	46.93	0.40	"	10.007	2
25	251.4	47.22	0.12	16.3	16.30	2

Determination of H_1.

From (220), page 46, remembering that the terms containing N_x and F_x are to be omitted, we have for vertical loads

$$H_1 = \frac{\sum_0^{l/2} \frac{K' y \Delta s}{\theta_x} - \frac{\sum_0^{l/2} \frac{K' \Delta s}{\theta_x}}{\sum_0^{l/2} \frac{\Delta s}{\theta_x}} \sum_0^{l/2} \frac{y \Delta s}{\theta_x}}{2\left\{ \sum_0^{l/2} \frac{y^2 \Delta s}{\theta_x} - \frac{\left(\sum_0^{l/2} \frac{y \Delta s}{\theta_x}\right)^2}{\sum_0^{l/2} \frac{\Delta s}{\theta_x}} \right\}},$$

in which for a load $P =$ unity

$$\sum_0^{l/2} \frac{K' y \Delta s}{\theta_x} = \sum_0^{l/2} \frac{x y \Delta s}{\theta_x} - \left(\sum_a^{l/2} \frac{x y \Delta s}{\theta_x} - a \sum_a^{l/2} \frac{y \Delta s}{\theta_x} \right) \quad . \quad . \quad (222)$$

and

$$\sum_0^{l/2} \frac{K' \Delta s}{\theta_x} = \sum_0^{l/2} \frac{x \Delta s}{\theta_x} - \left(\sum_a^{l/2} \frac{x \Delta s}{\theta_x} - a \sum_a^{l/2} \frac{\Delta s}{\theta_x} \right). \quad . \quad . \quad (223)$$

We note that only the quantities enclosed in the parentheses in (222) and (223) vary with a change in the location of the load. We will first compute the terms which are constant.

$$\sum_0^{l/2} y^2 \frac{\Delta s}{\theta_x} = 156{,}868.7;$$

$$\left(\sum_0^{l/2} y \frac{\Delta s}{\theta_x} \right)^2 = (4110.8)^2;$$

$$\sum_0^{l/2} \frac{\Delta s}{\theta_x} = 125.0.$$

Combining these quantities and multiplying the product by 2, we have for the value of the denominator 43324.1.

$$\sum_0^{l/2} x y \frac{\Delta s}{\theta_x} = 669{,}403.3$$

$$\sum_0^{l/2} x \frac{\Delta s}{\theta_x} = 16{,}863.4;$$

$$\frac{\sum_0^{l/2} \frac{y \Delta s}{\theta_x}}{\sum_0^{l/2} \frac{\Delta s}{\theta_x}} = \frac{4110.8}{125.0} = 32.88.$$

For our purpose it will not be necessary to compute H_1 for a load at each point of division. We have selected points 2, 4, 6, 10, 15, 20, 23, and 25.

The following tables show the method of procedure in the determination of H_1 for each point designated.

FIRST TERM OF NUMERATOR.

Load at Point No.	$\sum_0^{l/2} \frac{y \Delta s}{\theta_x}$	$\sum_a^{l/2} \frac{xy \Delta s}{\theta_x}$	$a \sum_a^{l/2} \frac{y \Delta s}{\theta_x}$	First Term of Numerator: $\sum_0^{l/2} \frac{K' y \Delta s}{\theta_x}$
2	669,403.3	668,910.9	74,606.5	75,099.1
4	"	666,152.5	153,035.0	156,285.8
6	"	656,225.8	222,170.7	235,348.2
10	"	611,473.5	322,514.9	380,444.7
15	"	495,454.3	353,462.5	527,411.6
20	"	303,881.1	260,155.3	625,677.5
23	"	152,699.9	141,462.2	658,165.5
25	"	0	0	669,403.3

SECOND TERM OF NUMERATOR.

Load at Point No.	$\sum_0^{l/2} \frac{x \Delta s}{\theta_x}$	$\sum_a^{l/2} \frac{x \Delta s}{\theta_x}$	$a \sum_a^{l/2} \frac{\Delta s}{\theta_x}$	Second Term of Numerator: $\sum_0^{l/2} \frac{K'' \Delta s}{\theta_x}(32.88)$
2	16,863.4	16,767.9	2,136.6	73,435.4
4	"	16,534.9	4,203.5	149,104.1
6	"	15,979.6	5,791.1	219,637.1
10	"	14,265.3	7,740.4	340,139.0
15	"	11,004.2	7,909.3	452,983.7
20	"	6,519.9	5,587.7	524,124.8
23	"	3,240.1	3,002.8	547,000.0
25	"	0	0	554,807.2

VALUE OF H_1.

Load at Point No.	Numerator.	Denominator.	H_1
2	1,663.7	43,324.1	0.039
4	7,181.6	"	0.165
6	15,711.1	"	0.362
10	40,305.8	"	0.930
15	74,427.9	"	1.7·7
20	101,542.7	"	2.343
23	111,163.5	"	2.565
25	114,596.1	"	2.645

If now, with the values of *a* as abscissas and the corresponding values of H_1 as ordinates, points be located on sectioned

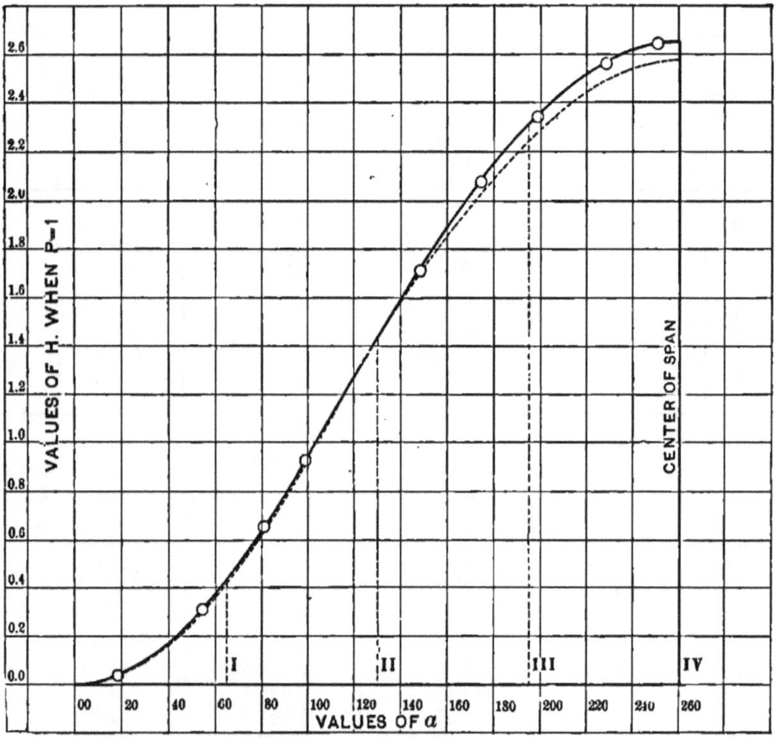

FIG. 60.

paper and a smooth curve drawn through them, the value of H_1 for any value of *a* can be readily and quite accurately determined.

For a uniform load of w per horizontal unit of span the total value of H_1 will be twice the area included between the above curve (extending from the support to the crown) and the axis of abscissas multiplied by w.

Such a curve is shown in Fig. 60. The full line represents the curve located by the above values of H_1. The broken line is located by values of H_1 which were obtained by another computation in which only one decimal place was employed in the data.

In the computations for the St. Louis arch uniform loads were assumed as follows:

For dead load 1.0 ton per lineal foot.
" live " 0.8 " " " "

In the history of the bridge the values of H_1 are given for a load extending from the support up to each of eight points of division. The corresponding points are marked I, II, III, etc., in Fig. 60.

The following table shows the relation between the values of H_1 given in the history of the bridge and those obtained from Fig. 60.

MOVING LOAD OF 0.8 TON PER LINEAL FOOT. VALUES OF H_1.

Load up to	History.	Fig. 60.	Difference.	Remarks.
	Tons.	Tons.		
I	8.10	8.04	0.06	The values in the third column were obtained from Fig. 60 by taking $\frac{8}{10}$ the area between the full line and the axis of a.
II	56.20	55.86	0.34	
III	155.20	154.78	0.42	
IV	286.60	286.56	0.04	
V	418.10	418.34	0.24	
VI	517.00	517.26	0.26	
VII	565.10	565.08	0.02	
over all	573.30	573.12	0.18	

The above table shows almost perfect agreement between the exact and approximate methods. The errors are of no practical importance. They exist only in the decimal figures, which are quite likely to be in error by either method.

For a load over all with $w = 0.8$ ton the area between the

broken line and the axis of a is 559.3, being in error 14 tons, or a little over 2 per cent. Even this is of no practical importance.

We will now show that the effect of the axial stress, which was neglected in the calculations made for the St. Louis Bridge, is very much larger than any error which is likely to be made by using the summation formula.

In (221) we have in the numerator

$$\sum_0^{\frac{1}{2}l} N_x \frac{\Delta x}{F_x} = \sum_0^{a} \frac{\Delta x \sin \phi}{F_x}.$$

Since we used only relative values for θ, it will be necessary to introduce a factor in the above expression. For the area 67, $\theta = 2\frac{67}{144}\cdot\left(\frac{12}{2}\right)^2 = 2(16.75)$ approximately. Therefore $2\frac{(16.75)}{2} = 2(8.37) =$ the factor required; then

$$\sum_0^{l/2} N_x \frac{\Delta x}{F_x} = 2\sum_0^{a} 8.37 \frac{\Delta x \sin \phi}{F_x}.$$

This is very small in comparison with the remaining terms in the numerator, and hence can be neglected without serious error.

In the denominator of the same equation we have

$$+\sum_0^{l/2} \frac{\Delta x \cos \phi}{F_x} \quad \text{or} \quad 2\sum_0^{l/2} 8.37 \frac{\Delta x \cos \phi}{F_x}.$$

We may replace $\Delta x \cos \phi$ by Δs nearly, and have

$$16.74 \sum_0^{l/2} \frac{\Delta s}{F_x} = 8300, \text{ about.}$$

Then the denominator, when the effect of the axial stress is considered, becomes

$$43,324 + 8,300 = 51,624.$$

$\frac{51624}{43324} = 1.19$, or the values of H_1 obtained above are *too large*, and should be divided by 1.19 to obtain the values which include the effect of the axial stress.

Deduction of M_1.

Neglecting the axial stress term, (225), page 47, becomes,

$$M_1 = \frac{\left\{\sum_0^l \frac{Kx\Delta s}{\theta_x}\right\}\sum_0^l \frac{x\Delta s}{\theta_x} - \sum_0^l \frac{K\Delta s}{\theta_x}\sum_0^l \frac{x^2\Delta s}{\theta_x}}{\left\{\sum_0^l \frac{\Delta s}{\theta_x}\sum_0^l \frac{x^2\Delta s}{\theta_x} - \left(\sum_0^l \frac{x\Delta s}{\theta_x}\right)^2\right\} = D},$$

where

$$\sum_0^l \frac{K\Delta xs}{\theta_x} = -H_1\sum_0^l \frac{xy\Delta s}{\theta_x} - \left(\sum_a^l \frac{x^2\Delta s}{\theta_x} - a\sum_a^l \frac{x\Delta s}{\theta_x}\right)P;$$

$$\sum_0^l \frac{K\Delta s}{\theta_x} = -H_1\sum_0^l \frac{y\Delta s}{\theta_x} - \left(\sum_a^l \frac{x\Delta s}{\theta_x} - a\sum_a^l \frac{\Delta s}{\theta_x}\right)P.$$

The values of the constant terms are as follows:

$$\sum_0^l \frac{\Delta s}{\theta_x} = 250 \qquad \sum_0^l \frac{x^2\Delta s}{\theta_x} = 22{,}035{,}617.$$

$$\sum_0^l \frac{x\Delta s}{\theta_x} = 64{,}887.$$

Hence the denominator $= 1{,}298{,}530{,}726$.

$$\frac{\sum_0^l \frac{x^2\Delta s}{\theta_x}}{D} = 0.0169696;$$

$$\sum_0^l \frac{x\Delta s}{\theta_x} = 0.000050046.$$

Then our equation becomes

$$M_1 = 0.000050046\sum_0^l \frac{Kx\Delta s}{\theta_x} - 0.0169696\sum_0^l \frac{K\Delta s}{\theta_x}.$$

The following tables contain the necessary quantities for substitution in the above equation, using the values of H_1 found above and the same points for the location of the loads.

FIRST TERM.

Load at Point No.	$H_1\sum_0^l \frac{xy\Delta s}{\theta_x}$	$\sum_a^l \frac{x^2\Delta s}{\theta_x}$	$a\sum_a^l \frac{x\Delta s}{\theta_x}$	$\sum_0^l \frac{Kx\Delta s}{\theta_x}$	First Term.
2	83,226.4	22,034,233.5	1,185,691.6	20,931,768	1047.551
4	352,111.5	22,026,302.9	2,472,605.5	19,905,809	996.205
6	772,511.3	21,996,451.0	3,731,410.1	19,037,552	952.750
10	1,985,268.8	21,851,139.9	6,123,034.1	17,713,374	886.483
15	3,664,093.9	21,427,952.6	8,754,026.2	16,338,020	817.652
20	4,999,983.7	20,623,541.6	10,816,260.6	14,807,264	741.044
23	5,473,733.7	19,906,568.2	11,703,816.8	13,676,485	684.453
25	5,644,454.5	19,107,367.4	12,073,221.3	12,678,600	634.513
25′	5,644,454.5	18,522,875.2	12,276,321.8	11,891,007	595.096
23′	5,473,733.7	17,704,577.3	12,502,906.4	10,675,404	534.261
20′	4,999,983.7	16,250,248.4	12,292,268.3	8,957,963	448.310
15′	3,664,093.9	13,141,610.2	10,927,427.6	5,878,276	294.184
10′	1,985,268.8	9,041,875.9	8,102,179.8	2,924,964	146.382
6′	772,511.3	4,928,737.8	4,623,750.2	1,077,498	53.924
4′	352,111.5	2,964,105.3	2,835,176.7	481,040	24.074
2′	83,226.4	1,237,593.8	1,209,345.9	111,514	5.580

SECOND TERM.

Load at Point No.	$H_1\sum_0^l \frac{y\Delta s}{\theta_x}$	$\sum_a^l \frac{x\Delta s}{\theta_x}$	$a\sum_a^l \frac{\Delta s}{\theta_x}$	$\sum_0^l \frac{K\Delta s}{\theta_x}$	Second Term
2	320.6	64,791.9	4,423.8	60,688	1029.864
4	1,356.5	64,558.9	8,990.4	56,925	965.994
6	2,976.2	64,003.6	13,077.6	53,902	914.697
10	7,648.5	62.289.3	20,026.3	49,911	846.978
15	14,116.4	59,029.2	26,444.2	46,701	792.502
20	19,263.1	54,544.9	30,371.8	43,436	737.094
23	21,088.3	51,265.1	31,536.4	40,816	692.646
25	21,726.0	48,023.9	31,420.7	38,329	650.431
25′	21,726.0	45,841.4	31,287.9	36,279	615.648
23′	21,088.3	42,980.1	31,075.4	32,993	559.877
20′	19,263.1	38,305.6	29,455.4	28,113	477.069
15′	14,116.4	29,461.9	24,698.2	18,880	320.386
10′	7,648.5	19,249.6	17,307.4	9.590	162.750
6′	2,976.2	10,032.0	9,425.4	3,582	60.798
4′	1,356.5	5,895.6	5,645.8	1,606	27.257
2′	320.6	2,414.3	2,359.2	375	6.375

In making the computations above *three* decimal places were used throughout. These have not been given, hence the last figures may not exactly check.

ALUES OF M_1.

Load at Point No.	Computed Values of M_1. Load Unity.	Load at Point No.	Computed Values of M_1. Load Unity.
2	− 17.7	25′	+ 20 6
4	− 30.2	23′	+ 25.6
6	− 38.1	20′	+ 28.8
10	− 39.5	15′	+ 26.2
15	− 25.2	10′	+ 16.4
20	− 4.0	6′	+ 6.9
23	+ 8.2	4′	+ 3.2
25	+ 15.9	2′	+ 0.8

With the values of a as abscissas and those of M_1 as ordinates the curve shown in Fig. 61, page 215, can be located. The following table shows the agreement between the values given in the History and those obtained from Fig. 61.

COMPARISON OF VALUES OF M_1. $w = 0.8$ TON.

Load Up to-	Values Given in History of Bridge. (1)	Values from Fig. 61. (2)	Difference. (3)	Percentage of Computed Values. (4)
I	− 1206	− 1244	+ 38	3.0
II	− 3114	− 3224	+ 110	3.4
III	− 4034	− 4226	+ 192	4.5
IV	− 3588	− 3848	+ 260	6.7
V	− 2235	− 2529	+ 294	11.6
VI	− 782	− 1123	+ 341	30.3
VII	+ 70	− 282	+ 352	124.8
VIII	+ 232	− 128	+ 360	281.2

COMPARISON OF VALUES OF M_2.

I	+ 161	+ 154	− 7	4.5
II	+ 1013	+ 985	− 28	2.8
III	+ 2466	+ 2401	− 65	2.7
IV	+ 3821	+ 3720	− 101	2.7
V	+ 4266	+ 4008	− 258	6.3
VI	+ 3346	+ 3096	− 250	8.0
VII	+ 1438	+ 1116	− 322	28.7
VIII	+ 232	− 128	+ 360	281.2

In column (3) the positive sign indicates that the values in column (2) are *too large.*

Here we see that the agreement in values is not as close as in the valves of H_1, but we also note that the greatest discrepancies occur in the small and non-important values.

The maximum negative value of M_1 is in error—but 4.5 per cent and the maximum positive value 6.3 per cent—errors which are of little importance.

The negative area in Fig. 61 is about 4.5 per cent *too large* and the positive about 6.3 per cent *too small.* Now since in this particular case the difference between these areas is small, we readily see why our discrepancy for a load over all is so large. The heavy broken line in Fig. 61 represents the correct curve.

For practical purposes our curve is quite exact, and will give results as near the truth as any of the common methods in their special cases.

Of course the particular advantage in the summation method is its adaptability to any case of the symmetrical arch.

Temperature.

Data.—$et^\circ = 0.000527$, where $t^\circ = 80^\circ$, $l = 520$, $E = 1944000$.

Value of H_t.—From (239),

$$H_t = \frac{Eet^\circ l}{D} = \frac{532734}{D}.$$

In this case the actual values of θ_x must be employed in the denominator, or

$$D = \frac{43324}{2 \times 8.37} = 2588;$$

$$\therefore H_t = \frac{532734}{2588} = 205.9.$$

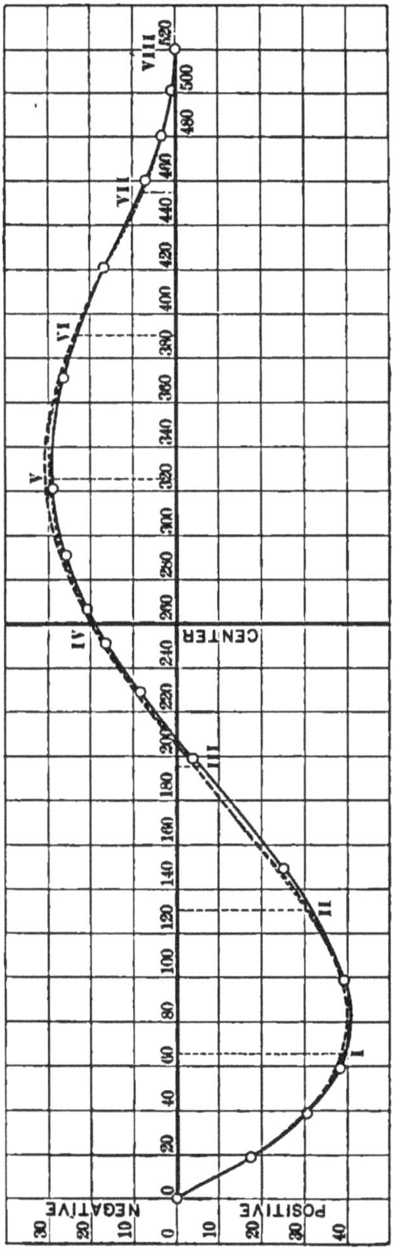
NEGATIVE
POSITIVE
CENTER
30
20
10
0
10
20
30
40
0
20
40
60
80
100
120
140
160
180
200
220
240
260
280
300
320
340
360
380
400
420
440
460
480
500
520
I
II
III
IV
V
VI
VII
VIII

Fig. 61.

From the history of the bridge,

$$H_t = 204.9.$$

$205.9 - 204.9 = 1$, or an error of about one half of 1 per cent.

Value of M_1.—From (240), page 49, we see that

$$M_1 = H_t \frac{\sum_0^l \frac{y\Delta s}{\theta_x}}{\sum_0^l \frac{\Delta s}{\theta_x}} = 205.9(32.88) = 6769.9.$$

From the history of the bridge,

$$M_1 = 6747.$$

$6769.9 - 6747 = 22.9$, or an error of about 3.4 per cent.

CHAPTER XI.

THE SPANDREL-BRACED ARCH.

THE so-called spandrel-braced arch usually consists of an arched bottom chord and a horizontal top chord connected by a system of web-bracing. Evidently the formulas of Chapters III and IV cannot be applied even approximately to this form of arch. The summation formulas, however, enable us to consider this type of arch either with or without hinges with comparatively little more labor than required for the ordinary form having a variable θ.

To illustrate the method to be pursued we will take the case of a proposed design for a bridge over the river Douro by Mr. Max Am Ende and Messrs. Handyside & Co.*

The form and general dimensions of the bridge are given in Fig. 62.

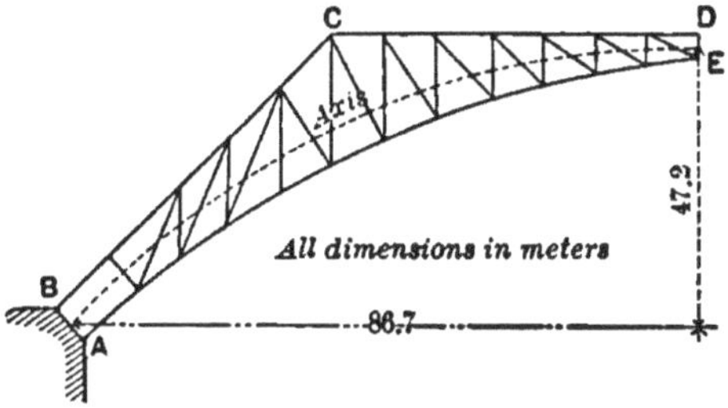

FIG. 62.

When the general form of the structure has been decided upon, the first step is to approximately determine the sections

* Design for a bridge over the river Douro by Mr. Max Am Ende and Messrs. Handyside & Co.; *Engineering*, London, 1881.

of the various members by the formulas of Chapter III or IV, using for the linear arch the parabola or circle which lies approximately midway between the two chords.

For the application of the summation formulas the linear arch is assumed to pass through the centres of gravity of each vertical section. (Of course in both cases mentioned above the linear arch must pass through the supports.) The method of procedure is now the same as already explained for the arch with a hinge at each support and the arch without hinges.

In computing the values of θ for each section the moments of inertia of the flange sections about an axis passing through their centres of gravity may be neglected and the moment of each flange be taken as $\frac{Fh^2}{4}$, where h is the distance centre to centre of the flanges.

Douro Spandrel-braced Arch.

Let *ABCDE*, Fig. 62, represent one half of the bridge, and suppose the approximate dimensions of members and the linear arch have been determined. We will divide the linear arch

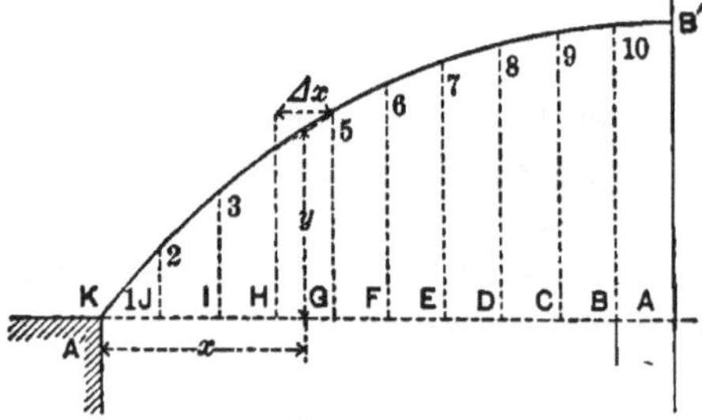

FIG. 63.

into twenty equal parts, measure the co-ordinates at the *centre* of each division, and take the moments of inertia at the same points.

Following are the data required for the determination of H_1:

DATA.

Division.	* Δs	* Δx	* Δy	x	y	* $\sin \phi$	* $\cos \phi$	* $\frac{\theta_x}{1000}$	* $\frac{F_x}{1000}$
1	10.2	7.0	7.4	3.5	3.7	0.735	0.68	20.73	5.60
2	"	7.0	7.3	10.5	11.1	0.720	0.70	11.23	4.70
3	"	7.6	7.1	17.8	18.2	0.685	0.73	21 64	5.00
4	"	7.8	6.6	25.5	25.1	0.630	0.77	40.41	4.80
5	"	8.2	6.2	33.5	31.5	0.600	0.80	87.07	4.80
6	"	9.0	4.8	42.1	37.0	0.480	0.88	157.57	4.36
7	"	9.5	3.9	51.4	41.4	0.367	0.93	101.91	4.30
8	"	10.0	2.3	61.2	44.4	0.235	0.97	51.96	4.18
9	"	10.1	1.2	71.2	46.2	0.122	0.99	29.27	4.00
10	"	10.2	0.4	81.2	47.0	0.045	1.00	17.84	4.04

DATA.

Division.	* $1000 \frac{\Delta s}{\theta_x}$	$1000\, y\frac{\Delta s}{\theta_x}$	$1000\, x\frac{\Delta s}{\theta_x}$	$1000\, y^2\frac{\Delta s}{\theta_x}$	$1000\, xy\frac{\Delta s}{\theta_x}$	* $1000 \frac{\Delta x \cos\phi}{F_x}$
1	0.492	1.820	1.722	6.734	6.371	2.525
2	0.892	9.901	9.366	109.901	103.962	2.499
3	0.471	8.572	8.384	156.010	152.589	2.320
4	0.252	6.325	6.426	156.757	161.292	2.055
5	0.117	3.685	3.919	116.077	123.448	1.817
6	0.065	2.405	2.736	88.985	101.232	1.367
7	0.100	4.140	5.140	171.396	212.796	1.251
8	0.196	8.702	11.995	386.368	532.578	1.109
9	0.348	16.077	24.777	742.757	1149.317	1.043
10	0.572	26.884	46.446	1263.548	2182.962	0.850
	3.505	88.511	120.911	3198.533	4726.547	16.836

From (221), page 46, we have, neglecting the axial stress term in the numerator,

$$H_1 = \frac{\sum_0^{l/2} \frac{K' y \Delta s}{\theta_x} - \frac{\sum_0^{l/2} \frac{K' \Delta s}{\theta_x}}{\sum_0^{l/2} \frac{\Delta s}{\theta_x}} \sum_0^{l/2} \frac{y \Delta s}{\theta_x}}{2\left[\sum_0^{l/2} \frac{y^2 \Delta s}{\theta_x} + \sum_0^{l/2} \frac{\Delta x \cos\phi}{F_x} - \frac{\left(\sum_0^{l/2} \frac{y\Delta s}{\theta_x}\right)^2}{\sum_0^{l/2} \frac{\Delta s}{\theta_x}}\right]}$$

* Data given by Mr. Max Am Ende.

or

$$H_1 = \frac{\sum\limits_0^{l/2}\frac{K'y\Delta s}{\theta_x} - 25.25\sum\limits_0^{l/2}\frac{K'\Delta s}{\theta_x}}{1961, \text{ say } 2000},$$

where

$$\sum_0^{l/2}\frac{K'y\phi s}{\theta_x} = \sum_0^{l/2} xy\frac{\Delta s}{\theta_x} - \left\{\sum_a^{l/2}\frac{xy\Delta s}{\theta_x} - a\sum_a^{l/2}\frac{y\Delta s}{\theta_x}\right\}$$

and

$$\sum_0^{l/2}\frac{K'\Delta s}{\theta_x} = \sum_0^{l/2}\frac{x\Delta s}{\theta_x} - \left\{\sum_a^{l/2}\frac{x\Delta s}{\theta_x} - a\sum_a^{l/2}\frac{\Delta s}{\theta_x}\right\},$$

for $P =$ unity.

For a load at 10, $a = 81.2$. Then

$$\sum_0^{l/2}\frac{K'y\Delta s}{\theta_x} = 4726.5 - \{0\} - 4726.5;$$

$$\sum_0^{l/2}\frac{K'\Delta s}{\theta_x} = 121 - \{0\} = 121.0.$$

$$\therefore H_1 = \frac{4726.5 - 121(25.25)}{2000} = 0.835.$$

For a load at 9, $a = 71.2$. Then

$$\sum_0^{l/2}\frac{K'y\Delta s}{\theta_x} = 4726.5 - \{2183 - 71.2(26.9)\} = 4458.7$$

and

$$\sum_0^{l/2}\frac{K'\Delta s}{\theta_x} = 121 - \{46.4 - 71.2(0.572)\} = 115.2.$$

$$\therefore H_1 = \frac{4458.7 - 115.2(25.25)}{2000} = 0.775.$$

In like manner the values of H_1 for loads at the other points are obtained. The following table contains the values of H_1 for each division, the interpolated values for the ends of the divisions, and the values given by Mr. Max Am Ende, who took his origin of co-ordinates at the crown and measured the x's and y's to the extremities of the divisions. He then substituted the proper quantities in * three equations, which he demonstrates, and eliminated all unknowns but H_1.

COMPARISON OF THE VALUES OF H_1

Division No.	H_1	H_1 at End of Divisions, Fig. 64.	H_1, Max Am Ende.
1	——	0.018	0.032
2	0.037	0.060	0.109
3	0.117	0.167	0.204
4	0.218	0.274	0.302
5	0.323	0.385	0.402
6	0.440	0.500	0.509
7	0.550	0.617	0.616
8	0.673	0.725	0.701
9	0.775	0.810	0.762
10	0.835	0.840	0.792 (?)

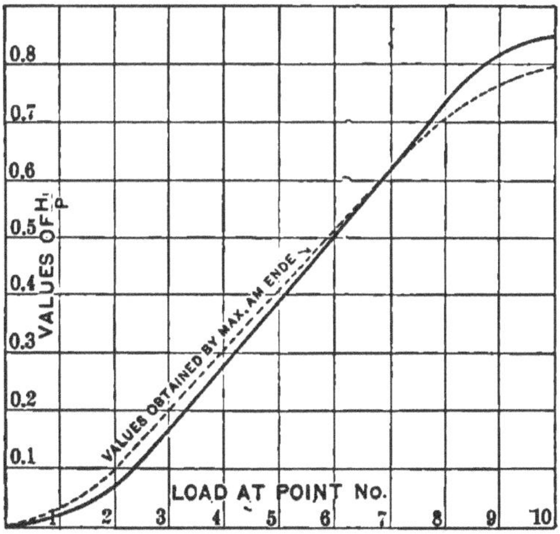

FIG. 64.

* As far as known by the author, Mr. Max Am Ende was the first to successfully treat the fixed arch with variable θ, using the summation formulas. By some manipulation his three formulas can be reduced to our general forms.

We see from Fig. 64 that our values lie above and below those given by Mr. Max Am Ende, and that the areas between the curves located by both series of values and the axis of a are very nearly equal, that is, for a uniform load over all the values of H_1 would be practically equal. We could not expect any closer agreement in values, considering the difference in method and the very approximate values of x and y which we used.

It will not be necessary to take up the deduction of M_1, V_1, etc., as the method of procedure is precisely the same as that employed for the St. Louis arch.

The more common form of the spandrel-braced arch is hinged at each support. The method of treatment is practically the same as outlined above; only the formulas for the hinged arch are, of course, used.

CHAPTER XII.

THE MASONRY ARCH.

UNDER this heading we will include arches constructed of stone, brick, and concrete having spans of at least twenty-five feet.

Before considering the many types of masonry arches we will first consider a type which is amenable for calculation by the formulas deduced for the elastic arch. This type consists of an arch-rib of masonry, with joints carefully made and as thin as practicable. At regular intervals this arch supports thin *lateral* walls, which in turn carry small arches or slabs which support the roadway. At the *abutments* the arch is protected from any horizontal pressures by retaining walls. The general features of this type are shown in Fig. 65.*

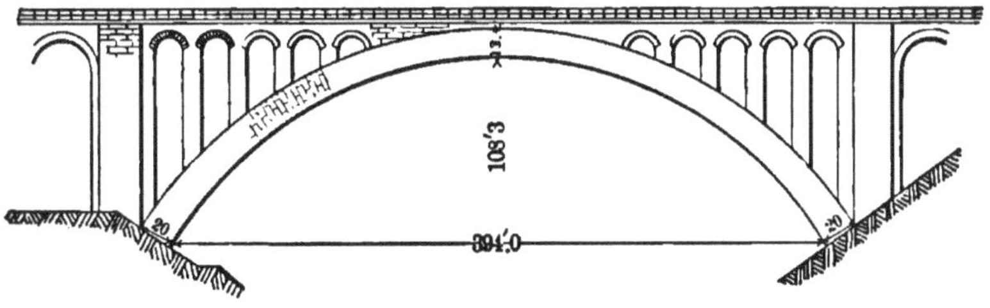

FIG. 65.

The dead weight of this style of bridge consists (1°) of the weight of the masonry in the rib proper, and (2°) the weight of

* See "Bericht des Gewölbe-Ausschusses. Sonderabdruck aus der Zeitschrift des Osterr. Ingenieur- und Architekten-Vereines," No. 20–34, 1895.

the material above the arch which is transmitted to the rib through the thin lateral walls. The forces acting upon the arch-ring are evidently *vertical.*

Now since any rectangular masonry joint will have the same kind of stress at all points when the resultant pressure upon the joint is applied within the middle third, our arch-ring will be in *compression throughout* if the equilibrium polygon lies within the middle third of each section. Then if the effect of the mortar joints be neglected the masonry rib will behave quite similarly to an elastic rib, and hence we may consider the *formulas already demonstrated as applicable in this case.*

If the skew-backs are well fitted and the abutments or piers supporting the arch practically immovable, then the masonry rib is *fixed* at the ends, or at least more nearly fixed than hinged, as long as the equilibrium polygon remains within the middle third of the section.

Since the arch-ring is necessarily made up of many pieces where either stone or brick is employed, it is practically impossible to so construct the arch-ring that there will not be more or less change in the position of the axis when the false-works or centring is removed. As a consequence the true position of the equilibrium polygon in the arch as constructed is somewhat uncertain.

To avoid this uncertainty in the location of the equilibrium polygon, it is advisable to place in three or more joints which divide the ring symmetrically some material, as lead, covering the middle third of the joint. This locates the polygon within the limits of the area of the lead plates, and hence the *maximum* possible thrusts at these joints can be determined.

After the falseworks are removed and the arch with its spandrels, etc., completed, these joints can be filled with cement, and become fixed at the ends for any additional loads.* This method is successfully followed by German engineers.

For arches of the above type all loads are considered vertical, aud the arch-rib is assumed to be without hinges for moving loads.

---

* See page 229.

As in all arch designs, the general dimensions must be* assumed, and then the corresponding loads computed and the equilibrium polygons† drawn to determine if they lie within the middle third of the arch-ring assumed, and further, to be sure that the intensity of the pressure at any point in the rib does not exceed the safe strength of the material and that frictional stability is not exceeded.

Having decided upon the shape of the arch, the span and rise of the axis being assumed, the next dimensions required are the thickness of the rib at the crown and that at the skew-backs. The assumption of these dimensions can be made with the aid of Table XXX.

THICKNESS OF ARCH-RING AT THE SKEW-BACK.

Theory (except in hinged arches), practice, and appearances demand that the depth of the arch-ring at the skew-back should be somewhat greater than at the crown. For vertical forces the horizontal thrust is constant throughout the arch, and hence the axial thrust increases as the secant of the angle of inclination of the axis. The thickness of the rib, however, should increase more rapidly than the secant of this angle, since it is seldom that the equilibrium polygons follow the centre of the arch-ring. As the polygon departs from the centre of the ring the maximum *intensity* of the pressure upon the joint changes quite rapidly, being *twice* the average intensity when the polygon passes through the third point of the joint.

Having decided upon the depths of the crown and the skew-backs, the arch-ring can be drawn to scale.

---

* See Alexander and Thomson's direct method for proportioning masonry arches, page 234.

† The graphic method is preferred for the preliminary investigations, being much shorter than the algebraic methods, and quite accurate enough.

EQUILIBRIUM POLYGON FOLLOWING THE AXIS OF THE ARCH-RING.

The ideal arch would be one in which the pressure over the area of each joint is uniform, or the resultant pressure would pass through the centre of each joint of the arch-ring. This, of course, is impossible when the loading is movable; but for the dead load of the structure the various parts can be so located that the equilibrium polygon will very nearly pass along the axis of the arch-ring.

Now since the dead load is usually much greater than the live load, if the arch be designed so that the equilibrium polygon follows the axis for the dead load and a live load over all, the ring will be safe usually for a variable moving load.

The loading necessary to make the equilibrium polygon follow the axis can be obtained approximately as follows:

Assume the dimensions of the arch-ring and draw it to scale as shown in Fig. 66. Determine the distance *mp* and the location of the points *a*, *b*, *c*, etc., where the lateral walls rest upon the arch-ring. Then in Fig. 66 let *abc*, etc., be these points of division; connect them by the straight lines *ab*, *bc*, *cd*, etc.; then *abcde* is one half of the equilibrium polygon which follows (nearly) the axis of the arch. We have now to determine the relative and actual magnitudes of $P_1P_2P_3$, etc., so that the points *a*, *b*, *c*, etc., will not be changed in position.

The load at the crown can be determined at once from the assumed dimensions and weights. Lay off *one half* of this load as shown in Fig. 66, and draw S_6 parallel to *ge* until it cuts the horizontal at P; draw S_4S_3, etc., parallel to *ed*, *dc*, etc., respectively: then the distances $P_6P_4P_3P_2$, etc., cut off on the vertical are the required values of the loads at *e*, *d*, *c*, *b*, etc. A few trials will place the material above the ring so that these values will very nearly obtain.

We have now all of the general dimensions of the structure from which the actual loads at *abc*, etc., can be computed.

Taking *abcde* as the axis of the arch, and assuming the above loads applied at the points *abcde*, the actual values of H_1, V_1, and M_1 can be found by means of the formulas already demonstrated, and the true equilibrium polygon drawn.

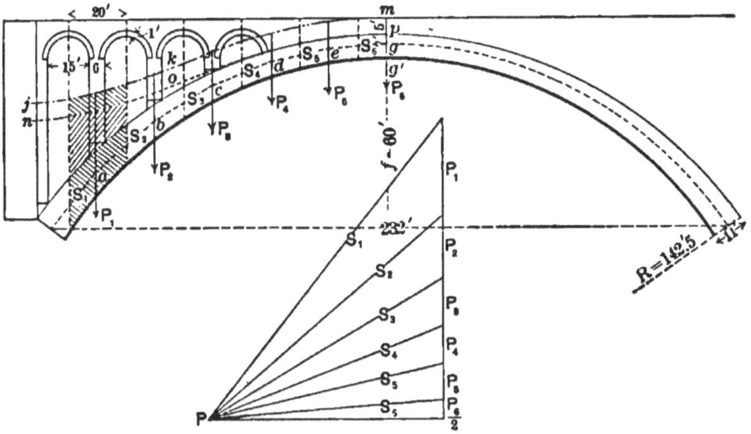

FIG. 66.

If lead joints are employed at the skew-backs and the crown, the values of H_1, V_1, etc., can be found under the assumption that the arch has three hinges, trials being made under the assumption that the hinges lie within the middle third of the arch-ring.

If an actual hinge is placed at the crown, the starting-point of the equilibrium polygon is fixed.

If no hinges are assumed, then the starting-point of the polygon must be determined in the same manner as for the metal arch.

Extent of Loading which will cause the Equilibrium Polygon to follow the Axis of the Arch.—In Fig. 66 let *mg′* represent the load at *g*; then at the joints *e, d, c*, etc., lay off upwards from the lower limit of the arch-ring the loads $P_5 P_4$, etc., and draw the curve *jkm*. This represents very nearly the upper limit of a homogeneous load corresponding to the polygon *abc*, etc. If now *nop* is drawn parallel to *jkm* at a distance *mp* below this curve, the shaded portion between the curve *nop* and the upper limit of the arch-ring represents the relative amount of material to be placed in the lateral walls.

If the live load over all is included with the dead load, the point *m* would be raised an amount proportional to the added live load measured in masonry units.

The axis of the arch shown in Fig. 66 is circular. If the angle at the centre had been larger and the curve *jkm* continued, we would have found the distance between it and the arch-ring increasing quite rapidly beyond an angle of 45° or 50° from the crown and becoming infinite for the semicircular arch.

For this reason it is customary to consider the arch-ring to act as an arch for only about 45° or 50° from the crown, the masonry in the abutments or piers being built solid in horizontal courses up to this point.

Moving Load.—There remains now to be determined the effect of the moving load. If the actual maxima stresses are desired, the best method of procedure is to determine the effect of each load or concentration independently and combine the results. In most cases, however, the effect of the moving load is small, and it is necessary to consider but two cases, namely, moving load over all and moving load extending from one support up to the crown.

Change in Dimensions.—If after trial it is found that some equilibrium polygon for *dead and live load combined* departs from the middle third of the ring, the depth of the ring may be changed; this need not necessitate a new calculation unless a great change is made, for the effect of the added material is likely to be very small, especially if the equilibrium polygon for the dead load follows the axis of the arch-ring. In case the equilibrium polygon lies outside of the middle third at any section, it does not necessarily make the structure unsafe unless the intensity of the pressure is sufficient to crush the material. The joints may open a little on the side farthest away from the polygon, so that it is not good policy to so design the ring that there is any such tendency.

Concrete and Brick Arches.—Evidently concrete and brick

arches can be designed in the manner outlined for the stone arch, using proper judgment as to the strengths of the materials.

The concrete arch may even be made lighter, since it has considerable strength in tension.

ARCHES WITH LEAD IN THE JOINTS AT THE SPRINGING AND THE CROWN.*

In order to reduce as much as possible the uncertainty of the location of the equilibrium polygon at the crown and the springing, and also to reduce to a certainty its location within limits, German engineers have placed lead in the middle thirds of the joints specified. Evidently the equilibrium polygon cannot lie far outside of the middle third at these joints, as the lead acts similarly to a hinge. After the false-works have been removed the masonry adjusts itself until every joint is in equilibrium. Nearly all, if not all, this adjustment takes place at the lead joints, which are compressed in thickness and expanded around the edges until the pressure per square inch does not exceed about 3500 pounds.

German engineers design these lead joints so that the maximum intensity of the pressure does not exceed about 1600 pounds per square inch, and have been very successful in their application of the method. After the structure is about completed and the entire dead weight is in place the joints at the springing are filled with cement and the arch becomes fixed at the ends for any additional loads.

---

* "Ponts en Maçonnerie avec Articulations à la clef et au joint de Rupture." Par M. G. La Rivière. *Annales des Ponts et Chaussées*, juin, 1891. Abstract, *Engineering News*, Oct. 24, 1891.

ARCHES WITH STEEL OR IRON PINS AT THE CROWN AND THE SKEW-BACKS.

Recently there has been constructed in Switzerland a concrete-arch bridge which has articulations at the springing-joints and at the crown composed of convex steel bearings resting in concave steel sockets or grooves. The entire depth of the arch-ring is reinforced with metal and the steel bearings placed at the centres of the joints.

This mode of construction definitely fixes the equilibrium polygon at the springing-joints and the crown.

EARTH-FILLED SPANDRELS

In small arches the spandrels are often filled with earth from the arch-ring up to the roadway.

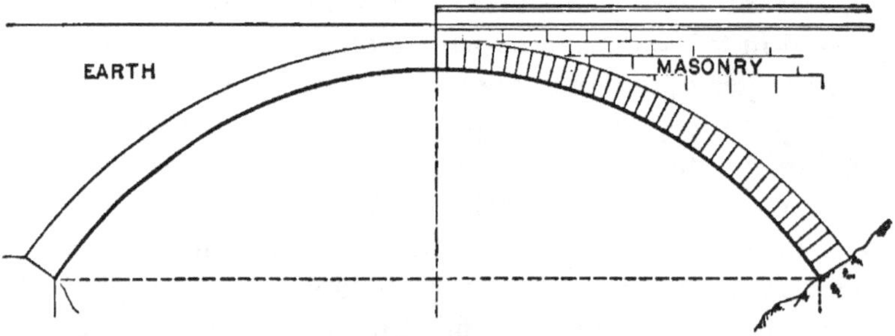

FIG. 67.

Assuming the earth to produce only the *vertical* pressures upon the arch-ring due to its weight, the determination of the equilibrium polygon offers no especial difficulties. But probably the earth causes other than vertical forces, and these are more or less indeterminate.

If the earth is assumed to be a homogeneous granular

* The Coulouvrenière Concrete-arch Bridge, Geneva, Switzerland. *Engineering News*, Aug. 6, 1896.

mass, then the pressure upon the arch-ring at any point can be fairly well determined from the Theory of Earth-pressure.*

If the arch is designed for this earth-pressure, it will support a very considerably increased load at the crown, owing to the resistance of the earth over the haunches against heaving.

Another feature which places the method of considering the earth-pressure acting against the ring as against a retaining-wall upon the safe side is that longitudinal side walls must be used to retain the earth in the spandrels. These walls undoubtedly relieve the arch-ring from the direct thrust of the earth.

If a retaining-wall is placed over the abutments, then the earth-filling may as well be treated as a vertical weight upon the arch-ring.

PART EARTH AND PART MASONRY SPANDREL-FILLING.

Under the assumption of *vertical loading*, it is found often that spandrels filled with earth alone are *too light* to cause the equilibrium polygon to follow the axis of the arch; then the spandrels are partially filled with masonry, as shown in Fig. 68.

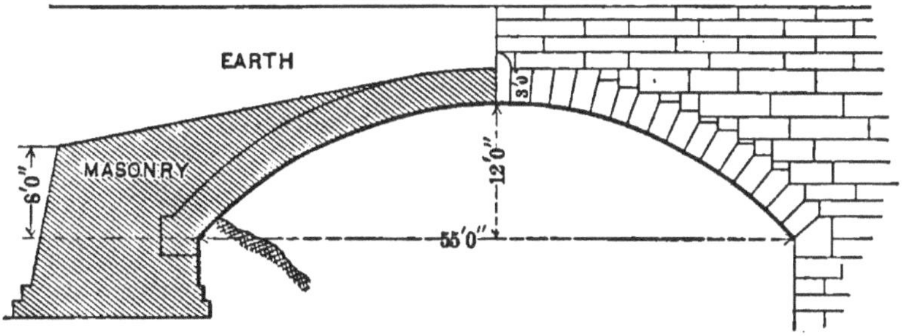

FIG. 68.

This masonry is usually concrete or rubble masonry. It is seldom of the same class as the arch-ring masonry.

As constructed, the upper limit of this masonry filling

---

* Retaining Walls for Earth, by M. A. Howe ; John Wiley & Sons, N. Y.

slopes very gradually from the crown towards the skew-backs; hence the horizontal thrust of the earth above is practically eliminated.

The exact action of this spandrel-filling upon the arch-ring is indeterminate. The assumption that it acts as vertical forces is on the safe side.

MASONRY SPANDRELS WITH LONGITUDINAL VOIDS.

Here the haunches are lightened by running longitudinal walls above the arch-rib and connecting them by arches or slabs immediately below the roadway, as shown in Fig. 69.

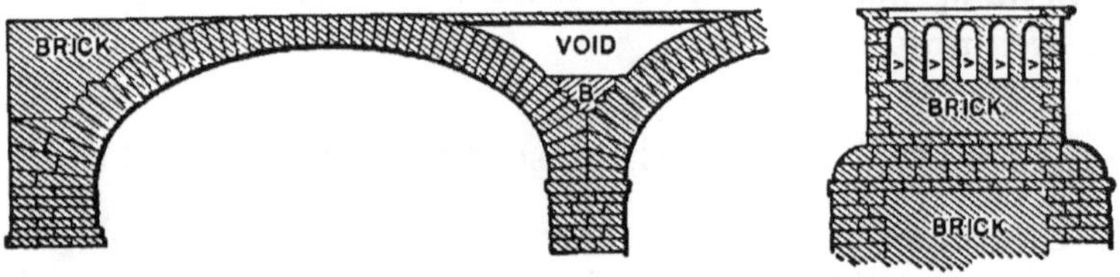

Fig. 69.

The amount of space to be left void can be found by the method outlined for *lateral voids*, but the masonry undoubtedly exerts a much less pressure upon the arch-ring than under the assumption of vertical loads. Just what the pressure is cannot be determined. Such arches seldom if ever fail at the haunches owing to the resistance offered by the solid longitudinal spandrel-walls.

If the arch-ring is designed for vertical loads the crown will not rise, as these walls cannot possibly exert a pressure equivalent to their weight. In fact good masonry can be stepped at an angle of at least 50° from the horizontal and be perfectly stable, provided the weight is balanced over the pier or abutment.

ARCHES HAVING SPANS LESS THAN TWENTY-FIVE FEET.

These can be proportioned in the manner outlined for larger arches, but usually the ring is made much deeper than necessary owing to the economy in using material of certain dimensions. Stone arches seldom have ring-stones less than one foot deep.

We have pointed out some of the difficulties which arise in the consistent designing of masonry arches of the usual type. The principal difficulty appears to be the determination of the magnitudes and directions of the forces due to the dead load. If these forces are assumed as acting vertically and in magnitude the weight of the material included between vertical planes then the arch can be designed by the formulas already deduced for elastic arches, or by the direct and very consistent method proposed by Alexander and Thomson, which we will explain in the following pages. For the assumptions made, this method is the most general and consistent which has been advanced up to the present time.

CHAPTER XIII.

ALEXANDER AND THOMSON'S METHOD FOR DESIGNING SEGMENTAL MASONRY ARCHES.*

"THE Transformed Catenary is shown by Rankine (Civil Engineering, Art. 131) to be the form of equilibrium for an ideal linear rib or chain under the uniform-vertical-load area between itself and a horizontal straight line. This curve has received considerable attention from early times because of its importance in designing arches, and is known best, perhaps, by engineers as the equilibrium curve.

"It seems to have been assumed that the transformed catenary, like the common catenary and the parabola, had its curvature continuously diminishing from the vertex outwards.

"In the following investigation it is shown that a very close resemblance exists between certain of these equilibrium curves and the circle—a fact important to engineers."

EQUATION OF THE COMMON CATENARY.

From Rankine's Civil Engineering, Art. 128,

$$y = \frac{m}{2}\left(\epsilon^{\frac{x}{m}} + \epsilon^{-\frac{x}{m}}\right), \quad . \quad . \quad . \quad . \quad . \quad . \quad . \quad (1)$$

---

* Transactions of the Royal Irish Academy, vol. XXIX, part III, 1888. On Two-nosed Catenaries and their Application to the Design of Segmental Arches. By T. Alexander, C.E., Professor of Engineering, Trinity College, Dublin; and A. W. Thomson, B.Sc., Assoc. Mem. Inst. C.E., Lecturer in the Glasgow and West of Scotland Technical College. This is an elaborate paper, containing many interesting things which are omitted here as not being essential for the mechanical method of designing arches.

where $y =$ the ordinate of any point;
$x =$ the abscissa of any point having the ordinate y;
$m =$ the parameter;
and $\epsilon =$ the base of the Naperian system of logarithms.

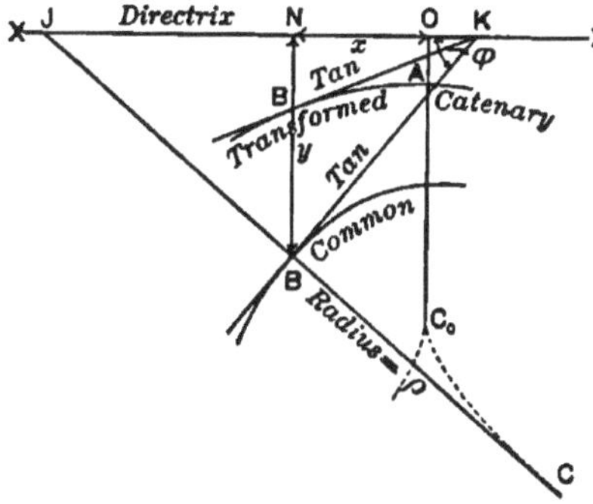

FIG. 70.

THE TRANSFORMED CATENARY.

The locus of a transformed catenary is obtained by increasing or decreasing all the ordinates of a common catenary by a given ratio r.

Then for the transformed catenary

$$y = r\frac{m}{2}\left(\epsilon^{\frac{x}{m}} + \epsilon^{-\frac{x}{m}}\right), \quad . \quad . \quad . \quad . \quad . \quad . \qquad \text{(II)}$$

$$\tan\phi = \frac{dy}{dx} = \frac{r}{2}\left(\epsilon^{\frac{x}{m}} - \epsilon^{-\frac{x}{m}}\right) = \frac{\sqrt{y^2 - y_0^2}}{m}, \quad . \quad . \quad . \qquad \text{(III)}$$

where y_0 is the value of y when $x = 0$.

$$\rho = \frac{(m^2 + y^2 - y_0^2)^{\frac{3}{2}}}{my}, \quad . \quad . \quad . \quad . \quad . \quad . \qquad \text{(IV)}$$

$$\sec^2\phi = (1 + \tan^2\phi)^{\frac{3}{2}} = \frac{\rho y}{m^2}, \quad . \quad . \quad . \quad . \qquad \text{(V)}$$

where ϕ is the slope at any point and ρ the radius of curvature.

For the crown (IV) becomes

$$\rho_0 = \frac{m^2}{y_0} = \frac{m}{r}; \quad . \quad . \quad . \quad . \quad . \quad . \quad . \quad \text{(VI)}$$

and hence (V) becomes

$$\sec^2 \phi = \frac{\rho y}{\rho_0 y_0}. \quad . \quad . \quad . \quad . \quad . \quad . \quad . \quad . \quad \text{(VII)}$$

THE TWO-NOSED CATENARY.

An investigation of (IV) for maxima and minima shows that for values of r *less* $\frac{1}{\sqrt{3}}$ there is a maximum radius of curvature ρ_0 at the crown and a minimum radius of curvature

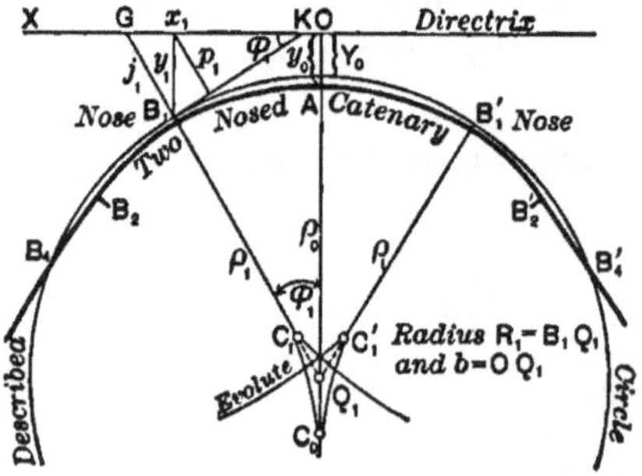

FIG. 71.

ρ_1 at a pair of points $(B_1 B_1')$ symmetrical about the crown, where

$$y_1 = \sqrt{\frac{m^2 - y_0^2}{2}} = m\sqrt{\frac{1 - r^2}{2}}. \quad . \quad . \quad . \quad \text{(VIII)}$$

Such catenaries are called *two-nosed.*

If m be assumed as unity and r be given values less than $\sqrt{\frac{1}{3}}$, the values of ρ_1, y_1, x_1, y_0, and ϕ_1 can be readily computed. A large number of these values are given in Table A.

As an aid in computing the ordinates, etc., of the two-nosed catenary, the general formulas may be put in the following forms:

Let

$$s = \frac{y_0}{\rho_0} = y_0 \div \frac{m^2}{y_0} = r^2. \quad . \quad . \quad . \quad . \qquad \text{(IX)}$$

Then, from (VIII),

$$y_1 = m\sqrt{\frac{1 - s}{2}}. \quad . \quad . \quad . \quad . \quad . \qquad \text{(X)}$$

From (III),

$$\tan \phi_1 = \sqrt{\frac{1 - 3s}{2}}; \quad . \quad . \quad . \quad . \quad . \qquad \text{(XI)}$$

$$\rho_1 = m\frac{3\sqrt{3}}{2}(1 - s). \quad . \quad . \quad . \quad . \qquad \text{(XII)}$$

From Rankine's Civil Engineering, Art. 131,

$$x_1 = m \log_e \frac{y_1 + \sqrt{y_1^2 - y_0^2}}{y_0}$$

$$= m \log_e \frac{\sqrt{1 - s} + \sqrt{1 - 3s}}{\sqrt{2s}}. \quad . \quad . \quad . \qquad \text{(XIII)}$$

It may be noted here that, given a certain value of s, all quantities are directly proportional to m excepting ϕ_1, which is *constant* for any given value of s, regardless of any change in m or ρ_1.

THE DESCRIBED CIRCLE.

In Fig. 71 if B_1C_1 be prolonged, it will cut AQ_1 in Q_1. If Q_1 be taken as a centre and B_1Q_1 as a radius, and a circle described, it will evidently lie wholly above the two-nosed catenary between the points B_1 and B_1'. This circle will also lie beyond the catenary curve for some distance beyond B_1 and B_1', cutting it finally in B_4 and B_4'.

Let R_1 be the radius of the described circle. Then, from Fig. 71,

$$R_1 = x_1 \operatorname{cosec} \phi_1, \quad . \quad . \quad . \quad . \quad . \quad . \quad . \quad . \quad (\text{XIV})$$

$$OQ_1 = b = y_1 + R_1 \cos \phi_1, \quad . \quad . \quad . \quad . \quad (\text{XV})$$

and

$$OK = Y_0 = b - R_1. \quad . \quad . \quad . \quad . \quad . \quad . \quad . \quad (\text{XVI})$$

The values of R_1, b, and Y_0 are given in Table A for the values of r which were used in computing the ordinates, etc., of the two-nosed catenary.

An examination of this table shows for $r = \sqrt{\frac{1}{8}}$, or $s = \frac{1}{8}$, that $y_0 = Y_0$, or the described circle touches the two-nosed catenary at the crown. That is, B_1B_1' and A and K coincide. Also, that between the values of $s = 0.027$ and $s = 0.0204$, Y_0 changes sign, indicating that the described circle cuts the directrix.

The distance apart of the described circle and the two-nosed catenary at the crown is

$$KA = \delta_0 = y_0 - Y_0. \quad . \quad . \quad . \quad . \quad . \quad . \quad (\text{XVII})$$

The values of δ_0 are given in Table A.

THE THREE-POINT CIRCLE.

Evidently for the two-nosed catenary there must be a point beyond B_1 which has the same radius of curvature as at the crown. There will be a similar point on the opposite side of the crown. A circle passed through these three points will

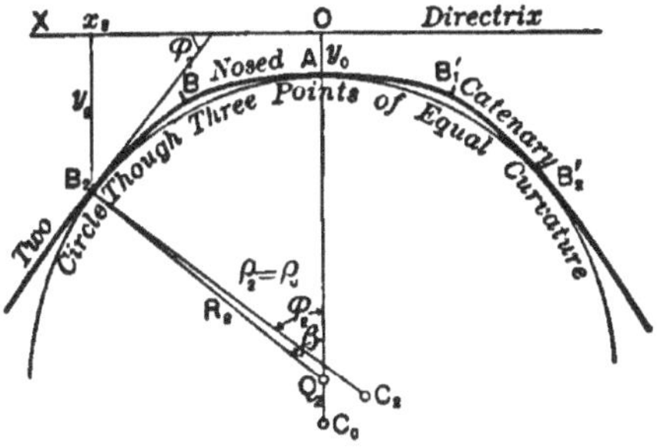

FIG. 72.

evidently lie below the catenary between B_2, A, and B_2'. This circle is called the three-point circle.

Let R_2 be the radius of the three-point circle which passes through the three points of equal curvature of the two-nosed catenary.

$$\rho_2 = \rho_0 = \frac{m_2}{y_0} = m\frac{1}{\sqrt{s}}. \quad \ldots \ldots \quad \text{(XVIII)}$$

$$y_2 = y_0 \sec^2 \phi_2 = m\sqrt{s}\sec^2 \phi_2. \quad \ldots \quad \text{(XIX)}$$

$$\sec^2 \phi_2 = \sqrt{\frac{1}{s} - \frac{3}{4}} - \frac{1}{2}. \quad \ldots \ldots \quad \text{(XX)}$$

$$\tan^2 \phi_2 = \sqrt{\frac{1}{s} - \frac{3}{4}} - \frac{3}{2}. \quad \ldots \ldots \quad \text{(XXI)}$$

$$x_2 = m \log_e\left(\sec^2 \phi_2 + \frac{\tan \phi_2}{\sqrt{s}}\right). \quad \ldots \quad \text{(XXII)}$$

$$R_2 = \frac{x_2}{2(y_2 - y_0)} + \frac{y_2 - y_0}{2}. \quad \ldots \ldots \quad \text{(XXIII)}$$

$$\tan\frac{\beta}{2} = \frac{y_2 - y_0}{x_2}. \quad \ldots \ldots \ldots \quad \text{(XXIV)}$$

The values of x_2, y_2, ϕ_2, β, ρ_2, and R_2 are tabulated in Table A, from which we see that ϕ_2 and β differ but little until $s = 0.02\dot{7}$, and then the difference is but 2° 28′, so that in many calculations one angle can be used for the other.

RELATIVE POSITIONS OF THE DESCRIBED AND THREE-POINT CIRCLES

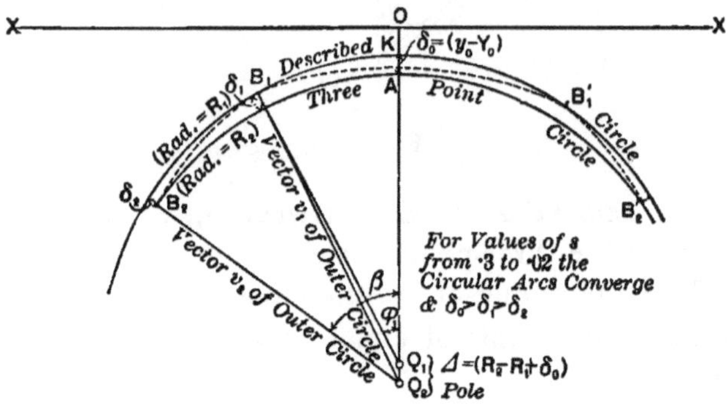

FIG. 73

From Table A the distance between the centres of the two circles is found to be very small when compared with the length of the radii, so that the angular distance of the nose B_2 from the crown is sensibly the same, whether measured on the described or three-point circle. Then

$$\delta_0 = y_0 - Y_0; \quad . \quad . \quad . \quad . \quad . \quad . \quad . \quad . \quad \text{(XXV)}$$

and approximately, for values of s from 0.333 to 0.01,

$$\delta_1 = R_1 - R_2 + \Delta \cos \phi_1 \quad . \quad . \quad . \quad \text{(XXVI)}$$

and

$$\delta_2 = R_1 - R_2 + \Delta \cos \beta, \quad . \quad . \quad . \quad \text{(XXVII)}$$

where

$$\Delta = R_2 - R_1 + \delta_0. \quad . \quad . \quad . \quad . \quad . \quad . \quad \text{(XXVIII)}$$

An examination of the values of δ_0, δ_1, and δ_2, given in Table A, shows that for values of s from 0.333 to 0.027, $\delta_0 > \delta_1 > \delta_2$, or the two circles approach each other as they leave the crown.

Between $s = 0.027$ and $s = 0.204$ ($s = \frac{1}{49}$), $\delta_0 = \delta_1 = \delta_2$, or the two circles are concentric. Beyond these values of s the circles diverge.

Then if through A a symmetrical circular arc is passed concentric with the described circle, the equilibrium polygon or two-nosed catenary will lie between the two out to the points of rupture B_2 and B_2'. Then much more will it lie between the described circle and one of a less radius than the concentric circle.

Table B.—In Table A are given the various co-ordinates of points on the described circle and the three-point circle for a modulus m = unity.

Suppose now we wish to base all of these quantities upon the radius of the described circle and take its value as unity, then it is necessary to divide each *linear* quantity in A by the corresponding value of R_1.

Table B is the result of such an operation. The object of this will appear from the following:

Suppose a circle of radius unity be drawn, and let this circle be taken as a described circle; then $R_1 = 1$.

Since s, ϕ_1, and ϕ_2 are independent of R_1, it is evident that all of the two-nosed catenaries in Table A can be constructed within this circle of unit-radius merely by changing each ordinate proportional to R_1. We have then to the scale unity (R_1 = unity) an exact representation of the relations between the several curves we have been considering.

If R_1 has any other value than unity, we have only to multiply these quantities by the new value of the radius.

In Fig. 73 let XX be the upper limit of masonry to be supported by an arch, and OA the depth at the crown when R_1 = unity; then, from Table B, $OA = y_0$ can have values from 0.23 to 0.05, R_2, ϕ_2, ϕ_1, etc., corresponding values, and for any particular case the equilibrium polygon will never be above B_1KB_1' between B_1 and B_1', and never below B_2AB_2' between B_2 and B_2'. Then if the portion of the masonry between $B_1'KB_2'$ and B_2AB_2' were cut into arch-stones the structure would be stable under the assumption that B_2AB_2' does not sensibly differ from $B_2B_1AB_1'B_2'$, which is the case within the limits of our values of s.

If this arch with R_1 = unity is in equilibrium, then any other arch of the *same proportions* would be in equilibrium.

But although we have equilibrium, we have not strength probably; and besides, in masonry the equilibrium polygon should not depart at any place from the middle third of the joints, and often it must follow more closely the centre line to obtain intensities of pressure consistent with the strength of the material employed.

The arch-ring specified above *decreases* in depth as it leaves the crown while the pressures upon the radial joints increase, indicating that the *lower boundary* of the ring should be changed so that the depth would increase as the stresses increase. This can be done after the depth at the crown is known. As this depends upon the material, we must determine the permissible intensities of the stresses, etc.

Horizontal Thrust.—According to Rankine's Civil Engineering, Art. 131,

$$H_1 = H_x = H = wm^2, \quad . \quad . \quad . \quad \text{(XXIX)}$$

where w is the weight per unit volume of the material taken as solid from the directrix down to the two-nosed catenary or the equilibrium polygon, or approximately the three-point circle.

If d is the depth at the crown from the directrix to the soffit,

$$H = \frac{d}{y_0} wm^2 = w\rho_0 d. \quad . \quad . \quad . \quad . \quad \text{(XXX)}$$

If T is the thrust at any point,

$$T = w\rho_0 d \sec \phi. \quad . \quad . \quad . \quad . \quad . \quad . \quad \text{(XXXI)}$$

Intensity of Pressure.—In Chapter I it was shown that if the resultant pressure on any rectangular joint was applied at the third point, the maximum intensity was twice the average intensity and the minimum intensity was zero.

Assuming, then, that the equilibrium polygon at the crown is to be applied at the *lower* third point of the keystone, if t_0 is the depth of the key,

$$t_0 = 3\delta_0. \quad . \quad . \quad . \quad . \quad . \quad . \quad . \quad \text{(XXXII)}$$

The average intensity of the pressure is

$$p_0 = \frac{H}{t_0} = \frac{w\rho_0 d}{t},$$

and the maximum intensity

$$p = 2p_0 = 2\frac{w\rho_0 d}{t_0}.$$

Let the safe strength of sandstone be 576008 pounds per square foot and $w = 140$ pounds per cubic foot; then

$$2\frac{w\rho_0 d}{t_0} \text{ must not } \textit{exceed } \frac{576008}{140}$$

or

$$\frac{\rho_0 d}{t_0} \text{ must not exceed } 205;$$

hence the *maximum* multiplier which can be used in Table B_1 for sandstone is $205 \div \frac{\rho_0 d}{t_0}$.

Table B_1 derived from Table B.—The values of s remain the same as in Table B.

$$t_0 = 3\delta_0.$$

$$d = y_0 + \delta_0.$$

$\frac{\rho_0 d}{t_0}$ is derived from taking ρ_0 from Table B and multiplying it by $\frac{d}{t_0}$.

The maximum multipliers are found as explained above, upon the supposition that t_0 is never less than *one foot* for stone masonry or one brick for brick masonry, nor greater than reasonable dimensions.

The remaining factors depend upon R, which is found as follows:

In order that equilibrium may still exist under the additional masonry due to increasing the depth at the crown by δ_0, and that the arch-ring may increase in depth as it departs from the crown, the limit of the soffit is made a three-point circle for a two-nosed catenary which lies the distance d below the directrix at the crown. Then entering Table A with $y_0 = d$, the corresponding value of R_2 is found for $m =$ unity. If this value of R_2 be divided by the corresponding value of R_1, the result will be the value of R given in Table B_1 on the line containing the assumed value of d.

In a manner quite similar other supplementary tables are formed.

ADVANTAGES OF THE METHOD.

One of the principal advantages of the above method is that any uniform load over the entire span can be added without changing the described circle which is the boundary of the kernel of the arch-ring, provided the added load is not sufficient in its equivalent of masonry to make $y_0 \div \rho_0$ greater than one third. The effect of the added load is merely a change in the angle ϕ_1, which grows smaller, and a change in δ_0, which also grows smaller, thus leaving the equilibrium polygon within the kernel designed for the original mass of masonry.

This is particularly advantageous in the case of moving loads.

It is to be noted that the depth of the key depends not only upon equilibrium, but upon the strength of the material, and that this depth is a function of y_0 and δ_0, thereby securing a keystone which varies consistently with different conditions.

Again, for given conditions there is a perfectly definite form and size of arch, which can be obtained without any of the usual cut-and-try methods.

Since for a solid load having the horizontal directrix for the upper limit the lower limit is a two-nosed catenary when equilibrium exists, and since within the limits of our tables an indefinite number of two-nosed catenaries can be constructed having different values of y_0, it must follow that the homogeneous material between any two two-nosed catenaries of the same family (that is, transformed from the same common catenary) must be in equilibrium, or any combination of such areas. For such conditions the weight of the material may be taken as the average, and then reduced by the ratio of the loaded area to the total area between the two-nosed catenary forming the soffit and the directrix.

The three-point circle can always be used for the limits of the loading since it differs so little from the two-nosed catenary within the limits of our tables.

Thus almost any kind of spandrel-filling can be used provided its upper limits are always three-point circles, members of the same family as the line of stress.

UNSYMMETRICAL MOVING LOAD.

Since the moving load is usually small in comparison with the dead load, unsymmetrical loading need not be considered.

For a moving load covering but one half of the span the equilibrium curve at the crown is raised a little, thus leaving a small distance between it and the dead-load curve from the unloaded portion. The effect of this is a couple tending to turn the key. This turning can be prevented by the masonry filling, as illustrated in Example 6.

CHAPTER XIV.

EXAMPLES ILLUSTRATING ALEXANDER AND THOMSON'S METHOD FOR DESIGNING SEGMENTAL MASONRY ARCHES.*

EX. 1°. Design of a sandstone segmental arch with vertical load: span 75 feet and depth of surcharge at crown about 1 foot 4 inches. The springing to be the joint of rupture.

Here $2c = 75$ and $d - t_0 = 1\frac{1}{3}$; their ratio is 56.[illegible]3. We find by trials on Table B_1 that $2c \div (d - t_0) = 53$ occurs on the line where $s = 0.05$, and the multiplier required on that line to make $2c$ into 75 is 50.07, about half the maximum multiplier given under sandstone in the table; so we shall have a factor of safety of about twice *ten*.

From Table B_1 we obtain the relative values given below, and multiplying them by 50.07 we obtain the absolute values.

s	*Mult.*	d	t_0	t_2	R	k	$2c$
.05	50.07	0.089	0.061	0.123	0.869	0.427	1.498
		4.46	3.05	6.15	43.5	21.4	75 feet.

The radius and rise of *soffit* are 43.5 and 21.4 feet; the thickness of arch-ring at the crown and springing, 3 feet 0 inches and 6 feet 2 inches; the surcharge being 1.41 feet or nearly 1 foot 4 inches, as required.

From Table B, for $s = 0.05$ we have

s	*Mult.*	R_1	ρ_0	Y_0	δ_0	δ_2
0.05	50.07	1	1.3634	0.0478	0.0478	0.0167
		50.07	68.265	2.393	1.02	0.836 feet.

* Examples 1 to 7 inclusive are from Alexander and Thomson's paper.

At the crown the thrust on the arch-ring per foot of breadth is $H = w\rho_0 d = 140 \times 68.265 \times 4.46 = 42,600$ pounds; the average intensity of the stress is $42,600 \div 3.05 = 14,000$ pounds; and hence the maximum intensity is $2 \times 14,000 = 28,000$ pounds per square foot, giving a factor of safety of $576,000 \div 28,000 = 20$.

At the springing $T = H \sec \phi_2 = 84,000$; the average stress is $84,000 \div 6.15 = 13,700$; and since the deviation of the stress is $\frac{1}{6}t_2 - \delta_2 = x_0 = 1.025 - 0.836 = 0.189$ above the centre of the joint, then from Chapter I, page 9,

$$\text{max. intensity} = p_0\left(1 + \frac{6x_0}{h}\right) = 13700\left(1 + \frac{6(0.189)}{6.15}\right)$$
$$= 16,200 \text{ pounds per square foot,}$$

which gives a factor of safety of $576,000 \div 16,200 = 35$.

2°. If a live load of 220 pounds per square foot of roadway be placed upon the structure (Ex. 1°), find the new line of stress in the arch-ring and the intensities of stress at the crown and springing.

The height of superstructure equivalent to this live load is $h = 220 \div 140 = 1.571$ feet of sandstone. Here we have to find a new two-nosed catenary still inscribed in the same circle, $R_1 = 50.07$, forming the upper boundary of the middle third of the arch-ring, as already designed (Ex. 1°), but to a directrix, h, higher than before. Adding h to the old value of Y_0 we get $2.393 + 1.571$ or 3.964, which, divided by $R_1 = 50.07$, gives us 0.0793 as a *new* relative value of Y_0, which is found in Table B at the line

s	ϕ_2	*mult.*	R_1	p_0	Y_0	δ_0	δ_2
0.075	55° 3′	50.07	1	1.2384	0.0793	0.0134	0.0097
			50.07	62.007	3.970	0.671	0.486 feet.

This is a new two-nosed catenary, of a different modulus and of a different family, so that the soffit already designed will not be mathematically the three-point circle of another member of the family of this line of stress, but it will sensibly be so. The joints of rupture have gone up to 55° 3′; but this is immaterial, as the line of stress is now *closer* to the upper boundary of the kernel, and will therefore be wholly in the kernel down to 59° 31′, the springing-joint.

At the crown now we have the thrust $H' = w\rho(d + h) =$ $140 \times 62.007(4.46 + 1.57) = 52,300$ pounds; average intensity = 17,200 and maximum intensity 22,600 pounds, being less than the maximum intensity for the dead load alone, because of the centre of stress being much nearer the centre of the joint.

At the springing $T' = H' \sec 59° 31' = 103,200$. $x_0 =$ 0.539 feet *above* the centre of the joint, and the maximum intensity of pressure is 25,600 pounds per square foot, giving for the live load on the structure a factor of safety of about 22.

3°. Let the live load of 2° cover but one half of the span; find the horizontal thrust to be balanced by the backing of the voussoirs.

For dead load alone,	$H = 42,600$	pounds;
For live and dead loads,	$H = 52,300$	"
Hence the required thrust is =	9,700	"

4°. Fig. 74. Suppose the arch-ring spandrels, etc., of 1° have by means of voids in the superstructure an average weight of 100 pounds per cubic foot. Find results corresponding to those of 1°.

For *stability*, and to give the required value of $d - t_0$, the dimensions in 1° are required, just as before, but the stresses will be altered in the ratio 140 : 100. H now becomes about 30,500 pounds and T 60,000 pounds, giving factors of safety of 28 at the crown and 49 at the springing.

The voids should be so arranged that their boundary may be roughly a member of the same family as the line of stress, by making the ordinates of the boundary a constant fraction of those of the soffit.

This should be done when the spandrels are partially filled with masonry and then the remainder with earth.

5°. Fig. 74. Let a live load of 157 pounds per square foot of roadway be over the whole span of the bridge

(Ex. 4°). Find the line of stress and the intensity of the stress at crown and springing.

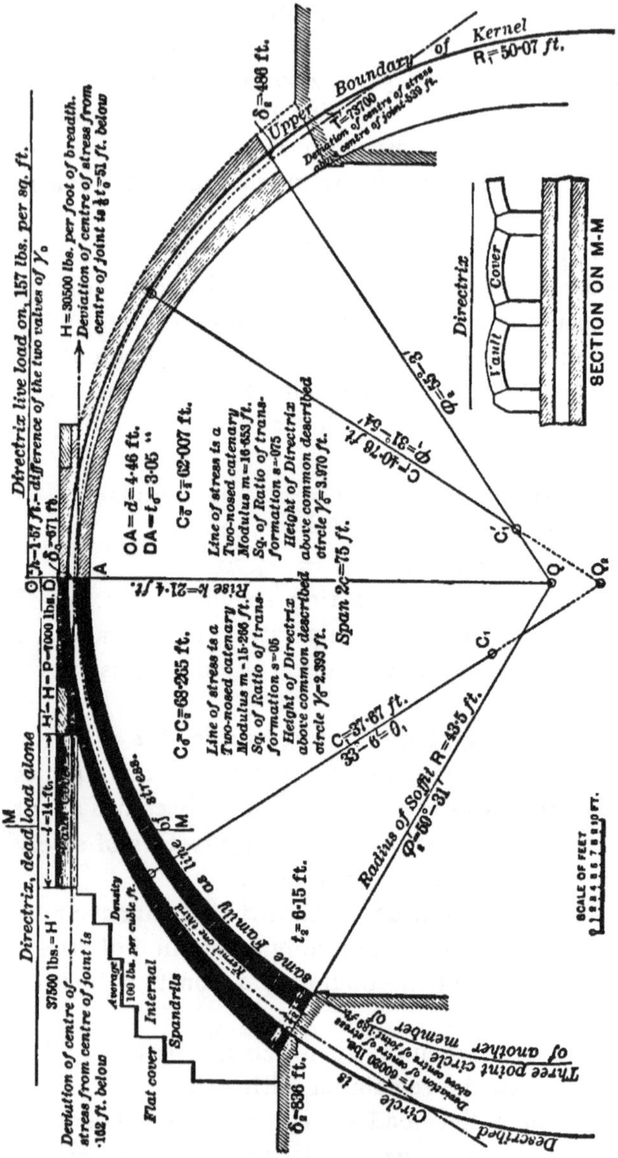

DESIGN OF A SEGMENTAL SANDSTONE ARCH.
FROM THE
TABLES OF THE PROPERTIES OF THE "TWO-NOSED CATENARIES."

FIG. 74.

The equivalent height of structure is 1.57 feet taking the

new density into account, so that the solution is the same as 2°, only we must alter the quantities in the ratio 140 : 100.

$H = 37,500$ nearly, and $T = 73,700$ nearly.

6°. Fig. 74. Let the live load in 5° be over only one half the span. Find the amount of horizontal thrust to be balanced by the frictional stability of the vault-covers butting against the higher voussoirs. Find also the distance back to which the vault-covers must extend to balance it.

The thrust $P = 37,500 - 30,500 = 7000$ pounds per foot of breadth. If the under side of the vault-covers come up to the level of the crown of the soffit, then the weight per foot of breadth of bridge on the spandrels due to the vault-covers, and the dead load over them alone, is $wdL = 140 \times 4.46L = 624L$. Taking the coefficient of friction as 0.7, then $0.7 \times 624L = 7000$, or $L =$ about 16 feet. The voussoirs near the keystone should have square-dressed side-joints until the sum of their vertical projection is t_0, so as to receive the horizontal thrust of the vault-covers truly; the spandrel walls must be built up to the height of the soffit for a distance equivalent to 16 feet of vault-covering, when they may be stepped down.

7°. Design of a semicircular arch-ring of common sandstone, the span to be 100 feet, and a surcharge of at least 1½ feet being required for the formation of the roadway, laying of gas-pipes, etc. The data are $R = 50$, and $R \div (d - t_0)$ not to be less than 33. On Table B_1 the lines above that with $s = .08$ (in order to make R into 50) require a multiplier greater than the maximum given for sandstone; these lines are therefore excluded on the question of strength, while the lines below that with $s = .05$ do not give $R \div (d - t_0)$ so great as 33, and are excluded by requirements of the roadway. Those two limiting lines give

s	*Mult.*	R	d	t^0	θ_2	Factor of Safety.
.08	53.6	50	5.9	2	54° 14′	$\frac{57 \times 10}{53.6} = 10.6$
.05	57.5	50	5.1	3.5	59° 31′	$\frac{104 \times 10}{57.5} = 18$

The upper gives *greatest economy* of material in *arch-ring*, which is only 2 feet at crown, but *less economy* of material in superstructure, as d is larger, and also *less economy* of solid backing, which has to be built to a joint 5° higher. Hence the line midway between them would be most suitable all round. For a single arch a line a little nearer the upper may be adopted; and for a series of arches a line nearer the lower, that is, in favor of a heavier arch-ring to withstand the shocks transmitted from arch to arch. The best lines then are

	s	*Mult.*	R	d	t_0	t_2	θ_2	Factor of Safety.
Single arch.....	.07	54.526	50	5.6	2.4	4.4	55° 53′	$\frac{69 \times 10}{54.5} = 12.7$
Series of arches.	.06	55.804	50	5.3	2.9	5.4	57° 38′	$\frac{85 \times 10}{55.8} = 15.2$

Compare Rankine's empirical rule, Civil Engineering, Art. 290, giving

$$t_0 = \sqrt{.12 \times 50} \quad \text{and} \quad \sqrt{.17 \times 50}$$
$$= \quad 2.45 \quad \text{"} \quad 2.92 \text{ respectively.}$$

The solid backing must be brought up to the point where the joint at θ_2 meets the back of the arch-ring, and below that joint the arch-ring may be of the uniform thickness t_2. The superstructure may readily be reduced by voids and the employment of material of less density than sandstone, till the average density of the whole is a fifth less than that of sandstone, which would raise the factors of safety at crown to 16 and 19. The factors of safety at joint of rupture are even greater as the centre of stress is nearer the centre of the joint, and $t_2 \div t_1 > \sec \theta_2$. By means of the values obtained for ρ_0, d, δ_2, the thrust at crown and joint of rupture and the centre of stress at joint of rupture are calculated, as in preceding example. A tangent from this last point enables a suitable abutment to be designed.

CHAPTER XV.

TESTS OF ARCHES.

RECENTLY the Austrian Society of Engineers and Architects have published a report of a series of tests made upon full-size arches. The publication contains 131 folio pages with 27 plates.* The experiments are minutely described and thoroughly discussed, and a comparison made between the results and those theoretically obtained.

The tests of greatest interest were those made upon five arches having a span of 75.4 feet, a clear rise of 15.1 feet, and a width of 6.56 feet. These arches were—

1°. Rough quarry-stone;
2°. Brick;
3°. Concrete;
4°. Concrete Monier type;
5°. Steel.

Rough Quarry-stone Arch.—This arch was constructed of rough quarry-stone laid in Portland cement-mortar composed of 1 part cement and 2.6 parts sand, the test being made 51 days after its completion.

The thickness at the crown was 23.6 inches and at the skew-backs 43.3 inches.

The loading was applied vertically at five points, dividing the half-span into five equal parts.

The ultimate load causing rupture was 660 pounds per square foot over one half of the span.

---

* Bericht des Gewölbe-Ausschusses des Oesterreichischen Ingenieur- und Architekten-Vereins. Vienna, 1895. See also *Eng. News*, Nov. 21, 1895, p. 351, and April 9, 1896.

The arch failed by radial cracks appearing on the extrados near the skew-backs on the loaded side and over the haunches on the unloaded side.

Brick Arch.—This arch was identical in dimensions with the stone arch, and failed in a similar manner under a load of 602 pounds per square foot over one half of the span.

Concrete Arch.—The thickness at the crown was 27.6 inches, and at the skew-backs 27.6 inches.

This arch was made up of segments of concrete composed of mixtures of different proportions, and at the skew-backs the joints between the arch-ring and the abutments were filled with asphalt about ½ inch thick.

The arch failed under a load of 742 pounds per square foot over one half of the span.

Monier Arch.—Here the general dimensions were the same as before, but the thickness at the crown was only 13.8 inches and at the skew-backs 23.6 inches.

The arch failed under a load of 1300 pounds per square foot over one half of the span, failing by cracking as follows:

1st. On the loaded side at the skew-back;

2d. On the unloaded side at the haunches; and,

3d. On the loaded side at the haunches.

Steel Arch.—Failure took place under a load of 1564 pounds per square foot over one half of the span, by the buckling of the unloaded portion near the haunches.

Deformations.—Throughout the tests careful measurements were made of all deformations caused by removing the falseworks, temperature changes, and the changes in loading.

The appearance of the first crack was noted, with the magnitude of the load causing it.

The arches were finally tested to destruction and the load causing failure carefully determined.

From these records a comparison was made with theoretical results.

Comparison with Theory.—It was found for the stone and brick arches that failure occurred in the joints, the mortar separating from the stone or brick. The adhesive strength of the mortar for the stone arch was found to be about 120 pounds per square inch and the value of E about 960,000.

In the brick arch the adhesive strength of the mortar was about 70 pounds per square inch, and the values of E varied from 340,000 to 470,000.

From the results of the tests of the concrete arch the average ultimate strength of the concrete was placed at 290 pounds per square inch, and the value of E at 1,430,000.

The Monier arch cannot be discussed theoretically, owing to the use of metal imbedded in the concrete.

The value of E as determined from the tests of the steel arch was about 26,000,000, which is a little smaller than the value obtained from the tests of small specimens.

Even in the masonry arches the deformations were proportional to the loads up to a certain point, showing that the material behaved the same in the arch as in small specimens for testing.

The measuring devices placed near the skew-backs indicated that on the loaded side the arch was practically fixed at the ends and on the unloaded side very nearly so. Of course the concrete arch with asphalt plates at the skew-backs must be excepted. This arch behaved neither as fixed nor hinged, and the theoretical results were taken as the mean of those obtained by considering the arch as fixed and then as hinged.

In all cases the arches failed at points which theory predicted.

CONCLUSIONS DRAWN FROM THE RESULTS OF THE FIVE EXPERIMENTS.

The very important conclusion drawn from these experiments was that the masonry arches behaved very nearly as *elastic arches fixed at the ends*, and hence the formulas for elastic arches were the only formulas which should be employed in designing such structures.

The close agreement with theory under the method of applying the loading employed in these experiments is a very strong argument in favor of the type of spandrel construction advocated in Chapter XII. A few very old bridges and some modern bridges have been constructed after this form. The only argument against this method is that in bridges of long span the effect of changes in temperature sometimes cracks the masonry above the small arches; but this can be avoided by making a vertical joint near the skew-backs, as was done in the Coulouvrenière bridge.

SPECIFICATIONS.

The following specifications are advocated in Chapter VII of the Austrian report.

All Large Arches must be designed according to the Elastic Theory.—Two cases of live loading may be considered: (1°) load covering entire span, and (2°) load covering but one half of the span.

For railway bridges the rails should be at least 3.28 feet above the crown of the arch, and this space filled with some cushioning material.

Brick and stone arches, where the ratio of the rise to the span lies between one half and one fifth, may have depths at the crown as specified below.

For spans of		30	metres, thickness of crown		=	1.1 m.
"	"	40	"	"	"	1.4 "
"	"	65	"	"	"	2.2 "
"	"	80	"	"	"	2.7 "
"	"	100	"	"	"	3.4 "
"	"	120	"	"	"	4.1 "

For segmental arches the thickness at the skew-backs may be $1\frac{1}{2}$ the thickness at the crown, and for semicircular arches 1.7 the thickness at the crown.

The width of the bridge at the crown should never be less than the following:

For spans of		30 metres,	width	=	2.4 m.
"	"	40	"	"	3.0 "
"	"	65	"	"	4.5 "
"	"	80	"	"	5.6 "
"	"	100	"	"	7.0 "
"	"	120	"	"	8.6 "

If the width at the crown is small, the width at the skew-backs should be one twentieth greater.

In all cases the falseworks should be as rigid as possible, and in order that the deformations should be symmetrical the arch should be constructed in symmetrical sections.

TEST OF SMALL ARCHES.

In connection with the tests mentioned above two small arches were tested.

Monier Arch.—This arch had a span of 32.8 feet, a rise of 3.28 feet, and a width of 13.92 feet. The thickness at the crown was 7.87 inches. The metal gridiron which was placed near the intrados of the arch was made of pieces 0.39 inch and 0.27 inch in thickness, the former running longitudinally.

The spandrels were filled even with the crown with concrete, which carried a single standard-gauge railway track.

The arch was first tested with locomotives covering one half of the span, then a uniform load of rails was placed upon one half of the span.

Cracks appeared near the springing on the loaded side under a load of 920 pounds per square foot. The arch failed under a load of 2010 pounds per square foot over one half of the span by the yielding of the abutments.

Concrete Arch.—A concrete arch of the same dimensions was tested six months after being built and no signs of failure appeared under a load of 2110 pounds per square foot over one half of the span.

TESTS OF FLOOR ARCHES.

Austrian Tests.—A synopsis of these tests was given by Prof. Merriman in *Engineering News*, April 9, 1896. This synopsis covers the ground so thoroughly that it is given below.

Seventeen arches, having spans of 4.43 feet and 8.86 feet, were tested to destruction by a uniform load. Of these four were common brick arches, five were of special forms of brick, three were concrete arches, three were Monier arches, one was of the Melan system, and two of corrugated plates. Most of these were built between rolled beams in the manner usual in floor construction, these beams being prevented from spreading by plates and channels at the ends, and also by a tie-rod at the middle. The space above the arch and between the beams was levelled up with earth, upon which a board floor was laid, and upon this pig iron was piled. The tests were made four months after the arches had been built. All these arches were designed for an allowable load of 123 pounds per square foot of load, besides their own weight, and were expected to rupture with about eight times this load, or, say, 1000 pounds per square foot.

Seven floor arches, with spans of 4.43 feet, were tested in this manner. Under a load of 1000 pounds per square foot none showed cracks or signs of failure. Under 1500 pounds per square foot the tests of two arches were discontinued on account of a deformation of the beams and their connections, although the arches themselves were intact. On the other arches the load was increased to about 1650 pounds per square foot, under which two failed and three remained unbroken. In each case the deflection of the crown of the arch was observed for different loads: under 1430 pounds per square foot, for example, this deflection varied between 0.39 and 0.98 inch, while for the two arches that failed the ultimate deflections were 1.0 and 1.65 inch.

The conclusions drawn from the tests of these small floor

arches are as follows: (1) That common brick arches 4.43 feet in span, with a rise of 0.44 foot, and laid with lime-mortar, show such slight deformations under a uniform load of 1430 pounds per square foot, that they afford ample security for all common buildings; (2) that ring-courses in brick arches are preferable to longitudinal courses; (3) that beton arches 3 inches thick, made of 1 part of Portland cement and 5 parts of sand, have about the same strength as brick arches 6 inches thick; (4) that flat arches give a much higher strength than expected (although the thrust upon the floor-beams is of course greater), and under careful construction they are of ample strength for all architectural purposes.

A second series of tests on arches of 8.86 feet span was conducted in a similar manner, except that the extra uniform load was applied only over one half the span. The following table gives the principal data regarding these arches, as also the load causing rupture:

Kind of Arch.	Rise, in.	Thickness, in.	Dead Load, lbs.	Applied Load, lbs., sq. ft.
Concrete	9.1	3.5	4430	1128
Monier, 1	10.2	2.0	3810	1218
Monier, 2	10.2	2.2	5410	1320
Brick, 1	9.8	5.5	4170	885
Brick, 2	5.3	3.9	3010	492
Corrugated plate, 1	9.8	....	2970	974
Corrugated plate, 2	10.2	....	2130	1100

The loads in all cases were applied gradually, and at each increment of 200 pounds per square foot the vertical and horizontal displacement of the crown of the arch was measured. The concrete arch fulfilled all expectations, and its deformation was less than one half of that for the brick arch. Of the two Monier arches, the first was built between rolled beams, while the second had solid concrete abutments, the effect of which was to greatly increase its strength. The first brick arch was of common brick, and the second of a patent brick of much less thickness; the test thus shows that brick less in thickness than the common kinds should not be employed. The first corrugated-plate arch simply butted against the floor-beams, while the second was provided at the ends with angle-irons; the deflections of these were greater than in any other arch except the brick ones.

A third series of tests on concrete and brick arches of 13.3 feet span was also undertaken, the abutments being made as nearly immovable as possible. A brick arch of 13.9 inches rise and 5.5 inches thickness was loaded over half the span. Under a load of 205 pounds per square foot the vertical deflection at the crown was 0.29 inch, and the horizontal movement was 0.11 inch. When the load reached 205 pounds per square foot, a small crack appeared on the unloaded extrados, and when it reached 275 pounds per square foot rupture occurred. A concrete arch, on the other hand, cracked at 410 pounds per square foot and ruptured at 663 pounds per square foot. A Monier arch, which is of beton built on an arched network of heavy wire or light round iron, cracked at 512 and ruptured at 894 pounds per square foot.

The Melan system, in which the beton or concrete is included, between arched I beams, was also tested, the span being 4 metres, or 13.1 feet, the rise 0.94 foot, the I beams 1 metre apart, and the thickness 3.15 inches. This arch was loaded on one side up to 1410 pounds per square foot, when the test was discontinued on account of lack of pig iron. Afterwards an area of 1 metre square over the second rib was loaded up to 3360 pounds per square foot, when failure occurred, large cracks having formed at 3100 pounds per square foot. This test shows, of course, the strength of the I rib rather than that of the total structure, yet there can be no doubt that this system is a highly efficient one, not only for floors, but for small bridges.

APPENDICES.

APPENDIX A.

INTEGRALS EMPLOYED IN THE DEDUCTION OF Δx FOR PARABOLIC ARCHES. EQUATION $p(79)$.

$$—— \int_0^x \Delta\phi dy. ——$$

Substituting the value of $\Delta\phi$ as given in $p(69)$,

$$\int_0^x \Delta\phi dy = y\Delta\phi_0 + \frac{1}{2A}\int_0^x \Big\{ 2M_1x + V_1x^2 - H_1\frac{3g-x}{3p}x^2$$

$$- \sum^x P(x-a)^2 + \sum^x Q\frac{1}{3p}[3g(x^2-a^2) - x^3 + a^3 - 6pb(x-a)] \Big\}.$$

Substituting the value of $dy = \frac{g-x}{p}dx$ [from $p(65)$],

$$\int_0^x \Delta\phi_0 dy_0 = y\Delta\phi_0 + \frac{1}{2Ap}\Big\{ \int_0^x [2M_1(g-x)x + V_1(g-x)x^2$$

$$- H_1\frac{3g-x}{3p}(g-x)x^2]dx$$

$$- \sum^x [P\int_a^x (x-a)^2(g-x)dx]$$

$$+ \frac{1}{3p}\sum^x \Big[Q\int_a^x [3g(x^2-a^2) - x^3 + a^3$$

$$- 6pb(x-a)](g-x)dx \Big] \Big\},$$

which reduces to

$$\int_0^x \Delta\phi_0 dy = y\Delta\phi_0 + \frac{x^2}{6Ap}\Big\{ M_1(3g - 2x) + V_1 x\Big(g - \frac{3x}{4}\Big)$$

$$- H_1\frac{x}{p}\Big(g^2 - gx + \frac{x^2}{5}\Big) - \sum^x P\frac{1}{4x^2}[(4g - 3x - a)(x - a)^3]$$

$$+ \frac{1}{10x^2 p}\sum^x Q[2x^5 - 10gx^4 + 10(g^2 + 2pb)x^3$$

$$+ (15ga^2 - 30pba - 5a^2)x^2 + 10(ga^3 - 3g^2a^2)x$$

$$- 30pbg(x-a)^2 + 3a^5 - 15a^4g + (20g^2 + 10pb)a^3]\Big\} \quad . \; p(71)$$

—— $et^\circ \int_0^x dx.$ ——

$$et^\circ \int_0^x dx = et^\circ x. \quad . \; . \; . \; . \; . \quad p(72)$$

—— $\frac{1}{E}\int_0^x \frac{N_x}{F_x}dx.$ ——

From (42),

$$N_x = V_x \sin\phi + H_x \cos\phi.$$

From $p(59)$,

$$\frac{1}{E} = \frac{\theta_x}{A}\frac{dx}{ds} \text{ since } \cos\phi = \frac{dx}{ds} \quad \text{and} \quad \sin\phi = \frac{dy}{ds}.$$

Then

$$\frac{1}{E}\int_0^x \frac{N_x}{F_x}dx = \int_0^x V_x \sin\phi \frac{\theta_x}{AF_x}\frac{d^2x}{ds} + \int_0^x H_x \cos\phi \frac{\theta_x}{AF_x}\frac{dx^2}{ds}$$

$$= \int_0^x V_x \frac{\theta_x}{AF_x}\frac{dx^2}{ds^2}dy + \int_0^x H_x \frac{\theta_x}{AF_x}\frac{dx^2}{ds^2}dx.$$

From $p(60)$, $m = \frac{\theta_x}{F_x}$; hence

$$\frac{1}{E}\int_0^x \frac{N_x}{F_x}dx = \frac{m}{A}\left\{\int_0^x V_x \cos^2\phi dy + \int_0^x H_x \cos^2\phi dx\right\}.$$

From (39) and (40),

$$H_x = H_1 - \overset{x}{\Sigma} Q \quad . \quad . \quad . \quad . \quad . \quad . \quad . \quad p(39)$$

and

$$V_x = V_1 - \overset{x}{\Sigma} P. \quad . \quad . \quad . \quad . \quad . \quad . \quad . \quad p(40)$$

Substituting these values in the above equation,

$$\frac{1}{E}\int_0 \frac{N_x}{F_x}dx = \frac{m}{A}\left\{\int_0^x\left[V_1 \cos^2\phi dy - \overset{x}{\Sigma}P\cos^2\phi dy\right]\right.$$

$$\left. + \int_0^x\left[H_1\cos^2\phi dx - \overset{x}{\Sigma}Q\cos^2\phi dx\right]\right\}$$

$$= \frac{m}{A}\left\{V_1\int_0^x \cos^2\phi dy + H_1\int_0^x \cos^2\phi dx\right.$$

$$\left. - \overset{x}{\Sigma}\left[P\int_a^x \cos^2\phi dy\right] - \overset{x}{\Sigma}\left[Q\int_a^x \cos^2\phi dx\right]\right\}. \quad p(73)$$

—— $\int_0^x \cos^2\phi dy.$ ——

Let $z = \tan\phi = \frac{g-x}{p} = \sqrt{2\frac{f-y}{p}}$; $\therefore \cos^2\phi = \frac{p^2}{p^2 + p^2z^2}$

$= \frac{1}{1+z^2}$, and $dx = -pdz$; hence $dy = -zpdz$, and we have

$$\int_0^x \cos^2\phi dy = \int_0^x - p\frac{z}{1+z^2}dz = -\frac{1}{2}p\log\frac{1+z^2}{1+z_0^2}.$$

Substituting the value of z,

$$\int_0^x \cos^2 \phi dy = -\frac{z}{2} p \log\left(1 - \frac{2y}{p+2f}\right).$$

In practice $\frac{f}{g}$ seldom exceeds $\frac{1}{5}$; then for this ratio $\frac{2y}{p+2f} = \frac{4}{29}$; when $y < f, \frac{2y}{p+2f} < \frac{4}{29}$. Then without sensible error we may take

$$\log\left(1 - \frac{2y}{p+2f}\right) = -\frac{2y}{p+2f}.$$

and

$$\int_0^x \cos^2 \phi dy = \frac{py}{p+2f}.* \quad . \quad . \quad . \quad . \quad p(74)$$

——— $\int_0^x \cos^2 \phi dx.$ ———

As before, let $z = \tan \phi = \frac{g-x}{p}$; then $\cos^2 \phi = \frac{1}{1+z^2}$ and $dx = -pdz$, and we have

$$\int_0^x \cos^2 \phi dx = \int_0^x \frac{-p}{1+z^2} dz = -p\Big[_0^z \tan^{-1} z = -p\Big[_{\phi_0}^{\phi} \phi.$$

Therefore

$$\int_0^x \cos^2 \phi dx = -p(\phi - \phi_0) = p(\phi_0 - \phi). \quad . \quad . \quad . \quad p(75)$$

——— $\int_a^x \cos^2 \phi dy.$ ———

* From demonstration by Prof. Weyrauch.

$$\int_a^x \cos^2\phi dy = \Big[_a^x \frac{py}{p+2f} = \Big[_a^x \frac{p}{p+2f}\frac{2g-x}{2p}x$$

$$= \frac{p}{p+2f}\left(\frac{2g-x}{2p}x - \frac{2g-a}{2p}a\right).$$

Therefore

$$\int_a \cos^2\phi dy = \frac{p}{p+2f}(y-b). \quad . \quad . \quad . \quad . \quad . \quad . \quad . \quad p(76)$$

—— $\int_a^x \cos^2 \phi dx.$ ——

$$\int_a^x \cos^2\phi dx = \Big[_a^x = -p\Big[_a \tan^{-1} z = -p\Big[_a^\phi \phi = -p(\phi - \alpha).$$

Therefore

$$\int_a^x \cos^2\phi dx = p(\alpha - \phi). \quad . \quad . \quad . \quad . \quad p(77)$$

Substituting $p(74)$, $p(75)$, $p(76)$, and $p(77)$ in $p(73)$, reducing and factoring, we obtain

$$\int_a^x \frac{N_x}{EF_x}dx = \frac{m}{A}\left\{\frac{p}{p+2f}\right\}\{V_1y + H_1(p+2f)(\phi_0 - \phi)$$

$$- \overset{x}{\Sigma}P(y-b) - \overset{x}{\Sigma}Q(p+2f)(\alpha-\phi)\}. \quad . \quad p(78)$$

Substituting $p(71)$, $p(72)$, and $p(78)$ in $p(70)$, we have

$$\Delta x = et^\circ x - y\Delta\phi_0 - \frac{x^2}{6Ap}\left\{M_1(3g-2x) + V_1x\left(g - \frac{3x}{4}\right)\right.$$

$$- H_1\frac{x}{p}\left(g^2 - gx + \frac{x^2}{5}\right) - \overset{x}{\Sigma}P\frac{1}{4x^2}[(4g-3x-a)(x-a)^3]$$

$$+\overset{x}{\Sigma}Q\frac{1}{10x^2p}[2x^5 - 10gx^4 + 10(g^2 + 2pb)x^3$$

$$+(15ga^2 - 30pba - 5a^3)x^2 + 10(ga^3 - 3g^2a^2)x$$

$$-30pbg(x-a)^2 + 3a^5 - 15a^4g + (20g^2 + 10pb)a^3]\Big\}$$

$$-\frac{m}{A}\frac{p}{p+2f}\{V_1y + H_1(p+2f)(\phi_0 - \phi) - \overset{x}{\Sigma}P(y-b)$$

$$-\overset{x}{\Sigma}Q(p+2f)(\alpha - \phi)\}. \quad \ldots\ldots\ldots\ldots \quad p(79)$$

APPENDIX B.

INTEGRALS EMPLOYED IN THE DEDUCTION OF Δy FOR PARABOLIC ARCHES. EQUATION $p(84)$.

$$\text{———} \quad et^{\circ}\int_0^x dy. \quad \text{———}$$

$$\epsilon t^{\circ}\int_0^x dy = et^{\circ}y. \quad . \quad . \quad . \quad . \quad . \quad . \quad . \qquad p(81)$$

$$\text{———} \quad \int_0^x \Delta\phi dx. \quad \text{———}$$

$$\int_0^x \Delta\phi dx = \int_0^x \Delta\phi_0 dx + \int_0^x \frac{dx}{2A}\Bigg\{ 2M_1x + V_1x^2 - H_1\frac{3g-x}{3p}x^2 - \sum^{x} P(x-a)^2 + \sum^{x} Q\frac{1}{3p}[3g(x^2-a^2) - x^3 + a^3 - 6pb(x-a)]\Bigg\}$$

or

$$\int_0^x \Delta\phi dx = x\Delta\phi_0 + \frac{x^2}{6A}\Bigg\{ 3M_1 + V_1x - H_1\frac{x}{p}\left(g - \frac{x}{4}\right) - \frac{1}{x^2}\sum^{x} P(x-a)^3 + \frac{1}{x^2p}\sum^{x} Q\Big[-\frac{x^4}{4} + gx^3 + (a^3 - 3ga^2)x - 3pb(x-a)^2 - \frac{3a^4}{4} + 2ga^3\Big]\Bigg\} . p(82)$$

$$—— \int_0^x \frac{N_x}{EF_x} dy. ——$$

$$\int_0^x \frac{N_x}{EF_x} dy = \int_0^x N_x \frac{\theta_x}{AF_x} \frac{dx}{ds} dy$$

$$= \int_0^x \frac{\theta_x}{AF_x} \frac{dx}{ds} dy (V_x \sin \phi + H_x \cos \phi)$$

$$= \int_0^x V_x \frac{\theta_x}{AF_x} \frac{dy^2}{ds^2} dx + \int_0^x H_x \frac{\theta_x}{AF_x} \frac{dx}{dy^2} dy$$

$$= \frac{m}{A} \left\{ V_x \int_0^x \sin^2 \phi dx + H_x \int_0^x \cos^2 \phi dy. \right.$$

From (39) and (40),

$$H_x = H_1 - \overset{x}{\Sigma} Q \quad . \quad . \quad . \quad . \quad . \quad . \quad . \quad (39)$$

and

$$V_x = V_1 - \overset{x}{\Sigma} P; \quad . \quad . \quad . \quad . \quad . \quad . \quad . \quad (40)$$

hence

$$\int_0^x \frac{N_x}{EF_x} dy = \frac{m}{A} \left\{ V_1 \int_0^x \sin^2 \phi dx + H_1 \int_0^x \cos^2 \phi dy \right.$$

$$\left. - \overset{x}{\Sigma} \left[P \int_a^x \sin^2 \phi dx \right] - \overset{x}{\Sigma} \left[Q \int_a^x \cos^2 \phi dy \right] \right\};$$

$$\int_0^x \sin^2 \phi dx = \int_0^x (1 - \cos^2 \phi) dx = x - p(\phi_0 - \phi);$$

$$\int_a^x \sin^2 \phi dx = \int_a^x (1 - \cos^2 \phi) dx = x - a - p(\alpha - \phi);$$

$$\int_0^x \cos \ \phi dy = \frac{py}{p + 2f};$$

$$\int_a^x \cos^2 \phi dy = \frac{p}{p + 2f} (y - b).$$

Therefore

$$\int_0^x \frac{N_x}{EF_x}dy = \frac{m}{A}\left\{ V_1[x - p(\phi_0 - \phi)] + H_1\frac{py}{p + 2f}\right.$$

$$\left. - \overset{x}{\Sigma} P[x - a - p(\alpha - \phi)] - \overset{x}{\Sigma} Q\frac{p}{p + 2f}(y - b)\right\}. \quad p(83)$$

Substituting $p(81)$, $p(82)$, and $p(83)$ in $p(80)$, we obtain $p(84)$.

APPENDIX C.

EFFECT OF THE AXIAL STRESS.

To illustrate the effect of the axial stress we will solve several examples by the common method and by the formulas which take into account the axial stress. A comparison of the results thus obtained will indicate the importance of this stress.

In the following examples let the form of the arch be *parabolic* and have a span of 100. Also, let the (radius of gyration)^2 $= 4 = m$.

ARCH WITH A HINGE AT EACH SUPPORT.

(a) *Vertical Loads.*

1°. Assume a single load on the crown of the arch, and let the rise be 10. Then $\phi_0 = 21° \, 48' = 0.38$, $p = 125$, and $k = \frac{1}{2}$.

From (64*a*),

$$H_1 = \frac{5}{8}\frac{100}{10}(0.3125)P = 1.9531P.$$

From (74),

$$H_1 = \frac{15}{85700}\{10416 - 34\}P = 1.8171P;$$

$$(1.9531 - 1.8171)P = 0.1360P;$$

$$\frac{0.1360}{1.9531} = 0.069 = \text{relative error.}$$

2°. Assume a single load acting at the crown of the arch, and let the rise be 25. Then $\phi_0 = 45° = 0.785$, $p = 50$, and $k = \frac{1}{2}$.

From (64*a*),

$$H_1 = \frac{5}{8}\frac{100}{25}(0.3125)P = 0.7813P.$$

From (74),

$$H_1 = \frac{15}{504712}\{26041 - 50\}P = 0.7724P;$$

$$(0.7813 - 0.7724)P = 0.0089P;$$

$$\frac{0.0089}{0.7813} = 0.0114 = \text{relative error.}$$

2°*a*. The same as 2°, with $k = \frac{1}{4}$.

From (64*a*),

$$H_1 = \frac{5}{8}\frac{100}{25}(0.2227)P = 0.5568P.$$

From (74),

$$H_1 = \frac{15}{504712}\{18560 - 37\}P = 0.5503P;$$

$$(0.5568 - 0.5503)P = 0.0065P;$$

$$\frac{0.0065}{0.5568} = 0.0116 = \text{relative error,}$$

which is practically the same relative error which was found in 2°.

3°. Assume a single load acting at the crown of the arch, and let the rise be 50. Then $\phi_0 = 63°\ 26' = 1.11$, $p = 25$, and $k = \frac{1}{2}$.

From (64*a*),

$$H_1 = \frac{5}{8} \cdot \frac{100}{50}(0.3125)P = 0.3904P.$$

From (74),

$$H = \frac{15}{2003330}\{52053 - 40\}P = 0.3894P;$$

$$(0.3904 - 0.3894)P = 0.001P;$$

$$\frac{0.001}{0.3904} = 0.0025 = \text{relative error.}$$

3°*a*. The same as 3°, with $k = \frac{1}{4}$.

From (64*a*),

$$H_1 = \frac{5}{8}\frac{100}{50}(0.2227)P = 0.2751P.$$

From (74),

$$H_1 = \frac{15}{2003330}\{36680 - 30\}P = 0.2744P;$$

$$(0.2751 - 0.2744)P = 0.0007P$$

$$\frac{0.0007}{0.2751} = 0.0025 = \text{relative error.}$$

The above results are tabulated below for convenience in comparison.

HINGED ARCH WITH VERTICAL LOADS.

f/l	Load at Crown, $k = \frac{1}{4}$.				Load at Quarter-point, $k = \frac{1}{4}$.			
	Values of H_1.				Values of H_1			
	(64*a*)	(74)	Diff.	Rel. error, %.	(64*a*)	(74)	Diff.	Rel. error, %.
0.10	1.9531	1.8171	0.1360	6.9				6.9
0.25	0.7813	0.7724	0.0089	1.1	0.5568	0.5503	0.0065	1.1
0.50	0.3904	0.3894	0.0010	0.2	0.2751	0.2744	0.0007	0.2

The results in the above table are probably not correct in the fourth decimal place, but for our purpose they are sufficiently exact.

The following conclusions may be drawn from the tabulated results:

1°. *The position of the load has little or no effect upon the magnitude of the relative error.*

2°. *The common method in general gives results which are too large.*

3°. *To obtain results which are not six or seven per cent too large, the formulas which consider the influence of the axial stress must be employed for flat arches.*

4°. *For arches having a rise equal to one fifth or more of the span the common formulas are sufficiently accurate.*

(*b*) *Horizontal Loads.*

A series of computations similar to those made for vertical loads indicated that for arches having a rise of one fifth or more of the span the common formulas can be employed. For flat arches the effect of the axial stress should be taken into account, as the results obtained by the common formulas are from six to ten per cent *too small* for arches having a *rise* of about *one tenth* the span.

ARCH WITHOUT HINGES.

(*a*) *Vertical Loads.*

1°. Assume a single load on the crown of the arch, and let the rise be 10. Then $\phi_0 = 21° \ 48' = 0.38$, $p = 125$, and $k = \frac{1}{2}$.

From (91*a*),

$$H_1 = \frac{150}{4}(0.0625)P = 2.3437P.$$

From (101),

$$H_1 = 0.2626\{6.25 - 0.103\}P = 1.6150P.$$

$$(2.3437 - 1.6150)P \qquad = 0.7287P.$$

$$\frac{0.7287}{2.3437} = 0.309 \qquad = \text{relative error}.$$

1°*a*. The same as 1°, with $k = \frac{1}{4}$.
From (91*a*),

$$H_1 = \frac{150}{4}(0.0351)P = 1.3162P.$$

From (101),

$$H_1 = 0.2626(3.51 - 0.077)P = 0.9007P.$$

$$(1.3162 - 0.9007)P \qquad = 0.4155P.$$

$$\frac{0.4155}{1.3162} = 31.5 \qquad = \text{relative error}$$

2°. Assume a single load on the crown of the arch and let the rise be 25. Then $\phi_0 = 45° = 0.785$, $p = 50$, and $k = \frac{1}{2}$.
From (91*a*),

$$H_1 = \frac{60}{4}(0.0625)P = 0.9375P.$$

From (101),

$$H_1 = 0.1419(6.25 - 0.06)P = 0.8784P.$$

$$(0.9375 - 0.8784)P = 0.0591P.$$

$$\frac{0.0591}{0.9375} = 0.063 = \text{relative error}.$$

2°*a*. The same as 2°, with $k = \frac{1}{4}$.
From (91*a*),

$$H_1 = 15(0.0351)P = 0.5265P.$$

From (101),

$$H_1 = 0.1419(3.51 - 0.04)P = 0.4910P.$$

$$P(0.5265 - 0.4910) = 0.0355P.$$

$$\frac{0.0355}{0.5265} = 0.067 = \text{relative error.}$$

3°. Assume a single load on the crown of the arch and let the rise be 50. Then $\phi_0 = 1.11$, $p = 25$, and $k = \frac{1}{2}$.

From (91a),

$$H_1 = \frac{15}{4}\frac{100}{50}(0.0625)P = 0.4687P.$$

From (101),

$$H_1 = 0.074(6.25 - 0.02)P = 0.4610P.$$

$$(0.4687 - 0.4610)P = 0.0077P.$$

$$\frac{0.0077}{0.4687} = 0.017 = \text{relative error.}$$

Collecting the above results for convenience, we have the following table:

ARCH WITHOUT HINGES—VERTICAL LOADS.

f/l	Load at Crown, $k = \frac{1}{2}$.				Load at Quarter-point, $k = \frac{1}{4}$.			
	Values of H_1.				Values of H_1			
	(91a)	(101)	Diff.	Relative Error, %.	(91a)	(101)	Diff.	Relative Error, %.
0.10	2.3437	1.6150	0.7287	30.9	1.3162	0.9007	0.4155	31.5
0.25	0.9375	0.8784	0.0591	6.3	0.5265	0.4910	0.0355	6.7
0.50	0.4687	0.4610	0.0077	1.7	......	......	......	1.7

From this table the following conclusions may be drawn:

1°. *The position of the load has little or no effect upon the magnitude of the relative error.*

2°. *The common method in general gives results which are too large.*

3°. *In arches which do not have a rise equal to at least one fourth the span, the effect of the axial stress is too great to be neglected. It amounts to about thirty per cent for arches having a rise equal to one tenth their span.*

(*b*) *Horizontal Loads.*

A series of computations similar to those made for vertical loads indicated that for loads near the crown of the arch the effect of the axial stress can be neglected.

For loads near the supports and the quarter-points the effect of the axial stress amounts to at least *six* per cent for arches having a rise of *one tenth* their spans, but decreases rapidly as the ratio increases.

Since horizontal loads are usually caused by wind, and the ratio of the wind stresses to the live and dead load stresses is small (ordinarily), the common method is probably sufficiently exact for practical purposes.

CIRCULAR ARCHES.

The above conclusions are based upon examples of parabolic arches. For flat circular arches (rise less than one fourth the span) we can safely predict that practically the same conclusions will obtain, since the parabola and circle so nearly coincide. We will solve a few examples, which will show the exact effect of the axial stress upon arches of greater rise.

CIRCULAR ARCH WITH HINGE AT EACH SUPPORT.

(*a*) *Vertical Loads.*

1°. Let $l = 100$ and $f = 25$. Then

$$R = 62.5 \quad \text{and} \quad k' = 62.5 - 25 = 37.5.$$

$$\tan \phi_0 = \frac{50}{37.5} = 1.333 \ldots \qquad \therefore \phi_0 = 53^\circ\, 7\tfrac{3}{4}'.$$

$$\frac{2\phi_0}{\pi} = \frac{106.25}{180} = 0.590.$$

From Table XVII, for $\alpha = 0$, or a load on the crown:

For $\dfrac{2\phi_0}{\pi} = 0.58$, $\quad \dfrac{\mathbf{A}}{\mathbf{B}} = 0.758.$

For $\dfrac{2\phi_0}{\pi} = 0.60$, $\quad \dfrac{\mathbf{A}}{\mathbf{B}} = 0.726.$

Then for $\dfrac{2\phi_0}{\pi} = 0.59$, $\quad \dfrac{\mathbf{A}}{\mathbf{B}} = \dfrac{0.758 + 0.726}{2} = 0.742.$

From (160),

$$H_1 = P\frac{\mathbf{A}}{\mathbf{B}} = 0.742P.$$

From (164),

$$H_1 = \mathfrak{H}\, \frac{1 - \dfrac{m}{2\mathbf{A}}(\sin^2 \phi_0 - \sin^2 \alpha)}{1 + \dfrac{m}{\mathbf{B}}(\phi_0 + \sin \phi_0 \cos \phi_0)},$$

where $\mathfrak{H} = 0.742P.$

From (153),

$$m = \frac{4}{(62.5)^2} = \frac{4}{3906}, \text{ say } \frac{4}{3900}, \quad = 0.00102.$$

From Table XVIII.

$$\text{for } \phi_0 = 53^\circ, \qquad \mathbf{B} = 0.1532454$$
$$\text{for } \phi_0 = 54^\circ, \qquad \mathbf{B} = 0.1671294$$

$$60 \,|\, 0.0138840$$

$$.00023140 = \text{diff. for } 1'$$
$$7\tfrac{3}{4}$$

$$0.00179335 = \text{diff. for } 7\tfrac{3}{4}'$$
$$0.1532454$$

$$\therefore \text{ for } \phi_0 = 53^\circ\ 7\tfrac{3}{4}', \qquad \mathbf{B} = 0.1550387$$

$$\frac{\mathbf{A}}{\mathbf{B}} = 0.742. \qquad \therefore \mathbf{A} = (0.742)(0.155) = 0.115.$$

From Table XIX, by interpolation,

$$(\phi_0 + \sin\phi_0 \cos\phi_0) = 1.407$$

Substituting these values,

$$H_1 = 0.742 \frac{1 - \frac{0.00102}{0.230}(0.64 - 0)}{1 + \frac{0.00102}{0.155}(1.407)} P$$

$$= 0.742 \frac{0.997170}{1.00926} P = (0.742)(0.988)P$$

or

$$H_1 = 0.733P;$$

$$(0.742 - 0.733)P = 0.009P;$$

$$\frac{0.009}{0.742} = 0.012 = \text{relative error};$$

or for a load at the crown the results by the common method formula (160), are about 1.2 per cent TOO LARGE, which is practically the same as found for parabolic arches of the same rise.

1°*a*. Let a load be placed at the quarter-point in 1°. Then

$$k = \tfrac{1}{4}, \quad \sin\alpha = \frac{50 - 25}{62.5} = 0.4.$$

$$\therefore \quad \alpha = 23^\circ\ 35' = 23^\circ.583, \quad \frac{\alpha}{\phi_0} = \frac{23.583}{53.125} = 0.443.$$

Interpolating in Table XVII, $\frac{\mathbf{A}}{\mathbf{B}} = 0.570.$

From example 1°,

$$\mathbf{B} = 0.155$$
$$\therefore \quad \mathbf{A} = (0.155)(0.570) = 0.08835.$$

From (160),

$$H_1 = 0.570P.$$

From *c*(116), which is (164) in another form,

$$H_1 = P\frac{2\mathbf{A} - m(\sin^2\phi_0 - \sin^2\alpha)}{2\mathbf{B} + 2m(\phi_0 + \sin\phi_0\cos\phi_0)}$$

or

$$H_1 = P\frac{0.1767 - 0.00102(0.64 - 0.16)}{0.31287} = 0.563P$$

$$(0.570 - 0.563)P = 0.007P;$$

$$\frac{0.007}{0.570} = 0.012 = \text{relative error},$$

which is the same relative error obtained for a load on the crown.

2°. Assume a vertical load on the crown of a semicircular arch. $l = 100$, $f = 50$, and $k = \tfrac{1}{2}$. Let (radius gyration)$^2 = 4$, $\phi_0 = 90^\circ$, and $\alpha = 0$. Then $\frac{2\phi_0}{\pi} = 1$ and $\frac{\alpha}{\phi_0} = 0.$

From Table XVII,

$$\text{for } \frac{2\phi_0}{\pi} = 1 \quad \text{and} \quad \frac{\alpha}{\phi_0} = 0, \quad \frac{\text{A}}{\text{B}} = 0.318.$$

From (160),

$$H_1 = 0.318P.$$

From (153),

$$m = \frac{4}{2500} = 0.0016.$$

Then from (164),

$$H_1 = 0.318\frac{1 - 0.0016}{1 + 0.0016}P = 0.317P;$$

$$(0.318 - 0.317)P = 0.001P;$$

$$\frac{0.001}{0.318} = 0.003 = \text{relative error,}$$

which is too small to be of any practical importance.

APPROXIMATE FORMULAS.

for parabolic arches

Very close approximate formulas can be formed by applying correction factors to the common formulas for *vertical loads.*

Arch with a Pin at Each Support.

Let $\mathfrak{H}$ = the horizontal thrust as given by the common method; then

$$H_1 = \mathfrak{H}(1 - \epsilon),$$

where ϵ is the relative error.

Computing the value of ϵ for several problems and plotting these results, the following table can be made by means of a curve drawn through the plotted points.

Arch without Hinges.

$$H_1 = \mathfrak{H}(1 - \epsilon').$$

Results obtained by the use of the approximate formulas will be sufficiently accurate for the ordinary problems met with in practice.

VALUES OF $1 - \epsilon$ AND $1 - \epsilon'$ IN THE APPROXIMATE FORMULAS FOR H_1.

t/l	Hinged.		Fixed.	
	ϵ	$1 - \epsilon$	ϵ'	$1 - \epsilon'$
0.10	.0690	.9310	.310	.690
0.11	.0570	.9430	.265	.735
0.12	.0480	.9520	.230	.770
0.13	.0420	.9580	.207	.793
0.14	.0370	.9620	.186	.814
0.15	.0330	.9670	.170	.820
0.16	.0295	.9705	.153	.847
0.17	.0265	.9735	.140	.860
0.18	.0240	.9760	.126	.874
0.19	.0215	.9785	.115	.885
0.20	.0195	.9805	.104	.896
0.21	.0170	.9820	.094	.906
0.22	.0153	.9847	.086	.914
0.23	.0140	.9850	.078	.922
0.24	.0125	.9875	.070	.930
0.25	.0115	.9885	.065	.935
0.26	.0100	.9900	.060	.940
0.27	.0095	.9905	.058	.942
0.28	.0085	.9915	.053	.947
0.29	.0080	.9920	.050	.950
0.30	.0074	.9926	.047	.953
0.31	.0068	.9932	.043	.957
0.32	.0063	.9937	.040	.960
0.33	.0060	.9940	.038	.962
0.34	.0056	.9944	.035	.965
0.35	.0052	.9948	.034	.966
0.50	.0020	.9980	.020	.980

The above table has been computed for a span of 1: and a (radius of gyration)^2 = 4; for any other v…
of l and m the tabular quantities must be multiplied by the quantity 2500 $\frac{m}{l^2}$.

APPENDIX D.

SPECIAL CASE—SEMICIRCULAR ARCH.

$$\frac{2E\theta}{R} = \text{a constant.}$$

SINCE semicircular arches are sometimes employed for large roof-supports, we will give the necessary formulas for determining the outer forces. *The effect of the axial stress will be omitted, as its effect can be neglected in practice.* See Appendix C.

ARCH WITH FIXED ENDS.

(*a*) *Vertical Loads.*

Since $\phi_0 = \frac{\pi}{2}$, $\sin \phi_0 = 1$, and $\cos \phi_0 = 0$.

Then from $c(133)$,

$$H_1 = \tfrac{1}{2}\Sigma P \left\{ \frac{2(\cos \alpha + \alpha \sin \alpha) - \frac{\pi}{2} - \frac{\pi}{2}\sin^2 \alpha}{\frac{\pi^2}{4} - 2} \right\}$$

and

$$M_1 = \frac{2H_1R}{\pi} + \frac{\overset{l}{\Sigma} PR}{\pi}\left\{ \begin{array}{l} \sin \alpha\left(\frac{\pi}{2} - \cos \alpha\right) - \alpha \\ + \cos \alpha + \alpha \sin \alpha - \frac{\pi}{2} \end{array} \right\}.$$

By making α negative (hence the $\sin \alpha$ will be negative) in the value for M_1 we have

$$M_2 = \frac{2H_1R}{\pi} + \frac{\overset{l}{\Sigma}PR}{\pi}\left\{\begin{array}{l}\sin\alpha\left(\cos\alpha - \frac{\pi}{2}\right) + \alpha \\ + \cos\alpha + \alpha\sin\alpha - \frac{\pi}{2}\end{array}\right\}.$$

For any single load we have, from (51),

$$y_1 = \frac{M_1}{H_1}.$$

Substituting the values of M_1 and H_1 found above, we obtain after reduction

$$y = \frac{2R}{\pi} + \frac{\left(\frac{\pi^2}{4} - 2\right)\left[\begin{array}{l}\sin\alpha\left(\frac{\pi}{2} - \cos\alpha\right) - \alpha \\ + \cos\alpha + \alpha\sin\alpha - \frac{\pi}{2}\end{array}\right]R}{\frac{\pi}{2}\left(\pi - 2\cos\alpha - 2\alpha\sin\alpha + \frac{\pi}{2}\sin^2\alpha\right)};$$

and by making α negative in the expression for y_1,

$$y_2 = \frac{2R}{\pi} + \frac{\left(\frac{\pi^2}{4} - 2\right)\left[\begin{array}{l}\alpha - \sin\alpha\left(\frac{\pi}{2} - \cos\alpha\right) \\ + \cos\alpha + \alpha\sin\alpha - \frac{\pi}{2}\end{array}\right]R}{\frac{\pi}{2}\left(\pi - 2\cos\alpha - 2\alpha\sin\alpha + \frac{\pi}{2}\sin^2\alpha\right)}.$$

From (50),

$$y_0 = \frac{M_1 + V_1a}{H_1}$$

Substituting the value of V_1 from (47) and reducing, this becomes

$$y_0 = \frac{1}{2}(1 + \sin\phi)y_1 + \frac{1}{2}(1 - \sin\phi)y_2 + \frac{PR}{2H_1}\cos^2\alpha.$$

The values of $\cos\alpha + \alpha\sin\alpha$ can be found from Table XXII.

(*b*) *Horizontal Loads.*

From *c*(154),

$$H_1 = \frac{\overset{l}{\Sigma} Q}{2}\left\{1 + \frac{\frac{\pi}{2}(\sin\alpha\cos\alpha - \alpha) + 2(\sin\alpha - \alpha\cos\alpha)}{2 - \frac{\pi^2}{4}}\right\}.$$

From *c*(156),

$$M_1 = \frac{2H_1R}{\pi} + \frac{\overset{l}{\Sigma} QR}{\pi}\left\{\begin{array}{l}\alpha\cos\alpha - \sin\alpha - 1\\ + \frac{\pi}{2}\cos\alpha - \cos^2\alpha\end{array}\right\}.$$

By making α negative in the above equations, they become

$$H_2 = \frac{\overset{l}{\Sigma} Q}{2}\left\{1 + \frac{\frac{\pi}{2}(\alpha - \sin\alpha\cos\alpha) + 2(\alpha\cos\alpha - \sin\alpha)}{2 - \frac{\pi^2}{4}}\right\}$$

and

$$M_2 = \frac{2H_2R}{\pi} + \frac{\overset{l}{\Sigma} QR}{\pi}\left\{\begin{array}{l}\sin\alpha - \alpha\cos\alpha - 1\\ + \frac{\pi}{2}\cos\alpha - \cos^2\alpha\end{array}\right\}$$

From (47),

$$V_1 = \frac{1}{l}\left\{M_2 - M_1 + \overset{l}{\Sigma} Qb\right\}.$$

The values of x_1, y_1, y_2, and x_2 can be found from (51) and (54).

(c) Effect of a Change in Temperature.

From $c(159)$,

$$H_1 = \frac{2A\frac{\pi}{2}let^\circ}{4R^2\left(\frac{\pi^2}{4}-2\right)} = \frac{A\frac{\pi}{2}et^\circ}{R\left(\frac{\pi^2}{4}-2\right)}.$$

From $c(161)$,

$$M_1 = \frac{2H_1R}{\pi} = \frac{Aet^\circ}{\frac{\pi^2}{4} - 2}.$$

From (51),

$$y_1 = \frac{M_1}{H_1} = \frac{2R}{\pi} = 0.632R.$$

ARCH WITH A HINGE AT EACH SUPPORT.

(a) Vertical Loads.

From $c(108)$,

$$H_1 = \overset{l}{\Sigma}P\left\{\frac{\frac{1}{2}\cos^2\alpha}{\frac{\pi}{2}}\right\} = \overset{l}{\Sigma}P\frac{\cos^2\alpha}{\pi}.$$

From $c(112)$,

$$V_1 = \overset{l}{\Sigma}P\frac{1 + \sin\alpha}{2}.$$

From (50),

$$y_0 = \frac{V_1}{H_1}a = \frac{1 + \sin\alpha}{2}\,\frac{\pi}{\cos^2\alpha}R(1 - \sin\alpha)$$

or

$$y_0 = \tfrac{1}{2}\pi R.$$

(*b*) *Horizontal Loads.*

From $c(120)$

$$H_1 = \frac{1}{2}\sum^{l} Q\left\{1 + \frac{\alpha - \sin\alpha\cos\alpha}{\frac{\pi}{2}}\right\}.$$

From $c(121)$,

$$V_1 = \sum^{l} Q\frac{\cos\alpha}{2}.$$

From $c(123)$,

$$x_0 = R\left\{1 + \frac{\alpha - \sin\alpha\cos\alpha}{\frac{\pi}{2}}\right\}.$$

The values of $\alpha - \sin\alpha\cos\alpha$ can be found from Table XIX.

(*c*) *Effect of a Change in Temperature.*

From $c(128)$,

$$H_1 = \frac{et^\circ A}{R}\cdot\frac{2}{\pi}.$$

APPENDIX E.

DEDUCTION OF FORMULAS FOR SPECIAL CASES FROM THE GENERAL FORMULAS OF CHAPTER V.

SYMMETRICAL PARABOLIC ARCH.

(a) Arch without Hinges. Special Case where
$\theta \cos \phi =$ a constant.

LET $\theta \cos \phi = A =$ a constant, and neglect the terms containing F_x; then for a *single vertical load* we have from $g(90)$, page 117,

Value of H_1.—

$$H_1 = \frac{\left[(1 - k\int_0^a yxdx + k\int_a^l y(l-x)dx - \dfrac{(1 - k\int_0^a xdx + k\int_a^l (l-x)dx}{\int_0^l dx} \int_0^l ydx \right]}{\int_0^l y^2dx - \dfrac{\left(\int_0^l ydx\right)^2}{\int_0^l dx}} P.$$

Substituting the value of y in terms of x, the following values of the respective integrals are easily obtained:*

$$(1 - k)\int_0^a yxdx = \frac{1-k}{3} fl^2(4k^3 - 3k^4);$$

$$k\int_a^l (l - x)ydx = \frac{1}{3} fl^2k(1 - 6k^2 + 8k^3 - 3k^4);$$

* The general formulas for the parabola are given on pages 52 and 53.

$$(1 - k)\int_0^a x dx = \tfrac{1}{2}l^2k^2(1 - k);$$

$$k\int_a^l (l - x)dx = \tfrac{1}{2}l^2k(1 - 2k + k^2);$$

$$\int_0^l dx = l; \qquad \int_0^l y dx = \tfrac{2}{3}fl; \qquad \int_0^l y^2 dx = \tfrac{8}{15}f^2 l.$$

Substituting these integrations in the expression for H_1 and reducing,

$$H_1 = \frac{\frac{1}{3}fl^2(k - 2k^3 + k^4) - \frac{1}{3}fl^2(k - k^2)}{\frac{12}{135}f^2 l}P$$

or

$$H_1 = \frac{15}{4}\frac{l}{f}Pk^2(1 - k)^2. \qquad \qquad (91)$$

Value of M_1.—For a *single vertical load*, from *g*(101), page 119, we have

$$M_1 = \frac{\begin{matrix} -H_1\int_0^l yxdx\int_0^l xdx - P\int_a^l (x - a)xdx\int_0^l xdx \\ + H_1\int_0^l ydx\int_0^l x^2dx + P\int_a^l (x - a)dx\int_0^l x^2dx \end{matrix}}{\int_0^l dx\int_0^l x^2dx - \left(\int_0^l xdx\right)^2}.$$

Replacing y in terms of x, the following integrations are easily obtained:

$$\int_0^l yxdx = \tfrac{1}{3}l^2f; \qquad \int_0^l xdx = \tfrac{1}{2}l^2;$$

$$\int_a^l (x - a)xdx = \tfrac{1}{6}l^3(2 - 3k + k^3);$$

$$\int_0^l ydx = \tfrac{2}{3}fl; \qquad \int_0^l x^2dx = \tfrac{1}{3}l^3;$$

$$\int_a^l (x - a)dx = \tfrac{1}{2}l^2(1 - k)^2.$$

Substituting these values, and that for H_1 given above, we obtain

$$M_1 = \frac{\frac{l^6}{24}(-2k+9k^2-12k^3+5k^4)}{\frac{1}{12}l^4}P$$

or

$$M_1 = \frac{P}{2}lk(1-k)^2(5k-2). \quad . \quad . \quad . \quad . \quad . \quad . \quad (92)$$

Value of H_1.—For a *single horizontal load* we have, from $g(95)$, page 118,

$$H_1 = \frac{Q}{2}\left\{1 - \frac{\int_{a_2}^{a_1}(y-b)ydx - \frac{\int_{a_2}^{a_1}(y-b)dx}{l}\frac{2}{3}fl}{\frac{12}{135}f^2l}\right\}.$$

We have introduced the values of the integrals which have been determined above.

The values of the two remaining integrals are as follows:

$$\int_{a_2}^{a_1}(y-b)ydx = \frac{8f^2l}{15}(-1+5k-5k^2-10k^3+20k^4-8k^5),$$

$$\frac{2}{3}f\int_{a_2}^{a_1}(y-b)dx = \frac{4f^2l}{9}(-1+6k-12k^2+8k^3).$$

Hence

$$H_1 = Q[1+k^2(-15+50k-60k^2+24k^3)]. \quad . \quad (115)$$

Value of M_1.—For a *single horizontal load*, from $g(106)$, page 120, we have

$$M_1 = \frac{\left[\begin{array}{l}\left(-H_1\int_0^l yxdx + Q\int_a^l(y-b)xdx\right)\int_0^l xdx \\ \left(+H_1\int_0^l ydx - Q\int_a^l(y-b)dx\right)\int_0^l x^2dx\end{array}\right]}{\left(\int_0^l dx\int_0^l xdx - \int_0^l xdx\int_0^l xdx\right) = \frac{1}{12}l^4}$$

Performing the integrations indicated,

$$\int_0^l yxdx \int_0^l xdx = \tfrac{1}{6}l^4f;$$

$$\int_0^l ydx \int_0^l x^2dx = \tfrac{2}{9}l^4f;$$

$$\int_a^l (y - b)xdx = \frac{fl^2}{3}(1 - 6k + 6k^2 + 2k^3 - 3k^4);$$

$$\int_a^l (y - b)dx = \frac{2fl}{3}(1 - 6k + 9k^2 - 4k^3).$$

Therefore

$$M_1 = Qf\{2k(1 - k)^2(2 - 7k + 8k^2)\}. \quad . \quad . \quad (117)$$

The value of H_1 is given by (115).

(b) Arch with a Hinge at each Support.

As for the arch without hinges let $\theta \cos \phi =$ a constant, and neglect the terms containing F_x.

Value of H_1.—For a *single vertical load* we have, from g(131), page 135,

$$H = \frac{(1 - k)\int_0^l xydx - \int_a^l (x - a)ydx}{\int_0^l y^2dx}P.$$

The values of the integrals are

$$(1 - k)\int_0^l xydx = \frac{1 - k}{3}l^2f, \quad \int_0^l y^2dx = \frac{8}{15}f^2l,$$

$$\int_a^l (x - a)ydx = \frac{fl^2}{3}(1 - 2k + 2k^3 - k^4).$$

Hence

$$H_1 = \frac{5}{8}\frac{l}{f}Pk(1 - 2k^2 + k^3). \quad . \quad . \quad . \quad (64)$$

Value of H_1.—For a *single horizontal load* we have, from $g(140)$, page 127,

$$H_1 = \frac{Q}{2}\left\{1 + \frac{\int_{a_1}^{a_2}(y - b)y\,dx}{\int_0^l y^2dx}\right\}.$$

The value of the integral in the numerator of the fraction is given above in the deduction of (115); the denominator equals $\frac{8}{15}f^2l$; hence

$$H_1 = Q\left\{1 - \frac{k}{2}(5[1 - k - 2k^2 + 4k^3] - 8k^4)\right\}. \quad . \quad (77)$$

SYMMETRICAL CIRCULAR ARCH.

(*a*) *Arch without Hinges—Special Case where* $\frac{2E}{R}\theta = a$ *constant.*

Value of H_1.—For a *single vertical load* we have, from $g(90)$, page 117, neglecting the terms containing F_x,

$$H_1 = \frac{P\left[-(1-k)\int_0^a xy\,d\phi - k\int_a^l y(l-x)d\phi \atop - \frac{-(1-k)\int_0^a x\,d\phi - k\int_a^l (l-x)d\phi}{-\int_0^l d\phi}\left(-\int_0^l y\,d\phi\right)\right]}{-\int_0^l y^2d\phi - \frac{\left(-\int_0^l y\,d\phi\right)^2}{-\int_0^l d\phi}}.$$

Performing the integrations indicated, we have, remembering that $a = kl$,

$$-\int_0^a xyd\phi = -\tfrac{1}{2}k'l\phi_0 + \tfrac{1}{2}k'l\alpha + \frac{a^2}{2} + bk';$$

$$-(1-k)\int_0^a xyd\phi = -\tfrac{1}{2}k'l\phi_0 + \tfrac{1}{2}k'l\alpha + \frac{a^2}{2} + bk' + \tfrac{1}{2}k'a\phi_0$$
$$-\tfrac{1}{2}k'a\alpha - \frac{a^2}{2l} - \frac{ab}{l}k';$$

$$-k\int_a^l y(l-x)d\phi = \tfrac{1}{2}al - a^2 + \frac{a^2}{2l} - \tfrac{1}{2}ak'\phi_0$$
$$-\tfrac{1}{2}ak'\alpha + \frac{ab}{l}k'.$$

Combining these values we obtain

$$\tfrac{1}{2}al - \tfrac{1}{2}k'l\phi_0 + \tfrac{1}{2}k'l\alpha + bk' - ak'\alpha - \frac{a^2}{2}.$$

$$-(1-k)\int_0^a xd\phi = -\tfrac{1}{2}l\alpha + \tfrac{1}{2}l\phi_0 - b + \tfrac{1}{2}a\alpha - \tfrac{1}{2}a\phi_0 + \frac{ab}{l};$$
$$-k\int_0^l (l-x)d\phi = \tfrac{1}{2}a\phi_0 + \tfrac{1}{2}a\alpha - \frac{ab}{l}.$$

Combining these two values, we have

$$a\alpha - \tfrac{1}{2}l\alpha + \tfrac{1}{2}l\phi_0 - b.$$

$$-\int_0^l yd\phi = l - 2k'\phi_0;$$

$$-\int_0^l d\phi \quad = 2\phi_0;$$

$$-\int_0^l y^2d\phi = \tfrac{1}{2}(4k'^2\phi_0 + 2R^2\phi_0 - 3k'l);$$

$$\left(-\int_0^l yd\phi\right)^2 = l^2 - 4k'l\phi_0 + 4k'^2\phi_0^2.$$

Substituting the values found above for the integrals indicated in the expression for H_1 it becomes, after reduction,

$$H_1 = P\frac{2bl - l(l-2a)(\phi_0 - \alpha) - 2a^2\phi_0}{2(ld\phi_0 + k'l\phi_0 - l^2)}, \quad . \quad c(132)$$

which readily reduces to (192).

Value of M_1.—For a *single vertical load* we have, from $g(101)$, page 119,

$$M_1 = \frac{+\left[-H_1\left\{-\int_0^l yxd\phi\right\} - P\left\{-\int_a^l (x-a)xd\phi\right\}\right]\left(-\int_0^l xd\phi\right) + \left[+H_1\left\{-\int_0^l yd\phi\right\} + P\left\{-\int_a^l (x-a)d\phi\right\}\right]\left(-\int_0^l x^2d\phi\right)}{-\int_0^l d\phi\left\{-\int_0^l x^2d\phi\right\} - \left\{-\int_0^l xd\phi\right\}}.$$

The values of the integrals not already found above are

$$-\int_0^l yxd\phi = \tfrac{1}{2}l^2 - k'l\phi_0;$$

$$-\int_0^l xd\phi = l\phi_0;$$

$$-\int_0^l x^2d\phi = \tfrac{1}{2}l^2\phi_0 - k'l + R^2\phi_0.$$

Then the terms containing H_1 reduce to

$$H_1\tfrac{1}{2}(l[d-k'][l-2k'\phi_0]),$$

where $\qquad d = \dfrac{2R^2\phi_0}{l}.$

$$-\int_a^l (x-a)xd\phi = \tfrac{1}{4}l^2\phi_0 + \tfrac{1}{4}l^2\alpha - \tfrac{1}{2}k'l - \tfrac{1}{2}al\phi_0 - \tfrac{1}{2}al\alpha$$
$$+ bk' + \tfrac{3}{4}bl + \tfrac{1}{2}ak' - \tfrac{1}{2}ab + \frac{R^2}{2}\phi_0 + \frac{R^2\alpha}{2};$$

$$-\int_0^l x d\phi = l\phi_0;$$

$$-\int_a^l (x-a)d\phi = \tfrac{1}{2}l\phi_0 + \tfrac{1}{2}l\alpha - a\phi_0 - a\alpha + b;$$

$$-\int_0^l x^2 d\phi = \tfrac{1}{2}l^2\phi_0 - \tfrac{1}{2}k'l + R^2\phi_0.$$

Then for the terms containing P we have

$$\frac{Pl}{2}\{(l-2a)(b-d)\phi_0 + 2R^2\alpha\phi_0 + (k'-d)(2b-2a\alpha-l\phi_0+l\alpha)\}.$$

The denominator becomes $l\phi_0(d-k')$. Hence

$$M_1 = H_1\left(\frac{l}{2\phi_0} - k'\right) + P\frac{\begin{array}{c}(l-2a)(b-d)\phi_0 + 2R^2\alpha\phi_0 \\ +(k'-d)(2b-2a\alpha-l\phi_0+l\alpha)\end{array}}{4\phi_0(k'-d)}.$$

which reduces to (196).

Value of H_1.—For a *single horizontal load* we have, from g(95), page 118, remembering that $R(-d\phi) = ds$,

$$H_1 = \frac{Q}{2}\left\{1 - \frac{-\int_{a_1}^{a_2}(y-b)yd\phi - \dfrac{-\int_{a_1}^{a_2}(y-b)d\phi}{-\int_0^l d\phi}\left(-\int_0^l yd\phi\right)}{-\int_0^l y^2 d\phi - \dfrac{\left(-\int_0^l yd\phi\right)^2}{-\int_0^l d\phi}}\right\}.$$

From integrals already evaluated the denominator of this at once becomes

$$-\frac{1}{2\phi_0}(l^2 - k'l\phi_0 - 2R^2\phi_0^2).$$

$$-\int_{a_1}^{a_2}(y-b)yd\phi = -2\int_{a}^{\frac{1}{2}l}(y-b)yd\phi =$$

$$-\tfrac{3}{4}k'l - \tfrac{1}{2}bl + 3ak' + ab + R^2\alpha + 2k'^2\alpha + 2bk'\alpha\,;$$

$$-\int_{a_1}^{a_2}(y-b)d\phi = -2\int_{a}^{\frac{1}{2}l}(y-b)d\phi = l - 2a - 2k'\alpha - 2b\alpha.$$

Making the proper substitutions, we obtain

$$H_1 = \frac{l^2 - al - lk'\alpha - lb\alpha - ak'\phi_0 + \tfrac{1}{2}bl\phi_0 - ab\phi_0 - R^2\phi_0\alpha - R^2\phi_0^2}{l^2 - k'l\phi_0 - 2R^2\phi_0},$$

which reduces to (207).

Value of M_1.—For a *single horizontal load* we have, from g(106), page 120,

$$M_1 = \frac{+\left\{-H_1\int_0^l xyd\phi + Q\int_a^l(y-b)xd\phi\right\}\int_0^l xd\phi + \left\{H_1\int_0^l yd\phi - Q\int_a^l(y-b)d\phi\right\}\int_0^l x^2d\phi}{\int_0^l d\phi\int_0^l x^2d\phi - \left(\int_0^l xd\phi\right)^2}$$

From integrals already evaluated the denominator becomes

$$-l\phi_0(k'-d),$$

where $$d = \frac{2R^2\phi_0}{l}.$$

$$\int_a^l(y-b)xd\phi = -\tfrac{1}{2}(l^2 - a^2 - (k'+b)[l(\phi_0+\alpha)+2b]);$$

$$\int_a^l(y-b)d\phi = -(l - a - b\alpha - k'\alpha - k'\phi_0).$$

The integrals in the terms containing H_1 have been evaluated above.

Substituting the values determined above, we have

$$M_1 = H_1\left(\frac{l}{2\phi_0} - k'\right) + \frac{Q}{2(k'-d)}\{a^2 - al + 3bk' - bd + 2b^2\}$$
$$- \frac{Q}{2\phi_0}\{l - a - b\alpha - k'(\phi_0 + \alpha)\}, \quad . \quad . \quad c(155)$$

which reduces to (212).

(*b*) *Arch with a Hinge at each Support—Special Case where* $\frac{2E}{R}\theta = a$ *constant.*

Value of H_1.—For a single vertical load we have, from $g(131)$, page 125,

$$H_1 = \frac{(1-k)\left\{-\int_0^l xyd\phi\right\} + \int_a^l (x-a)yd\phi}{-\int_0^l y^2 d\phi} P;$$

$$-\int_0^l xyd\phi = \tfrac{1}{2}l^2 - k'l\phi_0;$$

$$\int_a^l (x-a)yd\phi = -\left(\tfrac{1}{2}l^2 - \tfrac{1}{2}k'l\phi_0 + \frac{a^2}{2} - al + ak'\phi_0\right.$$
$$\left. - \tfrac{1}{2}k'l\alpha - bk' + ak'\alpha\right).$$

Therefore

$$\text{Numerator} = \tfrac{1}{2}a(l - a - 2k'\alpha) - \tfrac{1}{2}k'(l\phi_0 - l\alpha - 2b).$$

$$-\int_0^l y^2 d\phi = \tfrac{1}{2}(4k'^2\phi_0 + 2R^2\phi_0 - 3k'l).$$

Hence

$$H_1 = P\frac{a(l - a - 2k'\alpha) - k'(l\phi_0 - l\alpha - 2b)}{4k'^2\phi_0 + 2R^2\phi_0 - 3k'l}, \qquad c(106)$$

which reduces to (160).

Value of H_1.—For a *single horizontal* load we have, from $g(140)$, page 127,

$$H_1 = Q\tfrac{1}{2}\left\{1 + \frac{+\left\{-\int_{a_1}^{a_2}(y-b)y\,d\phi\right\}}{-\int_0^l y^2\,d\phi}\right\}.$$

$$-\int_{a_1}^{a_2}(y-b)y\,d\phi = -2\int_a^{\frac{1}{2}l}(y-b)y\,d\phi$$

$$= -R^3(-\alpha + \sin\alpha\cos\alpha + 2\cos\phi_0[\sin\alpha - \alpha\cos\alpha]).$$

$$-\int_0^l y^2\,d\phi = R^3(\phi_0 - 3\cos\phi_0\sin\phi_0 + 2\cos^2\phi_0\phi_0).$$

Hence

$$H_1 = \frac{Q}{2}\left\{1 + \frac{\alpha - \sin\alpha\cos\alpha - 2\cos\phi_0(\sin\alpha - \alpha\cos\alpha)}{\phi_0 - 3\cos\phi_0\sin\phi_0 + 2\cos^2\phi_0\phi_0}\right\}. \quad (172)$$

APPENDIX F.

EFFECT OF A COUPLE UPON A SYMMETRICAL ARCH.

(a) *Arch with a Hinge at each Support.*

Value of H_1.—Let M_a be any couple applied at any point a of the arch: then

$$V_1 l = M_a \quad \text{or} \quad V_1 = \frac{M_a}{l}.$$

Evidently V_2 is numerically equal to V_1, but acting in the opposite direction, and $H_1 = H_2$.

If another couple equal and symmetrical to M_a be placed upon the arch,

$$V_1 = 0 = V_2 \quad \text{and} \quad \mathfrak{H}_1 = 2H_1 = \left\{\begin{array}{l}\text{the horizontal thrust at the}\\ \text{left support.}\end{array}\right.$$

If the arch be assumed free to slide upon the supports, the change in the length of the span due to a horizontal load Q' applied at each support is given by g(116), page 122, or

$$\Delta' l = -\frac{Q'}{E}\int_0^l \frac{y^2 ds}{\theta_x} - \frac{Q'}{E}\int_0^l \frac{dx \cos\phi}{F_x}. \quad . \quad g(116)$$

Now let the loads Q' be removed and two equal and symmetrical moments be applied to the arch; the corresponding change in the length of the span will be

$$\Delta'' l = \frac{1}{E}\int_0^l \frac{M_x}{\theta_x} y ds - \frac{1}{E}\int_0^l \frac{N_x}{F_x} dx, \quad . \quad . \quad . \quad g(117)$$

where M_x is the resultant moment at any section x.

If $\mathfrak{H}_1$ represents the magnitude of the horizontal thrust necessary to cause a change in the length of the span of the loaded arch of $\Delta''l$, we have

$$\Delta'l : \Delta''l :: Q' : \mathfrak{H}_1 = \frac{\Delta''l}{\Delta'l}Q'.$$

Substituting the values of $\Delta'l$ and $\Delta''l$, we have

$$\mathfrak{H}_1 = \frac{\int_0^l \frac{M_x}{\theta_x} y\,ds + \int_0^l \frac{N_x}{\theta_x} dx}{\int_0^l \frac{y^2 ds}{\theta_x} + \int_0^l \frac{dx}{\theta_x} \cos\phi}.$$

$N_x = 0$ and $M_x = M_a$ from $x = a_1$ to $x = a_2$.

Then since $H_1 = \frac{1}{2}\mathfrak{H}_1$,

$$H_1 = \frac{M_a \int_a^{\frac{1}{2}l} \frac{y\,ds}{\theta_x}}{2\left\{\int_0^{\frac{1}{2}l} \frac{y^2 ds}{\theta_x} + \int_0^{\frac{1}{2}l} \frac{dx}{\theta_x} \cos\phi\right\}}.$$

This equation is perfectly general for any symmetrical arch having a pin at each support.

(*b*) *Arch without Hinges.*

Value of M_1.—Let a couple M_a be applied at any point a on the arch; then the moment at any point x is

$$M_x = M_1 + V_1 x - H_1 y + M_a \ldots x > a.$$

Since the arch is fixed at the ends $\Delta\phi_0 = \Delta\phi_l$, and as it is symmetrical $\Delta c = 0$. Substituting the value of M_x in $g(62)$ and $g(64)$, page 111, we obtain, neglecting the axial stress term,

$$M_1 \int_0^l \frac{ds}{\theta_x} + V_1 \int_0^l \frac{x\,ds}{\theta_x} - H_1 \int_0^l \frac{y\,ds}{\theta_x} + M_a \int_0^l \frac{ds}{\theta_x} = 0$$

and

$$M_1\int_0^l\frac{xds}{\theta_x}+V_1\int_0^l\frac{x^2ds}{\theta_x}-H_1\int_0^l\frac{xds}{\theta_x}+M_a\int_0^l\frac{xds}{\theta_x}=0.$$

Eliminating V_1 and solving for M_1, we have

$$M_1=\frac{+\left\{M_a\int_0^l\frac{ds}{\theta_x}-H\int_0^l\frac{yds}{\theta_x}\right\}\int_0^l\frac{x^2ds}{\theta_x}-\left\{M_a\int_0^l\frac{xds}{\theta_x}-H_1\int_0^l\frac{xds}{\theta_x}\right\}\int_0^l\frac{xds}{\theta_x}}{\int_0^l\frac{ds}{\theta_x}\int_0^l\frac{x^2ds}{\theta_x}-\left(\int_0^l\frac{xds}{\theta_x}\right)^2}.$$

in which everything is known excepting H_1, which can be determined as follows:

Value of H_1.—Let two equal and symmetrical couples act upon the arch, and assume the arch free to slide upon the supports. Also assume that there are equal and symmetrical moments applied at the supports. Then from g(62) we have

$$\int_0^l\frac{M_xds}{\theta_x}=0.$$

Since the arch is free to slide upon the supports, $H_1=0$; and since the applied couples are equal and symmetrical, $V_1=0$. Therefore

$$M_x=M_1+K',$$

where K' is the additional moment at the section x caused by the action of the applied couples.

Substituting the value of M_x and solving for M_1

$$M_1=-\frac{\int_0^l\frac{K'ds}{\theta_x}}{\int_0^l\frac{ds}{\theta_x}}.$$

The change in the length of the span due to our couples is (see $g(80)$)

$$\Delta' l = \frac{1}{E}\frac{\int_0^l \frac{K' ds}{\theta_x}}{\int_0^l \frac{ds}{\theta_x}}\int_0^l \frac{y ds}{\theta_x} - \frac{1}{E}\int_0^l \frac{K' y ds}{\theta_x} - \frac{1}{E}\int_0^l \frac{N_x dx}{F_x}.$$

Now suppose the arch unloaded and free to slide as before. Let two equal and symmetrical moments $Q'z$ be applied at the supports; then the corresponding change in the length of the span is given by $g(86)$, page 115,

$$\Delta'' l = -\frac{Q'}{E}\left\{\int_0^l \frac{y^2 ds}{\theta_x} + \int_0^l \frac{dx \cos\phi}{F_x} - \frac{\left(\int_0^l \frac{y ds}{\theta_x}\right)^2}{\int_0^l \frac{ds}{\theta_x}}\right\}. \quad g(86)$$

Let $\mathfrak{H}_1$ be the horizontal thrust necessary to cause a change in the length of span of $\Delta' l$; then

$$\Delta'' l : \Delta' l :: Q' : \mathfrak{H} = Q'\frac{\Delta' l}{\Delta'' l}$$

or

$$\mathfrak{H}_1 = \frac{\int_0^l \frac{K' y ds}{\theta_x} + \int_0^l \frac{N_x dx}{F_x} - \frac{\int_0^l \frac{K' ds}{\theta_x}}{\int_0^l \frac{ds}{\theta_x}}\int^l \frac{y ds}{\theta_x}}{\int_0^l \frac{y^2 ds}{\theta_x} + \int_0^l \frac{dx \cos\phi}{F_x} - \frac{\left(\int_0^l \frac{y ds}{\theta_x}\right)^2}{\int_0^l \frac{ds}{\theta_x}}}$$

$$N_x = 0.$$

For $x = 0$ to $x = a_1$, $K' = 0$;
For $x = a_1$ to $x = a_2$, $K' = M_a$;
For $x = a_2$ to $x = l$, $K' = 0$.

Therefore since H_1 for a single couple equals $\frac{1}{2}\mathfrak{H}_1$, we have

$$H_1 = \frac{M_a \int_a^{l/2} \frac{yds}{\theta_x} - \frac{M_a \int_a^{l/2} \frac{ds}{\theta_x}}{\int_0^{l/2} \frac{ds}{\theta_x}} \int_0^{l/2} \frac{yds}{\theta_x}}{2\left\{ \int_0^{l/2} \frac{y^2 ds}{\theta_x} + \int_0^{l/2} \frac{ax \cos\phi}{F_x} - \frac{\left(\int_0^{l/2} \frac{yds}{\theta_x}\right)^2}{\int_0^{l/2} \frac{ds}{\theta_x}} \right\}}.$$

This equation is perfectly general for any symmetrical arch which has no hinges.

PARABOLIC ARCH.

$$\theta \cos \phi = \text{a constant.}$$

(a) *Arch with a Hinge at each Support.*

Value of H_1.—Our general expression immediately becomes, neglecting the axial stress term,

$$H_1 = \frac{M_a \int_a^{l/2} ydx}{2\int_0^{l/2} y_2 dx}$$

$$\int_a^{l/2} ydx = \frac{1}{3} fl(1 - 6k^2 + 4k^3)$$

$$2\int_0^{l/2} y^2 dx = \frac{8}{15} f^2 l.$$

Therefore $$H_1 = M_a \frac{5}{8f}(1 - 6k^2 + 4k^3).$$

(*b*) *Arch without Hinges.*

Value of H.—Neglecting the term which contains F_x, we have

$$H_1 = \frac{M_a\int_a^{l/2} y dx - \dfrac{M_a\int_a^{l/2} dx}{\int_0^{l/2} dx}\int_0^{l/2} y dx}{2\left\{\int_0^{l/2} y^2 dx - \dfrac{\left(\int_0^{l/2} y dx\right)^2}{\int_0^{l/2} dx}\right\}}$$

From Appendix E, page 290, the denominator is found to be $\frac{12}{135} f^2 l$.

$$\int_a^{l/2} y dx = \frac{1}{3} fl(1 - 6k^2 + 4k^3):$$

$$\int_a^{l/2} dx = \frac{l}{2}(1 - 2k ;$$

$$\int_0^{l/2} y dx = \frac{1}{3} fl;$$

$$\int_0^{l/2} dx = \frac{1}{2} l.$$

Therefore $$H_1 = \frac{\frac{2}{3} fl(k - 3k^2 + 2k^3)}{\frac{12}{135} f^2 l}$$

or

$$H_1 = \frac{15}{2f} M_a(1 - 3k + 2k^2)k.$$

Value of M_1.—Our general equation becomes

$$M_1 = \frac{\left\{ M_a \int_0^l dx - H_1 \int_0^l y dx \right\} \int_0^l x^2 dx - \left\{ M_a \int_0^l x dx - H_1 \int_0^l x dx \right\} \int_0^l x dx}{\int_0^l dx \int_0^l x^2 dx - \left(\int_0^l x dx \right)^2}$$

The denominator becomes $\frac{1}{12} l^4$.

$$\int_0^l y dx = \frac{2}{3} f l ; \qquad \int_0^l x^2 dx = \frac{1}{3} l^3 ;$$

$$\int_0^l x dx = \frac{1}{2} l^2$$

Hence

$$M_1 = M_a + \frac{1}{3} (9 - 8f) H .$$

APPENDIX G.

SPECIAL CASE WHERE $\theta =$ A CONSTANT.

PARABOLIC ARCH WITH A HINGE AT EACH SUPPORT—VERTICAL LOAD.

In practice it sometimes happens that the arch-rib has a constant moment of inertia, especially in large arches. The formulas already deduced do not apply to such a condition, though they may be considered as approximately correct.

This case has been very thoroughly considered in two* papers by M. Belliard in the *Annales des Ponts et Chaussées.* The principal results are given below.

According to the assumption that $\theta \cos \phi =$ a constant, the general equation for H_1 becomes, from $g(131)$,

$$H_1 = \frac{P(1-k)\left\{\int_0^l xydx + \int_0^l \frac{dx}{F_x}\sin\phi\right\} - P\left\{\int_a^l y(x-a)dx + \int_a^l \frac{dx}{F_x}\sin\phi\right\}}{\int_0^l y^2dx + \int_0^l \frac{dx}{F_x}\cos\phi};$$

while for $\theta =$ a constant

$$H_1 = \frac{P(1-k)\left\{\int_0^l xyds + \int_0^l \frac{dx}{F_x}\sin\phi\right\} - P\left\{\int_a^l x)-a)yds + \int_a^l \frac{dx}{F_x}\sin\phi\right\}}{\int_0^l y^2ds + \int_0^l \frac{dx}{F_x}\cos\phi}.$$

---

* Note sur L'erreur relative que l'on commet en substituant dx a ds dans la Formule de Navier. April, 1893.

Mémoire sur le calcul de la Résistance des arcs paraboliques a grande flèche. November, 1893.

These two equations are the same in form, and their only difference is, in the second ds replaces dx in the first in all terms excepting those containing F_x.

Although the integration of the second equation offers no serious difficulty, yet the final results are long, and their application in practice tedious without special tables.

The equation for the common method is very simple and easy in application. Since the location of the load does not affect the relation between the results obtained by using the equation containing dx and that containing ds, the relative error between the results can be found, and the results obtained by the common method corrected to correspond with those which would have been obtained by the application of the correct formula containing ds.

M. Belliard found that the relative error depended only upon the ratio of the rise to the span. For $\frac{f}{l} = 0.50$ he found that the common formula $H_1 = \frac{5}{8}\frac{l}{f}Pk(1 - k)(1 + k - k^2)$, *which neglects the axial stress*, gave a result 3.3 per cent larger than that given by the exact formula. For $\frac{f}{l} = 0.25$ the result was 1.7 per cent larger, or practically one half that (per cent) for $\frac{f}{l} = 50$. This being the case, it is a very simple matter to find the percentage for any ratio of $\frac{f}{l}$ by interpolation.

The magnitudes of the above errors are too small to be of much practical importance, and since the formulas of the common method give results which are too large, an additional factor of safety is introduced by using them.

TABLES.

TABLES.

(*The tables that follow are arranged according to the scheme here given.*)

A. Tabulated Properties of the Two-nosed Catenary.

B. A Series of Two-nosed Catenaries inscribed in the Circle of Radius Unity

B_1. Arch-rings with the Line of Stress lying within the *middle Third.*

I. $k(1 - 2k^2 + k^3) = \Delta_1$.

II. $\frac{8}{5}\frac{1}{1 + k - k^2} = \Delta_2$.

III. $1 - \frac{k}{2}[5(1 - k - 2k^2 + 4k^3) - 8k^4] = \Delta_3$.

IV. $1 - 2k(2 - 5k + 5k^2) + 3k^4 = \Delta_4$.

V. k^2, $k(1 - k)$, and $(1 + k - k^2)$.

VI. $k^2(1 - k)(3 - 5k) = \Delta_6$.

VII. $(1 - k)^2(1 + 2k) = \Delta_7$.

VIII. $\frac{6}{5}\frac{5k - 2}{9k} = \Delta_8$.

IX. $\frac{k(5k - 2)}{2(1 + 2k)} = \Delta_9$.

X. $2k - 3k^2 + k^3 = \Delta_{10}$.

XI. $k^2(1 - k)^2 = \Delta_{11}$.

XII. $1 + k^2(-15 + 50k - 60k^2 + 24k^3) = \Delta_{12}$.

XIII. $2k(1 - k)^2(2 - 7k + 8k^2) = \Delta_{13}$.

XIV. $$\frac{2k(1-k)^2(2-7k+8k^2)}{1+k^2(-15+50k-60k^2+24k^3)}=\Delta_{14}.$$

XV. $$\frac{2-7k+8k^2}{6k}=\Delta_{15}.$$

XVI. $$3-12k+24k^2-16k^3=\Delta_{16}.$$

XVII. $$\frac{\frac{1}{2}(\sin^2\phi_0-\sin^2\alpha) + \cos\phi_0(\cos\alpha+\alpha\sin\alpha-\cos\phi_0-\phi_0\sin\phi_0)}{2\phi_0\cos^2\phi_0-3\sin\phi_0\cos\phi_0+\phi_0}=\Delta_{17}.$$

XVIII. $$2x^\circ\cos^2x^\circ-3\sin x^\circ\cos x^\circ+x^\circ=\Delta_{18}$$

XIX. $$x^\circ+\sin x^\circ\cos x^\circ=\Delta_{19}.$$

$$x^\circ-\sin x^\circ\cos x^\circ=\beta_{19}.$$

$$\sin x^\circ-x^\circ\cos x^\circ=\Delta\Delta_{19}.$$

XX. $$x^{\circ 2}+x^\circ\sin x^\circ\cos x^\circ-2\sin^2x^\circ=\Delta_{20}.$$

XXI. $$2\sin x^\circ\cos x^\circ+x^\circ\sin^2x^\circ=\Delta_{21}.$$

XXII. $$\cos x^\circ+x^\circ\sin x^\circ=\Delta_{22}.$$

XXIII. $$x^{\circ 2}-x^\circ\sin x^\circ\cos x^\circ=\Delta_{23}.$$

XXIV. $$x^{\circ 2}+x^\circ\sin x^\circ\cos x^\circ=\Delta_{24}.$$

XXV. Arc x° and $(\text{arc } x^\circ)^2$.

XXVI. $\sin x^\circ$, $\cos x^\circ$, and $1-\cos x^\circ$.

XXVII. $\sin^2 x^\circ$, $\cos^2 x^\circ$, $\sin x^\circ\cos x^\circ$;
$\sin^3 x^\circ$ and $\cos^3 x^\circ$.

XXVIII. $x^\circ\sin x^\circ$, $x^\circ\cos x^\circ$, and $\dfrac{\sin x^\circ}{x^\circ}$.

XXIX. $x^\circ\sin^2 x^\circ$, $x^\circ\cos^2 x^\circ$, and $x^\circ\sin x\ \cos x^\circ$.

XXX. General Dimensions of Masonry Arches constructed at Different Periods.

XXXI. Dimensions of a few Cast-iron Arches.

XXXII. Dimensions of a few Wrought-iron or Steel Arches.

XXXIII. Dimensions of a few Wrought-iron or Steel Roof-truss Arches.

TABLE A.—TABULATED PROPERTIES

Modulus of family, or parameter of common catenary from which members are transformed, is taken as unity.

				Described Circle.					
m	s	r	y_0	x_1	y_1	ϕ_1	ρ_1	R_1	Y_0
I	.3̂	$\sqrt{\frac{1}{3}}$	.5774	0.0000	.5774	0° 0.00'	1.7321	1.7321	+.5774
I	.32	—	.5657	.2475	.5831	8.03	1.7667	1.7672	.5657
I	.30	—	.5477	.3977	.5916	12.36	1.8187	1.8224	.5477
	.28	—	.5292	.5119	.6000	15.48	1.8706	1.8807	.5290
	.26	—	.5099	.6116	.6083	18.21	1.9226	1.9427	.5095
	.25	½	.5000	.6585	.6124	19.28	1.9485	1.9754	.4994
	.24	—	.4899	.7041	.6164	20.31	1.9745	2.0092	.4890
	.22	—	.4690	.7932	.6245	22.24	2.0265	2.0810	.4674
	.20	—	.4472	.8814	.6325	24.06	2.0785	2.1589	.4444
	.19	—	.4359	.9258	.6364	24.53	2.1044	2.2007	.4322
	.18	—	.4243	.9706	.6403	25.37	2.1304	2.2446	.4196
	.17	—	.4123	1.0163	.6442	26.20	2.1564	2.2910	.4065
	.16	—	.4000	1.0630	.6481	27.01	2.1824	2.3400	.3927
	.15	—	.3873	1.1110	.6519	27.40	2.2084	2.3922	.3783
	.14	—	.3742	1.1606	.6557	28.18	2.2344	2.4477	.3631
	.13	—	.3606	1.2122	.6600	28.55	2.2603	2.5075	.3471
	.12	—	.3464	1.2663	.6633	29.30	2.2863	2.5719	.3300
	.1̂	⅓	.3333	1.3170	.6667	30.00	2.3094	2.6339	.3138
	.11	—	.3317	1.3235	.6671	30.04	2.3123	2.6420	.3117
	.10	—	.3162	1.3843	.6708	30.37	2.3383	2.7188	.2920
	.09	—	.3000	1.4498	.6745	31.08	2.3643	2.8037	.2706
	.08	—	.2828	1.5211	.6782	31.39	2.3902	2.8988	.2471
	.07	—	.2646	1.5999	.6819	32.09	2.4162	3.0068	.2209
	.0625	¼	.2500	1.6655	.6847	32.31	2.4357	3.0987	.1990
	.06	—	.2450	1.6888	.6856	32.38	2.4422	3.1318	.1912
	.05	—	.2236	1.7914	.6892	33.06	2.4682	3.2801	.1569
	.04	⅕	.2000	1.9141	.6928	33.33	2.4942	3.4628	.1157
	.0357	—	.1889	1.9757	.6944	33.45	2.5053	3.5561	.0951
	.03	—	.1732	2.0688	.6964	34.00	2.5201	3.6994	.0639
	.027̂	⅙	.1667	2.1096	.6972	34.06	2.5259	3.7631	+.0503
		For a	value of	s here, sen	sibly the	next, dir	ectrix tou	ches descr	ibed circ
	.0204	$\frac{1}{7}$	.1429	2.2716	.6999	34.25	2.5451	4.0191	—.0037
	.02	—	.1414	2.2821	.7000	34.26	2.5461	4.0360	—.0072
	.01	$\frac{1}{10}$	.1000	2.6391	.7036	34.51	2.5721	4.6178	—.1249
	.005	—	.0707	2.9907	.7053	35.04	2.5851	5.2065	—.2394
	.0048	—	.0693	3.0114	.7054	35.04	2.5856	5.2410	—.2461
					For a va	lue of s	here, sen	sibly the l	ast, ϕ_2=
I	.002	—	.0447	3.4519	.7064	35.11	2.5929	5.9909	—.3881
I	.001	—	.0316	3.7994	.7068	35.14	2.5955	6.5872	—.4994
I	.000	—	.0000	∞	.7071	35.16	2.5981	∞	— ∞

For this table, the modulus of common catenary from which the members are transformed
two of the above values being given or assumed, the values of the others may be determined
for circular linear arches under vertical and conjugate horizontal loads has been calculated

OF THE "TWO-NOSED CATENARY."

Three-point Circle.										
x_2	y_2	i_0	ϕ_2	β	$\rho_0 = \rho_2$	R_2	$R_2 - R_1$	δ_0	δ_1	δ_2
		° ′	° ′	° ′						
0.0000	0.5774	0.00	0.00	0.00	1.7321	1.7321	.0000	.0000	.0000	.0000
.3511	.6009	10.30	11.28	11.28	1.7678	1.7673	.0001	.0000	.0000	.0000
.5671	.6382	16.39	18.08	18.08	1.8258	1.8228	.0003	.0000	.0000	.0000
.7336	.6780	—	22.58	22.57	1.8898	1.8818	.0010	.0001	.0001	.0001
.8808	.7208	—	27.00	26.56	1.9612	1.9446	.0019	.0004	.0003	.0001
.9507	.7435	26.36	28.49	28.44	2.0000	1.9777	.0023	.0006	.0004	.0002
1.0191	.7671	—	30.33	30.26	2.0412	2.0120	.0029	.0009	.0006	.0004
1.1538	.8174	—	33.48	33.36	2.1320	2.0846	.0037	.0017	.0013	.0008
1.2884	.8727	34.02	36.51	36.33	2.2361	2.1635	.0046	.0029	.0022	.0014
1.3567	.9025	—	38.19	37.57	2.2942	2.2057	.0050	.0037	.0029	.0018
1.4260	.9339	—	39.46	39.20	2.3570	2.2500	.0054	.0047	.0037	.0023
1.4969	.9672	—	41.11	40.41	2.4254	2.2965	.0055	.0059	.0047	.0030
1.5696	1.0026	—	42.36	42.00	2.5000	2.3456	.0055	.0073	.0059	.0039
1.6447	1.0404	40.24	44.00	43.19	2.5820	2.3975	.0053	.0090	.0074	.0050
1.7226	1.0809	—	45.24	44.37	2.6726	2.4527	.0050	.0111	.0091	.0063
1.8040	1.1247	—	46.49	45.55	2.7735	2.5116	.0041	.0135	.0113	.0079
1.8895	1.1721	—	48.14	47.13	2.8868	2.5748	.0029	.0165	.0139	.0100
1.9697	1.2179	—	49.31	48.22	3.0000	2.6353	.0013	.0195	.0167	.0122
1.9800	1.2240	—	49.41	48.31	3.0151	2.6429	.0009	.0200	.0171	.0126
2.0766	1.2812	46.13	51.09	49.51	3.1623	2.7170	−.0018	.0242	.0211	.0158
2.1808	1.3450	—	52.40	51.12	3.3333	2.7982	−.0055	.0294	.0259	.0199
2.2944	1.4170	—	54.14	52.36	3.5356	2.8879	−.0108	.0358	.0320	.0253
2.4202	1.4997	—	55.53	54.05	3.7796	2.9887	−.0181	.0437	.0397	.0324
2.5247	1.5709	—	57.11	55.14	4.0000	3.0732	−.0255	.0510	.0470	.0392
2.5619	1.5967	—	57.38	55.38	4.0825	3.1035	−.0283	.0537	.0497	.0418
2.7255	1.7139	51.43	59.31	57.20	4.4721	3.2374	−.0427	.0668	.0628	.0549
2.9210	1.8613	—	61.37	59.16	5.0000	3.3985	−.0642	.0843	.0809	.0738
3.0189	1.9382	—	62.36	60.11	5.2926	3.4796	−.0765	.0939	.0911	.0844
3.1667	2.0587	—	64.01	61.32	5.7733	3.6020	−.0974	.1093	.1073	.1026
3.2315	2.1131	—	64.36	62.08	6.0000	3.6557	−.1074	.1164	.1148	.1112
le; described and three-point circles are concentric; Y_0 changes sign.										
3.4875	2.3381	—	66.48	64.22	7.0000	3.8678	−.1512	.1465	.1473	.1494
3.5041	2.3532	—	66.56	64.31	7.0711	3.8816	−.1544	.1486	.1496	.1521
4.0639	2.9107	—	71.02	69.20	10.0000	4.3432	−.2746	.2249	.2337	.2582
4.6099	3.5526	—	74.17	74.08	14.1421	4.7926	−.4139	.3101	.3286	.3854
4.6418	3.5933	—	74.27	74.25	14.4338	4.8190	−.4220	.3154	.3344	.3924
β; $\rho_0 = \rho_2$; $(\phi_2 - \beta)$ changes sign.										
5.3190	4.5651	—	77.39	80.43	22.3607	5.3895	−.6013	.4328	.4628	.5659
5.8495	5.4874	—	79.40	86.01	31.6228	5.8637	−.7235	.5310	.5654	.6863
∞	∞	57.04	90.00	90.00	∞	∞	—	—	—	—

is taken as unity; all quantities except *s*, *r*, and angles are directly proportional to *m*. Any from the table. Intermediate values can be easily interpolated. Rankine's point of rupture for certain values of *s*; it is given in the column i_0.

TABLE B.

A Series of "TWO-NOSED CATENARIES" *inscribed in the Circle of Radius* (R_1) *Unity, and having Parallel Directrices at Graduated Distances* ($R_1 + Y_0$) *from its Centre from* (1 + .026) *to* (1 + .234).

This Table has for its purpose, in conjunction with supplementary tables, the designing of arch-rings, so as to secure the condition of the line of stress lying within the middle third, fifth, seventh, etc., of the arch-ring, as may be required to give strength *and* stability for every variation of proportion of parts and of the nature and distribution of load.

s	ϕ_1	ϕ_2	m	R_1	R_2	ρ_1	$\rho_0 = \rho_2$	y_0	Y_0	δ_0	δ_2
	° ′	° ′									
.230	21 29	32 13	.4890	1	1.0016	.9783	1.0201	.2345	.2340	.0007	.0003
.225	21 56	33 00	.4847	1	1.0017	.9761	1.0223	.2299	.2293	.0007	.0003
.220	22 24	33 48	.4805	1	1.0017	.9739	1.0245	.2254	.2247	.0008	.0004
.215	22 50	34 34	.4762		1.0019	.9713	1.0271	.2208	.2200	.0009	.0004
.210	23 16	35 21	.4719		1.0020	.9687	1.0296	.2163	.2153	.0010	.0005
.205	23 41	36 06	.4675		1.0021	.9658	1.0327	.2117	.2106	.0011	.0005
.200	24 06	36 51	.4632		1.0022	.9628	1.0358	.2072	.2059	.0013	.0006
.195	24 30	37 35	.4588		1.0022	.9596	1.0391	.2026	.2011	.0015	.0007
.190	24 53	38 19	.4544		1.0023	.9563	1.0425	.1981	.1964	.0017	.0008
.185	25 15	39 02	.4499		1.0023	.9527	1.0463	.1935	.1916	.0019	.0009
.180	25 37	39 46	.4455		1.0024	.9491	1.0501	.1890	.1869	.0021	.0010
.175	25 58	40 28	.4410		1.0024	.9452	1.0543	.1845	.1821	.0023	.0011
.170	26 20	41 11	.4365		1.0024	.9413	1.0586	.1800	.1774	.0026	.0013
.165	26 40	41 53	.4319		1.0024	.9370	1.0635	.1754	.1726	.0028	.0015
.160	27 01	42 36	.4273		1.0024	.9327	1.0684	.1709	.1678	.0031	.0017
.155	27 20	43 18	.4226		1.0023	.9279	1.0738	.1664	.1629	.0034	.0019
.150	27 40	44 00	.4180		1.0022	.9232	1.0793	.1619	.1581	.0038	.0021
.145	27 59	44 42	.4132		1.0021	.9180	1.0856	.1574	.1532	.0042	.0023
.140	28 18	45 24	.4085		1.0020	.9129	1.0919	.1529	.1484	.0045	.0026
.135	28 36	46 06	.4036		1.0019	.9071	1.0990	.1483	.1434	.0049	.0029
.130	28 55	46 49	.3988		1.0017	.9014	1.1061	.1438	.1384	.0054	.0032
.125	29 12	47 31	.3938		1.0014	.8952	1.1142	.1392	.1333	.0059	.0035
.120	29 30	48 14	.3888		1.0011	.8890	1.1224	.1347	.1283	.0064	.0039
.115	29 47	48 57	.3836		1.0007	.8821	1.1318	.1301	.1231	.0070	.0043
.110	30 04	49 41	.3785		1.0004	8752	1.1412	.1255	.1180	.0076	.0048
.105	30 20	50 25	.3731		0.9998	.8676	1.1521	.1209	.1127	.0082	.0053
.100	30 37	51 09	.3678		.9993	.8600	1.1631	.1163	.1074	.0089	.0058
.095	30 52	51 54	.3622		.9987	.8517	1.1760	.1116	1019	.0097	.0064
.090	31 08	52 40	.3567		.9981	.8433	1.1889	.1070	.0965	.0105	0071
.085	31 23	53 27	.3508		.9972	.8340	1.2043	.1023	.0908	.0114	.0079
.080	31 39	54 14	.3450		.9963	.8246	1.2197	.0976	0852	0124	0087
.075	31 54	55 03	.3388		.9951	.8141	1.2384	.0928	.0793	.0134	.0097
.070	32 09	55 53	.3326		.9940	.8036	1.2571	.0880	0735	.0145	.0108
.065	32 23	56 45	.3259		.9925	.7917	1.2803	.0831	.0673	.0158	.0121
.060	32 38	57 38	.3193		.9910	.7798	1.3036	.0782	.0611	.0172	0134
.055	34 52	58 34	.3121		.9890	.7661	1.3335	.0732	.0544	0188	.0150
.050	33 06	59 31	.3049	1	.9870	.7524	1.3634	.0682	.0478	.0204	.0167
.045	33 20	60 34	.2968	1	9842	.7363	1.4037	.0630	.0406	.0223	.0190
.040	33 33	61 37	.2888	1	.9815	.7203	1.4440	.0578	.0334	.0243	0213
.035	33 46	62 40	.2 08	1	.9788	.7040	1.4843	.0526	.0260	.0263	.0236
Independent of R_1.			Directly proportioned to R_1, and subject to any multiplier.								

(R_1 column, printed vertically:) The Radius of Described Circle is taken as unity. This circle forms the upper boundary of the "Kernel" of arch-ring, *i.e.* of its middle third, fifth, or seventh, etc.

For the values of s ending with 5, on this and on the Supplementary Table, the quantities are only interpolated as arithmetical means, and are correct to about 1 per cent.

SUPPLEMENTARY TABLE B_1.

Arch rings with the Two-nosed Catenary or Line of Stress lying within the middle third, and loaded from directrix to a circular soffit which is the three-point circle of another member of the same family of transformed catenaries as the line of stress.

Strong Brick. Average W't 112 lbs. p. cub. ft. Strength 154,000 lbs. per sq. ft.		**Sandstone.** Average W't 140 lbs. p. cub. ft. Strength 576,000 lbs. per sq. ft.		**Granite.** Average W't 164 lbs. p. cub. ft. Strength 1,350,000 lbs. per sq. ft.									
Max. mult.	Max. span.	Max. multiplier to give factor of safety 10.	Max. value of span $2c$ with that multiplier.	Max. mult.	Max. span.	Average intensity of stress at crown joint in lbs. per sq. ft.	Depth of load from extrados (directrix) to crown of soffit.	Thickness of arch-ring normal to soffit at crown.	Thickness of arch-ring normal to soffit at ϕ_2 joint of rupture.	Radius of the soffit.	Rise of segment of soffit subtending $2\phi_2$.	Span of segment of soffit subtending $2\phi_2$.	For reference.
						$\frac{p_0 d}{t_0}$	d	t_0	t_2	R	k	$2c$	s
	Feet.		Feet.		Feet.								
				50	72	8.25	.141	.019	.026	.973	.325	1.451	.120
				55	81	7.47	.137	.021	.029	.970	.333	1.462	.115
				61	90	6.70	.133	.023	.033	.966	.341	1.473	.110
				68	101	6.07	.129	.025	.037	.962	.349	1.482	.105
		38	56	76	112	5.45	.125	.027	.041	.957	.357	1.491	.100
		42	63	84	126	4.94	.121	.029	.045	.952	.364	1.497	.095
		46	70	92	140	4.44	.117	.032	.049	.946	.372	1.504	.090
		51	78	102	156	4.03	.113	.034	.055	.940	.380	1.509	.085
19	29	57	86	114	172	3.62	.110	.037	.062	.933	.388	1.514	.080
21	32	63	95	126	190	3.31	.106	.040	.071	.925	.395	1.516	.075
23	35	69	105	138	210	3.00	.103	.044	.081	.917	.403	1.518	.070
25	39	77	116	154	232	2.71	.099	.048	.089	.907	.409	1.516	.065
28	43	85	128	170	256	2.42	.095	.052	.097	.896	.416	1.514	.060
31	47	94	141	—	—	2.20	.092	.056	.110	.883	.421	1.506	.055
35	52	104	155	—	—	1.98	.089	.061	.123	.869	.427	1.498	.050
38	57	—	—	—	—	1.80	.085	.067	.141	.850	.431	1.480	.045
42	62	—	—	—	—	1.62	.082	.073	.159	.831	.436	1.462	.040
46	67	—	—	—	—	1.44	.078	.078	.177	.812	.441	1.444	.035
One third.	—	$205 + \frac{p_0 d}{t_0}$	—	Double	—	Directly prop. to R_1, and subject to any multiplier less than given max.							—

☞ Note that $\frac{k + d}{2c} = \frac{1}{3}$ nearly.

TABLE I.

VALUES OF $k(1 - 2k^2 + k^3) = \Delta_1$.

k	Δ_1	k	Δ_1	k	Δ_1	k	Δ_1	k	Δ_1
0	0	.21	0.1934	.42	0.3029	.63	0.2874	.84	0.1525
.01	0.0099	.22	.2010	.43	3052	.64	.2835	.85	.1438
.02	.0199	.23	.2085	.44	.3071	.65	.2793	.86	.1349
.03	.0299	.24	.2157	.45	.3088	.66	.2748	.87	.1259
.04	.0399	.25	.2227	.46	.3101	.67	.2699	.88	.1166
.05	.0498	.26	.2294	.47	.3112	.68	.2649	.89	.1075
.06	.0596	.27	.2359	.48	.3119	.69	.2597	.90	.0981
.07	.0693	.28	.2422	.49	.3124	.70	.2541	.91	.0886
.08	.0790	.29	.2483	.50	.3125	.71	.2483	.92	.0790
.09	.0886	.30	.2541	.51	.3124	.72	.2422	.93	.0693
.10	.0981	.31	.2597	.52	.3119	.73	.2359	.94	.0596
.11	.1070	.32	.2649	.53	.3112	.74	.2294	.95	.0498
.12	.1166	.33	.2699	.54	.3101	.75	.2227	.96	.0399
.13	.1259	.34	.2748	.55	.3088	.76	.2157	.97	.0299
.14	.1349	.35	.2793	.56	.3071	.77	.2085	.98	.0199
.15	.1438	.36	.2835	.57	.3052	.78	.2010	.99	.0099
.16	.1525	.37	.2874	.58	.3029	.79	.1934	1.00	0
.17	.1610	.38	.2911	.59	.3004	.80	.1856		
.18	.1694	.39	.2945	.60	.2976	.81	.1776		
.19	.1776	.40	.2976	.61	.2945	.82	.1694		
.20	.1856	.41	.3004	.62	.2911	.83	.1610		

TABLE II.

VALUES OF $\frac{8}{5} \frac{1}{1 + k - k^2} = \Delta_2$.

k	Δ_2	k	Δ_2	k	Δ_2	k	Δ_2	k	Δ_2
0	1.6000	.21	1.3723	.42	1.2866	.63	1.2975	.84	1.4104
.01	1.5843	.22	1.3657	.43	1.2850	.64	1.3004	.85	1.4191
.02	1.5692	.23	1.3593	.44	1.2837	.65	1.3035	.86	1.4280
.03	1.5548	.24	1.3532	.45	1.2826	.66	1.3068	.87	1.4374
.04	1.5408	.25	1.3474	.46	1.2816	.67	1.3103	.88	1.4472
.05	1.5274	.26	1.3418	.47	1.2809	.68	1.3141	.89	1.4573
.06	1.5146	.27	1.3366	.48	1.2804	.69	1.3181	.90	1.4679
.07	1.5022	.28	1.3316	.49	1.2801	.70	1.3223	.91	1.4789
.08	1.4903	.29	1.3268	.50	1.2800	.71	1.3268	.92	1.4903
.09	1.4789	.30	1.3223	.51	1.2801	.72	1.3316	.93	1.5022
.10	1.4679	.31	1.3181	.52	1.2804	.73	1.3366	.94	1.5146
.11	1.4573	.32	1.3141	.53	1.2809	.74	1.3418	.95	1.5274
.12	1.4472	.33	1.3103	.54	1.2816	.75	1.3474	.96	1.5408
.13	1.4374	.34	1.3068	.55	1.2826	.76	1.3532	.97	1.5548
.14	1.4280	.35	1.3035	.56	1.2837	.77	1.3593	.98	1.5692
.15	1.4191	.36	1.3004	.57	1.2850	.78	1.3657	.99	1.5843
.16	1.4104	.37	1.2975	.58	1.2866	.79	1.3723	1.00	1.6000
.17	1.4022	.38	1.2949	.59	1.2883	.80	1.3793		
.18	1.3942	.39	1.2925	.60	1.2903	.81	1.3866		
.19	1.3866	.40	1.2903	.61	1.2925	.82	1.3942		
.20	1.3793	.41	1.2883	.62	1.2949	.83	1.4022		

TABLE III.

VALUES OF $1 - \frac{k}{2}[5(1 - k - 2k^2 + 4k^3) - 8k^4] = \Delta_3$.

k	Δ_3	k	Δ_3	k	Δ_3	k	Δ_3	k	Δ_3
.00	1.0000	.21	.6137	.42	.5025	.63	.4892	.84	.3217
.01	.9753	.22	.6029	.43	.5017	.64	.4865	.85	.3066
.02	.9510	.23	.5927	.44	.5011	.65	.4834	.86	.2909
.03	.9274	.24	.5831	.45	.5006	.66	.4799	.87	.2745
.04	.9043	.25	.5742	.46	.5003	.67	.4760	.88	.2573
.05	.8818	.26	.5659	.47	.5001	.68	.4716	.89	.2395
.06	.8600	.27	.5583	.48	.5000	.69	.4667	.90	.2210
.07	.8387	.28	.5512	.49	.5000	.70	.4613	.91	.2017
.08	.8182	.29	.5447	.50	.5000	.71	.4553	.92	.1818
.09	.7983	.30	.5387	.51	.5000	.72	.4488	.93	.1613
.10	.7790	.31	.5333	.52	.5000	.73	.4417	.94	.1400
.11	.7605	.32	.5284	.53	.4999	.74	.4341	.95	.1182
.12	.7427	.33	5240	.54	.4997	.75	.4258	.96	.0957
.13	.7255	.34	.5201	.55	.4994	.76	.4169	.97	0726
.14	.7091	.35	.5166	.56	.4989	.77	.4073	.98	.0490
.15	.6934	.36	.5135	.57	.4983	.78	.3971	.99	.0247
.16	.6783	.37	.5108	.58	.4975	.79	.3863	1.00	.0
.17	.6640	.38	.5085	.59	.4964	.80	.3747		
.18	.6504	.39	.5066	.60	.4950	.81	.3625		
.19	.6375	.40	.5050	.61	.4934	.82	.3496		
.20	.6253	.41	.5036	.62	.4915	.83	.3360		

TABLE IV.

VALUES OF $1 - 2k(2 - 5k + 5k^2) + 3k^4 = \Delta_4$.

k	Δ_4	k	Δ_4	k	Δ_4	k	Δ_4	k	Δ_4
.00	1.0000	.21	.5142	.42	.4365	.63	.4211	.84	.2626
.01	.9610	.22	.5045	.43	.4365	.64	.4179	.85	.2498
.02	.9239	.23	.4957	.44	.4366	.65	.4143	.86	.2365
.03	.8887	.24	.4877	.45	.4368	.66	.4103	.87	.2227
.04	.8554	.25	.4805	.46	.4370	.67	.4059	.88	.2084
.05	.8238	.26	.4739	.47	.4372	.68	.4011	.89	.1936
.06	.7939	.27	.4681	.48	.4373	.69	.3959	.90	.1783
.07	.7656	.28	.4629	.49	.4375	.70	.3903	.91	.1625
.08	.7390	.29	.4583	.50	.4375	.71	.3842	.92	.1463
.09	.7139	.30	.4543	.51	4374	.72	.3777	.93	.1296
.10	.6903	.31	.4508	.52	.4372	.73	.3708	.94	.1124
.11	.6681	.32	.4478	.53	.4369	.74	.3634	.95	.0948
.12	.6473	.33	.4452	.54	.4365	.75	.3555	.96	.0767
.13	.6279	.34	.4431	.55	.4358	.76	.3471	.97	.0581
.14	.6097	.35	.4413	.56	.4349	.77	.3383	.98	.0392
.15	.5928	.36	.4398	.57	.4338	.78	.3289	.99	.0198
.16	.5770	.37	.4387	.58	.4324	.79	.3191	1.00	.0000
.17	.5624	.38	.4378	.59	.4307	.80	.3088		
.18	.5488	.39	.4372	.60	.4288	.81	.2980		
.19	.5363	.40	.4368	.61	4266	.82	.2867		
.20	.5248	.41	.4366	.62	.4240	.83	.2749		

TABLE V.

k	k^2	$k(1-k)$	$1+k-k^2$	$1-k$
0	0	0	1.0000	1.00
01	0.0001	0.0099	1.0099	.99
02	.0004	.0196	1.0196	.98
.03	.0009	.0291	1.0291	.97
.04	.0016	.0384	1.0384	.96
.05	.0025	.0475	1.0475	.95
.06	.0036	.0564	1.0564	.94
.07	.0049	.0651	1.0651	.93
.08	.0064	.0736	1.0736	.92
.09	.0081	.0819	1.0819	.91
.10	.0100	.0900	1.0900	.90
.11	.0121	.0979	1.0979	.89
.12	.0144	.1056	1.1056	.88
.13	.0169	.1131	1.1131	.87
.14	.0196	.1204	1.1204	.86
.15	.0225	.1275	1.1275	.85
.16	.0256	.1344	1.1344	.84
.17	.0289	.1411	1.1411	.83
.18	.0324	.1476	1.1476	.82
.19	.0361	.1539	1.1539	.81
.20	.0400	.1600	1.1600	.80
.21	.0441	.1659	1.1659	.79
.22	.0484	.1716	1.1716	.78
.23	.0529	.1771	1.1771	.77
.24	.0576	.1824	1.1824	.76
.25	.0625	.1875	1.1875	.75
.26	.0676	.1924	1.1924	.74
.27	.0729	.1971	1.1971	.73
.28	.0784	.2016	1.2016	.72
.29	.0841	.2059	1.2059	.71
.30	.0900	.2100	1.2100	.70
.31	.0961	.2139	1.2139	.69
.32	.1024	.2176	1.2176	.68
.33	.1089	.2211	1 2211	.67
.34	.1156	.2244	1.2244	.66
.35	.1225	.2275	1.2275	.65
.36	.1296	.2304	1.2304	.64
.37	.1369	.2331	1.2331	.63
.38	.1444	.2356	1.2356	.62
.39	.1521	.2379	1.2379	.61
.40	.1600	.2400	1.2400	.60
.41	.1681	.2419	1.2419	.59
.42	.1764	.2436	1.2436	.58
.43	.1849	.2451	1.2451	.57
.44	.1936	.2464	1.2464	.56
.45	.2025	.2475	1.2475	.55
.46	.2116	.2484	1.2484	.54
.47	.2209	.2491	1.2491	.53
.48	.2304	.2496	1.2496	.52
.49	.2401	.2499	1.2499	.51
.50	.2500	.2500	1.2500	.50
$1-k$	$(1-k)^2$	$k(1-k)$	$1+k-k^2$	k

TABLE VI.

VALUES OF $k^2(1 - k)(3 - 5k) = \Delta_6$.

k	Δ_6	k	Δ_6	k	Δ_6	k	Δ_6	k	Δ_6
.0	0.0000	.21	0.0679	.42	0.0920		Negative.		Negative.
.01	.0002	.22	.0717	.43	.0895	.62	0.0146	.82	0.1331
.02	.0011	.23	.0753	.44	.0867	.63	.0220	.83	.1346
.03	.0024	.24	.0787	.45	.0835	.64	.0294	.84	.1354
.04	.0043	.25	.0820	.46	.0799	.65	.0369	.85	.1354
.05	.0065	.26	.0850	.47	.0761	.66	.0444	.86	.1346
.06	.0091	.27	.0878	.48	.0718	.67	.0518	.87	.1328
.07	.0120	.28	.0903	.49	.0673	.68	.0591	.88	.1301
.08	.0153	.29	.0924	.50	.0625	.69	.0664	.89	.1263
.09	.0185	.30	.0945	.51	.0573	.70	.0735	.90	.1215
.10	.0225	.31	.0961	.52	.0519	.71	.0804	.91	.1155
.11	.0263	.32	.0974	.53	.0462	.72	.0871	.92	.1083
.12	.0304	.33	.0985	.54	.0402	.73	.0935	.93	.0998
.13	.0345	.34	.0991	.55	.0338	.74	.0996	.94	.0901
.14	.0373	.35	.0995	.56	.0275	.75	.1054	.95	.0785
.15	.0430	.36	.0995	.57	.0209	.76	.1109	.96	.0663
.16	.0473	.37	.0991	.58	.0141	.77	.1159	.97	.0522
.17	.0515	.38	.0984	.59	.0070	.78	.1204	.98	.0365
.18	.0557	.39	.0973	.60	.0000	.79	.1245	.99	.0191
.19	.0599	.40	.0960		Negative.	.80	.1280	1.00	.0000
.20	.0640	.41	.0942	.61	0.0068	.81	.1309		

TABLE VII.

VALUES OF $(1 - k)^2(1 + 2k) = \Delta_7$.

k	Δ_7	k	Δ_7	k	Δ_7	k	Δ_7	k	Δ_7
0	1.0000	.21	0.8862	.42	0.6189	.63	0.3093	.84	0.0686
.01	0.9997	.22	.8760	.43	.6043	.64	.2954	.85	.0607
.02	.9988	.23	.8656	.44	.5895	.65	.2817	.86	.0533
.03	.9973	.24	.8548	.45	.5747	.66	.2681	.87	.0463
.04	.9953	.25	.8437	.46	.5598	.67	.2548	.88	.0397
.05	.9927	.26	.8323	.47	.5449	.68	.2416	.89	.0336
.06	.9896	.27	.8206	.48	.5299	.69	.2287	.90	.0280
.07	.9859	.28	.8087	.49	.5149	.70	.2160	.91	.0228
.08	.9818	.29	.7964	.50	.5000	.71	.2035	.92	.0181
.09	.9771	.30	.7840	.51	.4850	.72	.1912	.93	.0140
.10	.9720	.31	.7712	.52	.4700	.73	.1793	.94	.0103
.11	.9663	.32	.7583	.53	.4550	.74	.1676	.95	.0072
.12	.9602	.33	.7451	.54	.4401	.75	.1562	.96	.0046
.13	.9536	.34	.7318	.55	.4252	.76	.1451	.97	.0026
.14	.9466	.35	.7182	.56	.4104	.77	.1343	.98	.0011
.15	.9392	.36	.7045	.57	.3956	.78	.1239	.99	.0002
.16	.9313	.37	.6906	.58	.3810	.79	.1137	1.00	.0000
.17	.9231	.38	.6765	.59	.3664	.80	.1040		
.18	.9144	.39	.6623	.60	.3520	.81	.0945		
.19	.9054	.40	.6480	.61	.3376	.82	.0855		
.20	.8960	.41	.6335	.62	.3234	.83	.0768		

TABLE VIII.

VALUES OF $\frac{6}{5}\,\frac{5k - 2}{9k} = \Delta_8$.

k	Δ_8	k	Δ_8	k	Δ_8	k	Δ_8	k	Δ_8
	Negative.		Negative.	.40	0	.61	0.2295	.82	0.3415
0	......	.20	0.6666	.41	0.0162	.62	.2365	.83	.3454
.01	26.000	.21	.6031	.42	.0317	.63	.2434	.84	.3492
.02	12.666	.22	.5454	.43	.0465	.64	.2500	.85	.3529
.03	8.222	.23	.4927	.44	.0606	.65	.2564	.86	.3566
.04	6.000	.24	.4444	.45	.0740	.66	.2626	.87	.3601
.05	4.667	.25	.4000	.46	.0869	.67	.2686	.88	.3636
.06	3.778	.26	.3589	.47	.0992	.68	.2745	.89	.3670
.07	3.143	.27	.3209	.48	.1111	.69	.2802	.90	.3704
.08	2.667	.28	.2857	.49	.1224	.70	.2857	.91	.3736
.09	2.296	.29	.2528	.50	.1333	.71	.2911	.92	.3768
.10	2.000	.30	.2222	.51	.1437	.72	.2963	.93	.3799
.11	1.757	.31	.1935	.52	.1538	.73	.3014	.94	.3830
.12	1.555	.32	.1666	.53	.1633	.74	.3063	.95	.3859
.13	1.384	.33	.1414	.54	.1728	.75	.3111	.96	.3889
.14	1.238	.34	.1176	.55	.1818	.76	.3158	.97	.3917
.15	1.111	.35	.0952	.56	.1904	.77	.3203	.98	.3945
.16	1.000	.36	.0740	.57	.1988	.78	.3248	.99	.3973
.17	0.9020	.37	.0540	.58	.2069	.79	.3291	1.00	.4000
.18	.8148	.38	.0350	.59	.2147	.80	.3333		
.19	.7368	.39	.0170	.60	.2222	.81	.3374		

TABLE IX.

VALUES OF $10\,\frac{k(5k - 2)}{2(1 + 2k)} = 10\Delta_9$.

k	Δ_9	k	Δ_9	k	Δ_9	k	Δ_9	k	Δ_9
	Negative.		Negative.	.40	0	.61	1.4425	.82	3.2617
0	0	.20	0.7143	.41	0.0563	.62	1.5222	.83	3.3595
.01	0.0955	.21	.7024	.42	.1141	.63	1.6029	.84	3.4480
.02	.1826	.22	.6875	.43	.1733	.64	1.6842	.85	3.5419
.03	.2617	.23	.6695	.44	.2340	.65	1.7663	.86	3.6365
.04	.3333	.24	.6486	.45	.2961	.66	1.8491	.87	3.7312
.05	.3977	.25	.6250	.46	.3594	.67	1.9327	.88	3.8266
.06	.4553	.26	.5987	.47	.4240	.68	2.0170	.89	3.9211
.07	.5063	.27	.5698	.48	.4898	.69	2.1020	.90	4.0182
.08	.5517	.28	.5385	.49	.5568	.70	2.1876	.91	4.1148
.09	.5911	.29	.5047	.50	.6262	.71	2.2740	.92	4.2113
.10	.6250	.30	.4687	.51	.6943	.72	2.3608	.93	4.3089
.11	.6537	.31	.4305	.52	.7647	.73	2.4484	.94	4.4067
.12	.6774	.32	.3902	.53	.8361	.74	2.5365	.95	4.5048
.13	.6964	.33	.3479	.54	.9086	.75	2.6250	.96	4.6032
.14	.7110	.34	.3036	.55	.9820	.76	2.7144	.97	4.7019
.15	.7212	.35	.2573	.56	1.0565	.77	2.8044	.98	4.8010
.16	.7273	.36	.2093	.57	1.1320	.78	2.8948	.99	4.9008
.17	.7295	.37	.1594	.58	1.2083	.79	2.9857	1.00	5.0000
.18	.7280	.38	.1079	.59	1.2625	.80	3.0771		
.19	.7229	.39	.0547	.60	1.3635	.81	3.1693		

TABLE X.

VALUES OF $2k - 3k^2 + k^3 = \Delta_{10}$.

k	Δ_{10}	k	Δ_{10}	k	Δ_{10}	k	Δ_{10}	k	Δ_{10}
0	0	.21	0.2969	.42	0.3848	.63	0.3193	.84	0.1559
.01	0.0197	.22	.3054	.43	.3848	.64	.3133	.85	.1466
.02	.0388	.23	.3134	.44	.3843	.65	.3071	.86	.1372
.03	.0573	.24	.3210	.45	.3836	.66	.3006	.87	.1278
.04	.0752	.25	.3281	.46	.3825	.67	.2940	.88	.1182
.05	.0926	.26	.3347	.47	.3811	.68	.2872	.89	.1086
.06	.1094	.27	.3409	.48	.3793	.69	.2802	.90	.0990
.07	.1256	.28	.3467	.49	.3773	.70	.2730	.91	.0892
.08	.1413	.29	.3520	.50	.3750	.71	.2656	.92	.0794
.09	.1564	.30	.3570	.51	.3723	.72	.2580	.93	.0696
.10	.1710	.31	.3614	.52	.3694	.73	.2503	.94	.0597
.11	.1850	.32	.3655	.53	.3661	.74	.2424	.95	.0498
.12	.1985	.33	.3692	.54	.3626	.75	.2343	.96	.0399
.13	.2114	.34	.3725	.55	.3588	.76	.2261	.97	.0299
.14	.2239	.35	.3753	.56	.3548	.77	.2178	.98	.0199
.15	.2358	.36	.3778	57	.3504	.78	.2093	.99	.0099
.16	.2472	.37	.3799	.58	.3459	.79	.2007	1.00	0
.17	.2582	.38	.3816	.59	.3410	.80	.1920		
.18	.2686	.39	.3830	.60	.3360	.81	.1831		
.19	.2785	.40	.3840	.61	.3306	.82	.1741		
.20	.2880	.41	.3846	.62	3251	.83	.1650		

TABLE XI.

VALUES OF $k^2(1 - k)^2 = \Delta_{11}$.

k	Δ_{11}	k	Δ_{11}	k	Δ_{11}	k	Δ_{11}	k	Δ_{11}
0	0.0000	.21	0.0275	.42	0.0593	.63	0.0543	.84	0.0180
.01	.0000	.22	.0294	.43	.0600	.64	.0530	.85	.0162
.02	.0003	.23	.0313	.44	.0607	.65	.0517	.86	.0144
.03	.0008	.24	.0332	.45	.0612	.66	.0503	.87	.0127
.04	.0014	.25	.0351	.46	.0617	.67	0488	.88	.0111
.05	.0022	.26	.0370	.47	.0620	.68	.0473	.89	.0095
.06	.0031	.27	.0388	.48	.0623	.69	.0457	.90	.0081
.07	.0042	.28	.0406	.49	.0624	.70	.0441	.91	.0067
.08	.0054	.29	.0423	.50	.0625	.71	.0423	.92	.0054
.09	.0067	.30	.0441	.51	.0624	.72	.0406	.93	.0042
.10	.0081	.31	.0457	.52	.0623	.73	.0388	.94	.0031
.11	.0095	.32	.0473	.53	.0620	.74	.0370	.95	.0022
.12	.0111	.33	.0488	.54	.0617	.75	.0351	.96	.0014
.13	.0127	.34	.0503	.55	.0612	.76	.0332	.97	.0008
.14	.0144	.35	.0517	.56	.0607	.77	.0313	.98	.0003
.15	.0162	.36	.0530	.57	.0600	.78	.0294	.99	.0000
.16	.0180	.37	.0543	.58	.0593	.79	.0275	1.00	.0000
.17	.0199	.38	.0555	.59	.0585	.80	.0256		
.18	.0217	.39	.0566	.60	.0576	.81	.0236		
.19	.0236	.40	.0576	.61	.0566	.82	.0217		
.20	.0256	.41	.0585	.62	.0555	.83	.0199		

TABLE XII.

VALUES OF $1 + k^2(-15 + 50k - 60k^2 + 24k^3) = \Delta_{12}$.

k	Δ_{12}	k	Δ_{12}	k	Δ_{12}	k	Δ_{12}	k	Δ_{12}
0	1.0000	.21	0.6947	.42	0.5051	.63	0.4790	.84	0.2161
.01	0.9986	.22	.6783	.43	.5034	.64	.4739	.85	.1973
.02	.9944	.23	.6625	.44	.5022	.65	.4681	.86	.1784
.03	.9879	.24	.6473	.45	.5013	.66	.4616	.87	.1600
.04	.9791	.25	.6329	.46	.5007	.67	.4543	.88	.1415
.05	.9684	.26	.6192	.47	.5003	.68	.4462	.89	.1234
.06	.9561	.27	.6063	.48	.5002	.69	.4374	.90	.1058
.07	.9493	.28	.5942	.49	.5001	.70	.4277	.91	.0888
.08	.9273	.29	.5829	.50	.5000	.71	.4172	.92	.0728
.09	.9113	.30	.5724	.51	.5000	.72	.4059	.93	.0578
.10	.8943	.31	.5627	.52	.5000	.73	.3939	.94	.0440
.11	.8767	.32	.5538	.53	.4999	.74	.3808	.95	.0316
.12	.8586	.33	.5458	.54	.4997	.75	.3672	.96	.0210
.13	.8401	.34	.5385	.55	.4988	.76	.3528	.97	.0122
.14	.8215	.35	.5320	.56	.4979	.77	.3376	.98	.0056
.15	.8028	.36	.5262	.57	.4967	.78	.3218	.99	.0015
.16	.7841	.37	.5212	.58	.4951	.79	.3054	1.00	0
.17	.7659	.38	.5167	.59	.4928	.80	.2884		
.18	.7472	.39	.5130	.60	.4903	.81	.2708		
.19	.7293	.40	.5098	.61	.4871	.82	.2530		
.20	.7117	.41	.5072	.62	.4834	.83	.2346		

TABLE XIII.

VALUES OF $2k(1 - k)^2(2 - 7k + 8k^2) = \Delta_{13}$.

k	Δ_{13}	k	Δ_{13}	k	Δ_{13}	k	Δ_{13}	k	Δ_{13}
0	0	.21	0.2314	.42	0.1331	.63	0.1319	.84	0.0758
.01	0.0073	.22	.2268	.43	.1311	.64	.1321	.85	.0699
.02	.0715	.23	.2217	.44	.1293	.65	.1321	.86	.0639
.03	.1011	.24	.2164	.45	.1279	.66	.1319	.87	.0577
.04	.1277	.25	.2109	.46	.1268	.67	.1315	.88	.0515
.05	.1506	.26	.2052	.47	.1260	.68	.1307	.89	.0453
.06	.1705	.27	.1994	.48	.1254	.69	.1298	.90	.0392
.07	.1874	.28	.1937	.49	.1250	.70	.1285	.91	.0331
.08	.2019	.29	.1879	.50	.1250	.71	.1269	.92	.0272
.09	.2138	.30	.1823	.51	.1251	.72	.1249	.93	.0219
.10	.2235	.31	.1767	.52	.1253	.73	.1226	.94	.0168
.11	.2311	.32	.1714	.53	.1257	.74	.1200	.95	.012207
.12	.2369	.33	.1661	.54	.1263	.75	.1171	.96	.008135
.13	.2410	.34	.1613	.55	.1269	.76	.1138	.97	.004806
.14	.2436	.35	.1567	.56	.1276	.77	.1101	.98	.002213
.15	.2448	.36	.1524	.57	.1283	.78	.1062	.99	.000576
.16	.2448	.37	.1483	.58	.1291	.79	.1018	1.00	0
.17	.2438	.38	.1446	.59	.1298	.80	.0972		
.18	.2418	.39	.1412	.60	.1305	.81	.0922		
.19	.2390	.40	.1382	.61	.1311	.82	.0870		
.20	.2355	.41	.1355	.62	.1316	.83	.0814		

TABLE XIV.

VALUES OF $\dfrac{2k(1-k)^2(2-7k+8k^2)}{1+k^2(-15+50k-60k^2+24k^3)} = \Delta_{14}.$

k	Δ_{14}	k	Δ_{14}	k	Δ_{14}	k	Δ_{14}	k	Δ_{14}
0	0	.21	0.3331	.42	0.2635	.63	0.2754	.84	0.3507
.01	0.0073	.22	.3344	.43	.2604	.64	.2787	.85	.3542
.02	.0719	.23	.3346	.44	.2575	.65	.2822	.86	.3581
.03	.1023	.24	.3343	.45	.2551	.66	.2857	.87	.3606
.04	.1304	.25	.3332	.46	.2532	.67	.2895	.88	.3639
.05	.1555	.26	.3314	.47	.2518	.68	.2929	.89	.3670
.06	.1783	.27	.3289	.48	.2507	.69	.2967	.90	.3705
.07	.1976	.28	.3260	.49	.2501	.70	.3004	.91	.3731
.08	.2177	.29	.3224	.50	.2500	.71	.3042	.92	.3748
.09	.2346	.30	.3185	.51	.2500	.72	.3077	.93	.3797
.10	.2499	.31	.3140	.52	.2500	.73	.3112	.94	.3833
.11	.2636	.32	.3095	.53	.2514	.74	.3151	.95	.3873
.12	.2759	.33	.3043	.54	.2527	.75	.3189	.96	.3882
.13	.2868	.34	.2995	.55	.2544	.76	.3225	.97	.3940
.14	.2965	.35	.2945	.56	.2563	.77	.3261	.98	.3944
.15	.3049	.36	.2896	.57	.2583	.78	.3300	.99	.3962
.16	.3122	.37	.2845	.58	.2608	.79	.3333	1.00	.4000
.17	.3183	.38	.2793	.59	.2633	.80	.3370		
.18	.3236	.39	.2752	.60	.2661	.81	.3405		
.19	.3277	.40	.2711	.61	.2691	.82	.3438		
.20	.3309	.41	.2671	.62	.2722	.83	.3470		

TABLE XV.

VALUES OF $\dfrac{2-7k+8k^2}{6k} = \Delta_{15}.$

k	Δ_{15}	k	Δ_{15}	k	Δ_{15}	k	Δ_{15}	k	Δ_{15}
0	∞	.21	0.7006	.42	0.1870	.63	0.2024	.84	0.3502
.01	32.1600	.22	.6418	.43	.1818	.64	.2075	.85	.3588
.02	15.5266	.23	.5892	.44	.1775	.65	.2128	.86	.3676
.03	9.9844	.24	.5422	.45	.1740	.66	.2189	.87	.3765
.04	7.2116	.25	.5000	.46	.1713	.67	.2242	.88	.3855
.05	5.5666	.26	.4620	.47	.1692	.68	.2302	.89	.3945
.06	4.4688	.27	.4279	.48	.1677	.69	.2364	.90	.4037
.07	3.6888	.28	.3947	.49	.1669	.70	.2428	.91	.4126
.08	3.1070	.29	.3698	.50	.1666	.71	.2495	.92	.4223
.09	2.6572	.30	.3444	.51	.1669	.72	.2563	.93	.4318
.10	2.3000	.31	.3219	.52	.1667	.73	.2633	.94	.4413
.11	2.0107	.32	.3016	.53	.1690	.74	.2705	.95	.4509
.12	1.7711	.33	.2834	.54	.1707	.75	.2778	.96	.4606
.13	1.5709	.34	.2670	.55	.1727	.76	.2853	.97	.4703
.14	1.4011	.35	.2523	.56	.1752	.77	.2929	.98	.4801
.15	1.2555	.36	.2392	.57	.1781	.78	.3007	.99	.4900
.16	1.1301	.37	.2275	.58	.1813	.79	.3086	1.00	.5000
.17	1.0208	.38	.2171	.59	.1849	.80	.3167		
.18	0.9251	.39	.2080	.60	.1889	.81	.3249		
.19	.8410	.40	.2000	.61	.1931	.82	.3332		
.20	.7666	.41	.1930	.62	.1976	.83	.3416		

TABLE XVI.

VALUES OF $3 - 12k + 24k^2 - 16k^3 = \Delta_{16}$.

k	Δ_{16}	k	Δ_{16}	k	Δ_{16}	k	Δ_{16}	k	Δ_{16}
0	3.0000	.21	1.3902	.42	1.0081	.63	0.9648	.84	0.3711
.01	2.8823	.22	1.3512	.43	1.0054	.64	.9560	.85	.3140
.02	2.7694	.23	1.3149	.44	1.0034	.65	.9460	.86	.2535
.03	2.6611	.24	1.2812	.45	1.0020	.66	.9344	.87	.1895
.04	2.5573	.25	1.2500	.46	1.0010	.67	.9213	.88	.1220
.05	2.4580	.26	1.2211	.47	1.0004	.68	.9066	.89	.0508
.06	2.3629	.27	1.1946	.48	1.0001	.69	.8902		Negative.
.07	2.2721	.28	1.1703	.49	1.0000	.70	.8720	.90	0.0240
.08	2.1854	.29	1.1481	.50	1.0000	.71	.8518	.91	.1027
.09	2.1027	.20	1.1280	.51	0.9999	.72	.8296	.92	.1854
.10	2.0240	.31	1.1097	.52	.9998	.73	.8053	.93	.2721
.11	1.9491	.32	1.0933	.53	.9995	.74	.7788	.94	.3629
.12	1.8779	.33	1.0786	.54	.9989	.75	.7500	.95	.4580
.13	1.8104	.34	1.0655	.55	.9980	.76	.7187	.96	.5573
.14	1.7464	.35	1.0540	.56	.9965	.77	.6850	.97	.6611
.15	1.6860	.36	1.0439	.57	.9945	.78	.6487	.98	.7694
.16	1.6288	.37	1.0351	.58	.9918	.79	.6097	.99	.8823
.17	1.5749	.38	1.0276	.59	.9883	.80	.5680	1.00	1.0000
.18	1.5242	.39	1.0212	.60	.9840	.81	.5233		
.19	1.4766	.40	1.0160	.61	.9787	.82	.4757		
.20	1.4320	.41	1.0116	.62	.9723	.83	.4250		

TABLE XVII (Bresse).

Values of $\frac{A}{B}$ in Equation (109) $H_1 = \sum^{l} P \frac{A}{B}$.

$\frac{2\phi_0}{\pi}$	Values of $\frac{a}{\phi_0}$.									
	0.00	0.05	0.10	0.15	0.20	0.25	0.30	0.35	0.40	0.45
0.12	4.125	4.112	4.075	4.012	3.926	3.816	3.682	3.526	3.348	3.149
13	3.804	3.793	3.758	3.700	3.621	3.519	3.396	3.251	3.087	2.903
.14	3.529	3.518	3.486	3.432	3.359	3.264	3.150	3.016	2.863	2.692
.15	3.291	3.281	3.251	3.200	3.132	3.043	2.936	2.811	2.669	2.509
.16	3.082	3.072	3.044	2.997	2.933	2.862	2.749	2.632	2.498	2.349
.17	2.897	2.888	2.862	2.817	2.757	2.679	2.584	2.474	2.348	2.207
.18	2.733	2.725	2.700	2.657	2.600	2.526	2.437	2.333	2.214	2.081
.19	2.586	2.578	2.554	2.514	2.460	2.390	2.305	2.206	2.094	1.968
.20	2.453	2.446	2.423	2.385	2.334	2.267	2.187	2.093	1.985	1.866
.21	2.333	2.326	2.304	2.268	2.219	2.156	2.079	1.989	1.887	1.774
.22	2 224	2.217	2.196	2.162	2.115	2.054	1.981	1.895	1.798	1.689
.23	2.124	2.117	2.098	2.064	2.019	1.961	1.891	1.809	1.716	1.612
.24	2.032	2.026	2.007	1.975	1.932	1.876	1.809	1.730	1.641	1.541
.25	1.947	1.941	1.923	1.893	1.851	1.798	1.733	1.658	1.572	1.476
.26	1.869	1.863	1.846	1.817	1.777	1.725	1.663	1.590	1.508	1.416
.27	1.797	1.791	1.774	1.746	1.707	1.658	1.598	1.528	1.448	1.360
.28	1.729	1.724	1.708	1.680	1.643	1.595	1.537	1.470	1.393	1.308
.29	1.666	1.661	1.645	1.619	1.583	1.537	1.481	1.415	1.341	1.259
.30	1.607	1.602	1.587	1.561	1.527	1.482	1.428	1.365	1.293	1.213
.31	1.552	1.547	1.533	1.508	1.474	1.431	1.378	1.317	1.248	1.170
.32	1.500	1.496	1.481	1.457	1.424	1.389	1.332	1.272	1.205	1.130
.33	1.452	1.447	1.433	1.410	1.378	1.337	1.288	1.230	1.165	1.092
.34	1.406	1.401	1.388	1.365	1.334	1.294	1.246	1.190	1.127	1.057
.35	1.362	1.358	1.344	1.322	1.292	1.254	1.207	1.153	1.091	1.023
.36	1.321	1.317	1.304	1.282	1.253	1.215	1.170	1.117	1.057	0.991
.37	1.282	1.278	1.265	1.244	1.216	1.179	1.135	1.083	1.025	.960
.38	1.245	1.241	1.228	1.208	1.180	1.144	1.101	1.051	0.994	.931
.39	1.209	1.205	1.194	1.174	1.146	1.111	1.069	1.021	.965	.904
.40	1.176	1.172	1.160	1.142	1.114	1.080	1.039	0.991	.937	.877
.42	1.113	1.109	1.098	1.080	1.054	1.022	0.983	.937	.885	.828
.44	1.056	1.052	1.042	1.024	0.999	0.968	.931	.887	.838	.783
.46	1.003	1.000	0.990	0.972	.949	.919	.883	.841	.794	.742
.48	0.955	0.951	.942	.925	.903	.874	.839	.799	.754	.704
.50	.910	.907	.897	.881	.859	.832	.798	.760	.716	.668
.52	.868	.865	.856	.840	.819	.793	.760	.723	.681	.635
.54	.829	.826	.817	.802	.782	.756	.725	.689	.648	.604
.56	.793	.790	.781	.767	.747	.722	.692	.657	.618	.575
.58	.758	.756	.747	.733	.714	.690	.661	.627	.589	.548
.60	.726	.723	.715	.702	.683	.659	.631	.599	.562	.522
.62	.696	.693	.685	.672	.654	.631	.603	.572	.536	.497
.64	.667	.665	.657	.644	.626	.607	.577	.546	.512	.474
.68	.614	.612	.604	.592	.575	.554	.528	.499	.467	.431
.72	.566	.564	.557	.545	.529	.508	.484	.456	.426	.392
.76	.522	.520	.516	.502	.486	.467	.444	.417	.388	.356
.80	.482	.480	.473	.462	.447	.429	.406	.381	.353	.323
.84	.445	.443	.436	.426	.411	.393	.372	.347	.320	.292
.88	.410	.408	.402	.391	.378	.360	.339	.316	.290	.262
.92	.378	.376	.370	.360	.346	.329	.309	.286	.261	.235
.96	.347	.345	.349	.329	.316	.300	.280	.258	.234	.209
1.00	.318	.316	.311	.301	.288	.272	.253	.231	.208	.184

TABLE XVII—*Continued.*

$\frac{2\phi_0}{\pi}$	Values of $\frac{a}{\phi_0}$.									
	0.50	0.55	0.60	0.65	0.70	0.75	0.80	0.85	0.90	0.95
0.12	2.931	2.694	2.441	2.171	1.888	1.592	1.286	0.972	0.651	0.327
.13	2.702	2.484	2.250	2.001	1.740	1.467	1.185	.895	.600	.301
.14	2.506	2.303	2.086	1.855	1.612	1.360	1.098	.830	.556	.279
.15	2.335	2.146	1.943	1.728	1.502	1.266	1.023	.772	.517	.259
.16	2.186	2.008	1.818	1.617	1.405	1.184	0.956	.722	.484	.242
.17	2.054	1.887	1.708	1.518	1.319	1.112	.898	.678	.454	.227
.18	1.936	1.778	1.610	1.431	1.243	1.048	.845	.638	.427	.214
.19	1.830	1.681	1.521	1.352	1.175	0.990	.799	.603	.403	.202
.20	1.735	1.594	1.442	1.281	1.112	.937	.756	.571	.382	.191
.21	1.649	1.514	1.370	1.217	1.057	.890	.718	.542	.362	.181
.22	1.571	1.442	1.304	1.159	1.006	.847	.683	.515	.344	.172
.23	1.499	1.376	1.244	1.105	0.959	.807	.651	.491	.328	.164
.24	1.433	1.315	1.189	1.056	.916	.771	.621	.468	.313	.157
.25	1.372	1.259	1.138	1.010	.876	.737	.594	.448	.299	.149
.26	1.315	1.207	1.091	0.968	.839	.706	.569	.428	.286	.143
.27	1.263	1.158	1.047	.929	.805	.677	.545	.411	.274	.137
.28	1.214	1.114	1.006	.892	.773	.650	.523	.394	.263	.131
.29	1.169	1.072	0.968	.858	.744	.625	.503	.379	.253	.126
.30	1.126	1.032	.932	.826	.716	.601	.484	.364	.243	.121
.31	1.086	0.995	.899	.796	.690	.579	.466	.350	.234	.116
.33	1.049	.961	.867	.768	.665	.558	.449	.337	.225	.112
.33	1.013	.928	.837	.742	.642	.539	.433	.325	.217	.108
.34	0.980	.897	.809	.716	.620	.520	.418	.314	.209	.104
.35	.948	.868	.782	.693	.599	.502	.403	.303	.202	.100
.36	.918	.840	.757	.670	.579	.486	.390	.292	.195	.097
.37	.890	.814	.733	.649	.560	.470	.377	.283	.188	.093
.38	.863	.789	.711	.628	.543	.454	.364	.273	.181	.090
.39	.837	.765	.689	.609	.526	.440	.353	.264	.175	.087
.40	.812	.742	.668	.590	.509	.426	.341	.256	.170	.084
.42	.766	.700	.629	.555	.479	.400	.320	.240	.159	.079
.44	.724	.661	.594	.524	.451	.377	.301	.225	.149	.074
.46	.685	.625	.561	.494	.425	.345	.283	.211	.140	.069
.48	.650	.592	.531	.467	.401	.334	.266	.198	.131	.065
.50	.616	.559	.502	.442	.379	.315	.251	.187	.123	.061
.52	.585	.532	.476	.418	.358	.297	.236	.176	.115	.057
.54	.556	.505	.451	.396	.339	.281	.223	.165	.108	.053
.56	.529	.480	.428	.375	.320	.265	.210	.155	.102	.050
.58	.503	.456	.406	.355	.303	.250	.198	.146	.096	.047
.60	.479	.433	.385	.336	.285	.236	.186	.137	.090	.044
.62	.456	.412	.366	.319	.271	.223	.175	.129	.084	.041
.64	.434	.391	.347	.302	.256	.210	.165	.121	.078	.038
.68	.393	.354	.313	.271	.228	.187	.146	.106	.068	.033
.72	.356	.319	.281	.242	.203	.165	.128	.092	.059	.028
.76	.322	.287	.251	.215	.180	.145	.111	.080	.050	.024
.80	.291	.258	.224	.191	.158	.126	.096	.068	.042	.019
.84	.261	.230	.199	.168	.137	.108	.081	.057	.035	.016
.88	.234	.204	.175	.146	.118	.092	.068	.046	.027	.012
.92	.208	.180	.152	.125	.100	.076	.055	.036	.021	.008
.96	.183	.157	.131	.106	.082	.061	.042	.027	.014	.005
.100	.159	.134	.110	.087	.066	.047	.030	.017	.008	.002

TABLE XVIII.

VALUES OF $2x^\circ \cos^2 x^\circ - 3 \sin x^\circ \cos x^\circ + x^\circ = \Delta_{18}$.

x°	Δ_{18}	x°	Δ_{18}	x°	Δ_{18}	x°	Δ_{18}
1		24	0.0033256	47	0.0870414	70	0.5433799
2		25	.0040687	48	.0961634	71	.5783850
3		26	.0049329	49	.1059980	72	.6149548
4		27	.0059379	50	.1165809	73	.6531228
5		28	.0071012	51	.1279485	74	.6929174
6		29	.0084352	52	.1401378	75	.7343691
7		30	.0099597	53	.1532454	76	.7775070
8		31	.0108680	54	.1671294	77	.8201831
9		32	.0136523	55	.1834638	78	.8689473
10		33	.0158626	56	.1978582	79	.9172998
11		34	.0183339	57	.2153204	80	.9674381
12	0.0001064	35	.0211197	58	.2336325	81	1.0193834
13	.0001586	36	.0242131	59	.2508080	82	1.0731551
14	.0002295	37	.0277103	60	.2717593	83	1.1287710
15	.0003238	38	.0314555	61	.2930503	84	1.18 From Diagram.
16	.0004463	39	.0356564	62	.3155469	85	1.25
17	.0006032	40	.0402812	63	.3391808	86	1.30
18	.0008006	41	.0453581	64	.3643046	87	1.37
19	.0010472	42	.0509171	65	.3906428	88	1.43
20	.0013503	43	.0569885	66	.4183343	89	1.50
21	.0017190	44	.0636040	67	.4474183	90	1.5707963
22	.0021640	45	.0707964	68	.4779307		
23	.0026954	46	.0785976	69	.5099063		

TABLE XIX.

VALUES OF $x^\circ + \sin x^\circ \cos x^\circ = \Delta_{19}$.
" " $x^\circ - \sin x^\circ \cos x^\circ = \beta_{19}$.

x°	Δ_{19}	β_{19}	x°	Δ_{19}	β_{19}
0	0	0	46	1.3025470	0.3031560
1	0.0349031	0.0000035	47	1.3190868	.3215226
2	.0697848	.0000284	48	1.3350190	.3404970
3	.1046243	.0000955	49	1.3503454	.3600772
4	.1393998	.0002266	50	1.3650685	.3802607
5	.1740906	.0004424	51	1.3791917	.4010441
6	.2086737	.0007659	52	1.3927190	.4224234
7	.2431339	.0012121	53	1.4056552	.4443938
8	.2774450	.0018076	54	1.4180060	.4669496
9	.3115881	.0025711	55	1.4297774	.4900848
10	.3455421	.0035219	56	1.4409764	.5137924
11	.3792895	.0046829	57	1.4516104	.5380650
12	.4128078	.0060712	58	1.4616881	.5628939
13	.4460784	.0077072	59	1.4714926	.5879960
14	.4790819	.0096103	60	1.4802103	.6141849
15	.5117994	.0117994	61	1.4886749	.6406267
16	.5442123	.0142931	62	1.4966229	.6675853
17	.5763024	.0171096	63	1.5040658	.6950490
18	.6080520	.0202666	64	1.5110162	.7230052
19	.6394434	.0237818	65	1.5174862	.7514418
20	.6704597	.0276721	66	1.5234897	.7803449
21	.7010844	.0319538	67	1.5290405	.8097007
22	.7313016	.0366432	68	1.5341531	.8394947
23	.7610956	.0417558	69	1.5388425	.8697119
24	.7904514	.0473066	70	1.5431243	.9003367
25	.8193545	.0533101	71	1.5470146	.9313530
26	.8477911	.0597801	72	1.5505298	.9627444
27	.8757473	.0667305	73	1.5536868	.9944940
28	.9032110	.0741734	74	1.5565032	1.0265840
29	.9301696	.0821214	75	1.5589969	1.0589969
30	.9566115	.0905861	76	1.5611860	1.0917144
31	.9828004	.0993038	77	1.5630891	1.1247179
32	1.0079025	.1091083	78	1.5647251	1.1579885
33	1.0327314	.1191860	79	1.5661134	1.1915068
34	1.0570039	.1298199	80	1.5672735	1.2252533
35	1.0807115	.1410189	81	1.5682252	1.2592082
36	1.1038467	.1527903	82	1.5689887	1.2933513
37	1.1264025	.1651411	83	1.5695842	1.3276624
38	1.1513729	.1810773	84	1.5700305	1.3621227
39	1.1697522	.1916046	85	1.5703540	1.3967058
40	1.1905356	.2057278	86	1.5705698	1.4313966
41	1.2107191	.2204509	87	1.5707008	1.4661720
42	1.2302993	.2357773	88	1.5707679	1.5010115
43	1.2492737	.2517095	89	1.5707928	1.5358932
44	1.2676404	.2682494	90	1.5707963	1.5707963
45	1.2853982	.2853982			

TABLE XIX—*Continued.*

VALUES OF $\sin x^\circ - x^\circ \cos x^\circ = \Delta\Delta_{19}$.

x°	Δ_{19}	x°	Δ_{19}	x°	Δ_{19}	x°	Δ_{19}
0	0	23	0.0212167	46	0.1616323	69	0.5020059
1		24	.0240716	47	.1719073	70	.5218362
2		25	.0271669	48	.1825754	71	.5420800
3		26	.0305115	49	.1936404	72	.5627342
4	0.0001134	27	.0341145	50	.2051066	73	.5837967
5	.0002203	28	.0379820	51	.2169764	74	.6052637
6	.0003829	29	.0421247	52	.2292541	75	.6271325
7	.0006071	30	.0465502	53	.2419419	76	.6493983
8	.0009047	31	.0512658	54	.2550423	77	.6720575
9	.0012888	32	.0562797	55	.2685580	78	.6951057
10	.0017668	33	.0615995	56	.2824910	79	.7185380
11	.0023501	34	.0672321	57	.2968431	80	.7423494
12	.0030490	35	.0731849	58	.3116145	81	.7665342
13	.0038735	36	.0794651	59	.3268099	82	.7910878
14	.0048339	37	.0860787	60	.3424266	83	.8160034
15	.0059402	38	.0930329	61	.3584668	84	.8412750
16	.0072019	39	.1003340	62	.3749305	85	.8668965
17	.0086304	40	.1079877	63	.3919339	86	.8928607
18	.0103238	41	.1159999	64	.4091287	87	.9191605
19	.0120222	42	.1243768	65	.4268617	88	.9457892
20	.0140056	43	.1331235	66	.4450185	89	.9927379
21	.0161930	44	.1422451	67	.4635955	90	1.0000000
22	.0185936	45	.1517464	68	.4825916		

TABLE XX.

VALUES OF $x^{\circ 2} + x^\circ \sin x^\circ \cos x^\circ - 2 \sin^2 x^\circ = \Delta_{20}$.

x°	Δ_{20}	x°	Δ_{20}	x°	Δ_{20}	x°	Δ_{20}
0	0.0000	23	0.0001818	46	0.0108522	69	0.1100482
1		24	.0002342	47	.0122966	70	.1191376
2		25	.0002985	48	.0138994	71	.1290242
3		26	.0003766	49	.0156577	72	.1385871
4		27	.0004688	50	.0175990	73	.1504997
5		28	.0005851	51	.0197320	74	.1622442
6		29	.0007206	52	.0220699	75	.1746965
7		30	.0008810	53	.0246290	76	.1878785
8		31	.0010693	54	.0274221	77	.1971387
9		32	.0012902	55	.0304683	78	.2166033
10		33	.0015486	56	.0337811	79	.2321884
11		34	.0018452	57	.0373797	80	.2486336
12		35	.0021893	58	.0412871	81	.2659695
13		36	.0025843	59	.0455069	82	.2837735
14		37	.0030365	60	.0500725	83	.3034412
15		38	.0035517	61	.0549993	84	.3236408
16		39	.0041368	62	.0603087	85	.3448606
17		40	.0047987	63	.0663576	86	.3671316
18		41	.0055458	64	.0721610	87	.3904870
19		42	.0063850	65	.0787504	88	.4149618
20		43	.0077259	66	.0858036	89	.4405888
21	0.0000776	44	.0083771	67	.0933561	90	.4674012
22	0.0001394	45	.0095494	68	.1014311		

TABLE XXI.

VALUES OF $2 \sin x^\circ \cos x^\circ + x^\circ \sin^2 x^\circ = \Delta_{21}$.

x°	Δ_{21}	x°	Δ_{21}	x°	Δ_{21}	x°	Δ_{21}
0	0	23	0.7806258	46	1.4148264	69	1.7187453
1	0.0349049	24	.8124424	47	1.4363274	70	1.7216027
2	.0697989	25	.8439761	48	1.4571858	71	1.7235986
3	.1046722	26	.8752146	49	1.4773852	72	1.7244246
4	.1385129	27	.9061424	50	1.4959084	73	1.7243722
5	.1743111	28	.9367471	51	1.5157396	74	1.7233366
6	.2090523	29	.9670128	52	1.5338617	75	1.7213107
7	.2437363	30	.9969252	53	1.5512591	76	1.7210540
8	.2783410	31	1.0270184	54	1.5679161	77	1.7142691
9	.3128610	32	1.0556305	55	1.5838162	78	1.7092456
10	.3472830	33	1.0843930	56	1.5989435	79	1.7032170
11	.3815964	34	1.1127420	57	1.6132827	80	1.6961813
12	.4157902	35	1.1406611	58	1.6268207	81	1.6885656
13	.4498526	36	1.1681353	59	1.6400866	82	1.6790869
14	.4837723	37	1.1951479	60	1.6514236	83	1.6690300
15	.5175372	38	1.2216839	61	1.6624632	84	1.6579655
16	.5501357	39	1.2477263	62	1.6726419	85	1.6459092
17	.5845556	40	1.2732589	63	1.6817446	86	1.6318529
18	.6177850	41	1.2982657	64	1.6903668	87	1.6188062
19	.6509107	42	1.3227295	65	1.6978876	88	1.6037755
20	.6836206	43	1.3466341	66	1.7044952	89	1.5877332
21	.7162018	44	1.3699635	67	1.7101816	90	1.5707963
22	.7485413	45	1.3926991	68	1.7149359		

TABLE XXII.

VALUES OF $\cos x^\circ + x^\circ \sin x^\circ = \Delta_{22}$.

x°	Δ_{22}	x°	Δ_{22}	x°	Δ_{22}	x°	Δ_{22}
0	1.0000	23	1.0773544	46	1.2721814	69	1.4826576
1	1.0001521	24	1.0839186	47	1.2819313	70	1.4900699
2	1.0006092	25	1.0907097	48	1.2917062	71	1.4972383
3	1.0013697	26	1.0977206	49	1.3014952	72	1.5042390
4	1.0024339	27	1.1049445	50	1.3112876	73	1.5107902
5	1.0038005	28	1.1123748	51	1.3210720	74	1.5171505
6	1.0054680	29	1.1200040	52	1.3308372	75	1.5232129
7	1.0074354	30	1.1278248	55	1.3405723	76	1.5289699
8	1.0097006	31	1.1358278	54	1.3502657	77	1.5344104
9	1.0122609	32	1.1440108	55	1.3599061	78	1.5395194
10	1.0151152	33	1.1523602	56	1.3694812	79	1.5442864
11	1.0182597	34	1.1608693	57	1.3789801	80	1.5486991
12	1.0216925	35	1.1695300	58	1.3883923	81	1.5527461
13	1.0254098	36	1.1783332	59	1.3977012	82	1.5564119
14	1.0294083	37	1.1872707	60	1.4068996	83	1.5596947
15	1.0336845	38	1.1963330	61	1.4159743	84	1.5625740
16	1.0382339	39	1.2055108	62	1.4249128	85	1.5650397
17	1.0430530	40	1.2147947	63	1.4337014	86	1.5670838
18	1.0481371	41	1.2241757	64	1.4423340	87	1.5686913
19	1.0534812	42	1.2336435	65	1.4507935	88	1.5698536
20	1.0590802	43	1.2431877	66	1.4590655	89	1.5705585
21	1.0649291	44	1.2527990	67	1.4671423	90	1.5707963
22	1.0710225	45	1.2624672	68	1.4750112		

TABLE XXIII

VALUES OF $x^{\circ 2} - x^\circ \sin x^\circ \cos x^\circ = \Delta_{23}$.

x°	Δ_{23}	x°	Δ_{23}	x°	Δ_{23}	x°	Δ_{23}
0		23	0.0167618	46	0.2433892	69	1.0473746
1		24	.0198158	47	.2637466	70	1.0998688
2		25	.0232609	48	.2852497	71	1.1541174
3		26	.0271274	49	.3079423	72	1.2106700
4		27	.0314472	50	.3318402	73	1.2670749
5		28	.0362481	51	.3569768	74	1.3258782
6		29	.0415654	52	.3833793	75	1.3862235
7	0.0001481	30	.0564302	53	.4110740	76	1.4481051
8	.0002521	31	.0538771	54	.4400895	77	1.5162235
9	.0004044	32	.0609376	55	.4704481	78	1.5764351
10	.0006148	33	.0686462	56	.5021727	79	1.6528610
11	.0008991	34	.0770368	57	.5352871	80	1.7107760
12	.0012617	35	.0861435	58	.5698155	81	1.7801637
13	.0017487	36	.0960009	59	.6057683	82	1.8514589
14	.0023472	37	.1066433	60	.6431729	83	1.9232828
15	.0030891	38	.1181055	61	.6820439	84	1.9969740
16	.0039914	39	.1304212	62	.7223967	85	2.0720558
17	.0050765	40	.1436249	63	.7637878	86	2.1485030
18	.0063670	41	.1577514	64	.8076048	87	2.2262890
19	.0078863	42	.1728338	65	.8424868	88	2.3053880
20	.0096600	43	.1885059	66	.8988928	89	2.3857684
21	.0117398	44	.2060007	67	.9468397	90	2.4674012
22	.0140700	45	.2241512	68	.9963335		

TABLE XXIV.

VALUES OF $x^{\circ 2} + x^{\circ} \sin x^{\circ} \cos x^{\circ} = \varDelta_{24}$.

x°	$\varDelta_{24}$	x°	$\varDelta_{24}$	x°	$\varDelta_{24}$	x°	$\varDelta_{24}$
0	0	23	0.3055234	46	1.0457516	69	1.8531934
1	0.0006092	24	.3311036	47	1.0820530	70	1.8851820
2	.0024360	25	.3575109	48	1.1184278	71	1.9170352
3	.0054782	26	.3847152	49	1.1548309	72	1.9476041
4	.0097319	27	.4126846	50	1.1912472	73	1.9795371
5	.0151922	28	.4413923	51	1.2276436	74	2.0102920
6	.0218524	29	.4708012	52	1.2639917	75	2.0407219
7	.0297045	30	.5008810	53	1.3002664	76	2.0708359
8	.0387385	31	.5315977	54	1.3364391	77	2.0959325
9	.0489436	32	.5629190	55	1.3724885	78	2.1301487
10	.0603086	33	.5948106	56	1.4083877	79	2.1593724
11	.0728185	34	.6272388	57	1.4441165	80	2.1883262
12	.0864681	35	.6601691	58	1.4796585	81	2.2170261
13	.1012119	36	.6935673	59	1.5149785	82	2.2450355
14	.1170628	37	.7273991	60	1.5500725	83	2.2737366
15	.1339887	38	.7616299	61	1.5849191	84	2.3017884
16	.1519728	39	.7962252	62	1.6195015	85	2.3296680
17	.1709923	40	.8311505	63	1.6531418	86	2.3573996
18	.1911252	41	.8663726	64	1.6878216	87	2.3850088
19	.2120475	42	.9018566	65	1.7215380	88	2.4125262
20	.2339556	43	.9379693	66	1.7549342	89	2.4399794
21	.2569324	44	.9734777	67	1.7880145	90	2.4674012
22	.2808096	45	1.0095494	68	1.8207711		

TABLE XXV (Winkler).

x	Arc *x*		(Arc *x*)²		
0	0	1.5707963	0	2.4674012	90
1	0.0174533	1.5533430	0.0003046	2.4128739	89
2	.0349066	1.5358897	.0012185	2.3589571	88
3	.0523599	1.5184364	.0027416	2.3056489	87
4	.0698132	1.5009832	.0048739	2.2529513	86
5	.0872665	1.4835299	.0076154	2.2008619	85
6	.1047198	1.4660766	.0109662	2.1493812	84
7	.1221730	1.4486233	.0149263	2.0985097	83
8	.1396263	1.4311700	.0194953	2.0482472	82
9	.1570796	1.4137167	.0246740	1.9985949	81
10	.1745329	1.3962634	.0304617	1.9495511	80
11	.1919862	1.3788101	.0368588	1.9011167	79
12	.2094395	1.3613568	.0438649	1.8532919	78
13	.2268928	1.3439035	.0514803	1.8060780	77
14	.2443461	1.3264502	.0597050	1.7594705	76
15	.2617994	1.3089969	.0685389	1.7134727	75
16	.2792527	1.2915436	.0779821	1.6680851	74
17	.2967060	1.2740904	.0880344	1.6233060	73
18	.3141593	1.2566371	.0986961	1.5791371	72
19	.3316126	1.2391838	.1099669	1.5355763	71
20	.3490659	1.2217305	.1218476	1.4925254	70
21	.3665191	1.2042772	.1343361	1.4502840	69
22	.3839724	1.1868239	.1474348	1.4085523	68
23	.4014257	1.1693706	.1611426	1.3674271	67
24	.4188790	1.1519173	.1754597	1.3269135	66
25	.4363323	1.1344640	.1903859	1.2870124	65
26	.4537856	1.1170107	.2059213	1.2477132	64
27	.4712389	1.0995574	.2220651	1.2084648	63
28	.4886922	1.0821041	.2388202	1.1709491	62
29	.5061455	1.0646508	.2561833	1.1334815	61
30	.5235988	1.0471976	.2741556	1.0966227	60
31	.5410521	1.0297443	.2927374	1.0603734	59
32	.5585054	1.0122910	.3119283	1.0247370	58
33	.5759587	0.9948377	.3317284	0.9897018	57
34	.5934119	0.9773844	.3521378	.9552802	56
35	.6108652	0.9599311	.3731563	.9214683	55
36	.6283185	0.9424778	.3947841	.8882643	54
37	.6457718	0.9250245	.4170212	.8556702	53
38	.6632251	0.9075712	.4398677	.8236855	52
39	.6806784	0.8901179	.4633232	.7923102	51
40	.6981317	0.8726646	.4873877	.7615437	50
41	.7155850	0.8552113	.5120620	.7313866	49
42	.7330383	0.8377580	.5373452	.7018386	48
43	.7504916	0.8203047	.5632376	.6728998	47
44	.7679449	0.8028515	.5897392	.6445704	46
45	.7853982	0.7853982	.6168503	.6168503	45
		Arc *x*		(Arc *x*)²	*x*

TABLE XXVI (WINKLER).

x	Sin x	Cos x	1 − cos x		
0	0	1	0	0	90
1	0.0174524	0.9998475	0.0001525	0.9825476	89
2	.0348995	.9993910	.0006090	.9651005	88
3	.0523359	.9986293	.0013707	.9476641	87
4	.0697565	.9975640	.0024360	.9302435	86
5	.0871547	.9961947	.0038053	.9128453	85
6	.1045287	.9945218	.0054782	.8954713	84
7	.1218694	.9925462	.0074538	.8781306	83
8	.1391731	.9902682	.0097318	.8608269	82
9	.1564345	.9876882	.0123118	.8435655	81
10	.1736482	.9848079	.0151921	.8263518	80
11	.1908090	.9816273	.0183727	.8091910	79
12	.2079117	.9781476	.0218524	.7920883	78
13	.2249510	.9743700	.0256300	.7750490	77
14	.2419219	.9702957	.0297043	.7580781	76
15	.2588190	.9659258	.0340742	.7411810	75
16	.2756374	.9612614	.0387386	.7243626	74
17	.2923717	.9563046	.0436954	.7076283	73
18	.3091070	.9510565	.0489435	.6908930	72
19	.3255681	.9455187	.0544813	.6744319	71
20	.3420202	.9396926	.0603074	.6579798	70
21	.3583680	.9335804	.0664196	.6416310	69
22	.3746066	.9271839	.0728161	.6253934	68
23	.3907311	.9205049	.0794951	.6092689	67
24	.4067366	.9135455	.0864545	.5932634	66
25	.4226183	.9063077	.0936923	.5773817	65
26	.4383712	.8987941	.1012059	.5616288	64
27	.4539905	.8910065	.1089935	.5460095	63
28	.4694717	.8829476	.1170524	.5305283	62
29	.4848096	.8746198	.1253802	.5151904	61
30	.5000000	.8660254	.1339746	.5000000	60
31	.5150380	.8571673	.1428327	.4849620	59
32	.5299192	.8480480	.1519520	.4700808	58
33	.5446391	.8386706	.1613294	.4553609	57
34	.5591929	.8290375	.1709625	.4408071	56
35	.5735764	.8191521	.1808479	.4264236	55
36	.5877853	.8090169	.1909831	.4122147	54
37	.6018150	.7986355	.2013645	.3981850	53
38	.6156615	.7880108	.2119892	.3843385	52
39	.6293204	.7771459	.2228541	.3706796	51
40	.6427876	.7660446	.2339554	.3572124	50
41	.6560589	.7547096	.2452904	.3439411	49
42	.6691306	.7431449	.2568551	.3308694	48
43	.6819983	.7313537	.2686463	.3180017	47
44	.6946584	.7193398	.2806602	.3053416	46
45	.7071068	.7071068	.2928932	.2928932	45
	cos x	sin x		1 − cos x	x

TABLE XXVII (Winkler).

x	$\sin^2 x$	$\cos^2 x$	$\sin x \cos x$	$\sin^3 x$	$\cos^3 x$	
0	0	1	0	0	1	90
1	0.0003046	0.9996953	0.0174498	0.0000053	0.9995428	89
2	.0012180	.9987822	.0348782	.0000425	.9981739	88
3	.0027391	.9972609	.0522644	.0001437	.9958942	87
4	.0048660	.9951340	.0695866	.0003394	.9927100	86
5	.0075961	.9924037	.0868241	.0006620	.9886273	85
6	.0109262	.9890738	.1039539	.0011421	.9836552	84
7	.0148521	.9851477	.1209609	.0018100	.9778047	83
8	.0193692	.9806310	.1378187	.0026957	.9710876	82
9	.0244717	.9755283	.1545085	.0038282	.9635178	81
10	.0301537	.9698463	.1710101	.0052361	.9551124	80
11	.0364080	.9635920	.1873033	.0069470	.9458883	79
12	.0432273	.9567727	.2033683	.0089875	.9358650	78
13	.0506030	.9493969	.2191856	.0113832	.9250638	77
14	.0585262	.9414737	.2347358	.0141588	.9135079	76
15	.0669873	.9330127	.2500000	.0173376	.9012213	75
16	.0759760	.9240239	.2649596	.0209418	.8882288	74
17	.0854812	.9145187	.2795964	.0249923	.8745788	73
18	.0954915	.9045085	.2938927	.0295085	.8602386	72
19	.1059947	.8940055	.3078308	.0345085	.8452989	71
20	.1169778	.8830222	.3213938	.0400088	.8297694	70
21	.1284274	.8715726	.3345653	.0465573	.8136828	69
22	.1403301	.8596700	.3473292	.0525686	.7970722	68
23	.1526708	.8473292	.3596699	.0596532	.7799707	67
24	.1654347	.8345653	.3715724	.0672884	.7624135	66
25	.1786062	.8213938	.3830222	.0754823	.7444355	65
26	.1921693	.8078308	.3940055	.0842415	.7260735	64
27	.2061079	.7938921	.4045084	.0935708	.7073636	63
28	.2204036	.7795964	.4145188	.1034732	.6883427	62
29	.2350403	.7649599	.4240241	,1139498	.6690480	61
30	.2500000	.7500000	.4330127	.1250000	.6495189	60
31	.2652642	.7347358	.4417483	.1366211	.6297915	59
32	.2808144	.7191857	.4493971	.1488090	.6099040	58
33	.2966310	.7033684	.4567727	.1615573	.5898943	57
34	.3126968	.6873033	.4635920	.1748579	.5698003	56
35	.3289899	.6710101	.4698463	.1887009	.5496592	55
36	.3454915	.6545085	.4755282	.2030748	.5295083	54
37	.3621813	.6378187	.4806307	.2179661	.5093846	53
38	.3790391	.6209609	.4851478	.2333598	.4893237	52
39	.3960442	.6039558	.4890738	.2492387	.4693619	51
40	.4131759	.5868241	.4924039	.2655844	.4495335	50
41	.4304134	.5695866	.4951341	.2823766	.4298725	49
42	.4477358	.5522642	.4972610	.2995937	.4104124	48
43	.4651217	.5348782	.4987821	.3172122	.3911853	47
44	.4825503	.5174497	.4996955	.3352075	.3722223	46
45	.5000000	.5000000	.5000000	.3535534	.3535534	45
	$\cos^2 x$	$\sin^2 x$	$\sin x \cos x$	$\cos^3 x$	$\sin^3 x$	x

TABLE XXVIII (WINKLER).

x	$x \sin x$		$x \cos x$		$\frac{\sin x}{x}$		
0	0	1.5707963	0	0	1	0.6366197	90
1	0.0003046	1.5531061	0.0174506	0.0271096	0.9999486	.6436748	89
2	.0012182	1.5349541	.0348853	.0536018	.9997967	.6506919	88
3	.0027404	1.5163554	.0522881	.0794688	.9995428	.6576697	87
4	.0048699	1.4973273	.0696431	.1047033	.9991875	.6646070	86
5	.0076058	1.4778850	.0869344	.1292982	.9987307	.6715028	85
6	.0109462	1.4580453	.1041458	.1532468	.9981757	.6783559	84
7	.0148892	1.4378255	.1212623	.1765428	.9975147	.6851651	83
8	.0194324	1.4172388	.1382684	.1991804	.9967479	.6919290	82
9	.0245727	1.3963116	.1551457	.2211540	.9958928	.6986465	81
10	.0303073	1.3750509	.1718814	.2424585	.9949309	.7053168	80
11	.0366327	1.3534774	.1884589	.2630893	.9938682	.7119380	79
12	.0435449	1.3316077	.2048627	.2830419	.9927055	.7185100	78
13	.0510398	1.3094594	.2210775	.3023125	.9914420	.7250297	77
14	.0591126	1.2870480	.2370880	.3208974	.9900789	.7314980	76
15	.0677587	1.2643939	.2528788	.3387933	.9886157	.7379131	75
16	.0769725	1.2415131	.2684355	.3559977	.9870534	.7442735	74
17	.0867484	1.2184185	.2837413	.3725079	.9853918	.7505785	73
18	.0970806	1.1951320	.2987832	.3883223	.9833316	.7568267	72
19	.1079625	1.1716702	.3135459	.4034387	.9817725	.7630174	71
20	.1193876	1.1480497	.3280146	.4178564	.9798155	.7691489	70
21	.1313487	1.1242896	.3421750	.4315745	.9777607	.7752204	69
22	.1438386	1.1004046	.3560130	.4445923	.9756082	.7812309	68
23	.1568495	1.0764112	.3695144	.4569094	.9733584	.7871798	67
24	.1703731	1.0523289	.3826650	.4685270	.9710122	.7930653	66
25	.1844020	1.0281752	.3954514	.4794460	.9685698	.7989851	65
26	.1989265	1.0039628	.4078597	.4896654	.9660318	.8046422	64
27	.2139380	0 9797109	.4198760	.4990726	.9634002	.8105204	63
28	.2294272	.9554411	.4314897	.5080171	.9606691	.8159544	62
29	.2453842	.9311647	.4426849	.5161530	.9578462	.8215086	61
30	.2617994	.9068996	.4534498	.5235988	.9549298	.8269933	60
31	.2786605	.8826632	.4637722	.5303574	.9519194	.8324079	59
32	.2959628	.8584731	.4736395	.5364335	.9488167	.8377498	58
33	.3136896	.8343410	.4830396	.5418275	.9456220	.8432167	57
34	.3318318	.8102883	.4919608	.5465465	.9424352	.8482206	56
35	.3503779	.7863297	.5003915	.5505941	.9389574	.8533443	55
36	.3693163	.7624804	.5083202	.5539746	.9354896	.8583937	54
37	.3886352	.7387573	.5157363	.5566936	.9319313	.8633671	53
38	.4083222	.7151757	.5226286	.5587567	.9282843	.8682632	52
39	.4283649	.6917516	.5289864	.5601695	.9245487	.8730820	51
40	.4487501	.6685000	.5347999	.5609380	.9207255	.8778223	50
41	.4694661	.6454363	.5400590	.5610692	.9168148	.8824831	49
42	.4904986	.6225756	.5447538	.5605695	.9128181	.8870639	48
43	.5118340	.5999330	.5488748	.5594464	.9087354	.8915636	47
44	.5334592	.5775230	.5524133	.5577075	.9045682	.8959812	46
45	.5553604	.5553604	.5553604	.5553604	.9003163	.9003163	45
		$x \sin x$		$x \cos x$		$\frac{\sin x}{x}$	x

TABLE XXIX (Winkler).

x	$x \sin^2 x$		$x \cos^2 x$		$x \sin x \cos x$		
0	0	1.5707963	0	0	0	0	90
1	0.0000053	1.5528336	0.0174480	0.0004731	0.0003046	0.0271055	89
2	.0000425	1.5340191	.0348641	.0018707	.0012175	.0535055	88
3	.0001434	1.5142774	.0522165	.0041591	.0027366	.0793599	87
4	.0003397	1.4936797	.0694735	.0073037	.0048580	.1044483	86
5	.0006629	1.4722610	.0866036	.0112691	.0075768	.1288061	85
6	.0011445	1.4500577	.1035753	.0160187	.0108862	.1524072	84
7	.0018145	1.4271082	.1203585	.0215152	.0147782	.1752269	83
8	.0027036	1.4034495	.1369227	.0277206	.0192432	.1967883	82
9	.0038440	1.3795486	.1532356	.0345961	.0242696	.2184312	81
10	.0052628	1.3541611	.1692701	.0421025	.0298469	.2387751	80
11	.0069898	1.3286104	.1849964	.0501998	.0359597	.2582557	79
12	.0090536	1.3025090	.2003859	.0588477	.0426032	.2768568	78
13	.0114814	1.2758979	.2154113	.0669182	.0497316	.2898545	77
14	.0143007	1.2515824	.2300454	.0776321	.0573578	.3113654	76
15	.0175372	1.2213107	.2442622	.0876861	.0654498	.3272492	75
16	.0212165	1.1934174	.2580362	.0981263	.0739907	.3422069	74
17	.0253628	1.1651794	.2713432	.1089108	.0829579	.3562311	73
18	.0299996	1.1366392	.2841597	.1199979	.0923291	.3684670	72
19	.0351491	1.1078370	.2964635	.1313468	.1020806	.3814589	71
20	.0408330	1.0788151	.3082329	.1429154	.1121876	.3926566	70
21	.0470712	1.0496147	.3194479	.1546625	.1225963	.4029094	69
22	.0538829	1.0202775	.3300896	.1665472	.1333648	.4122188	68
23	.0612860	0.9908418	.3401397	.1785287	.1443808	.4205874	67
24	.0692976	.9613504	.3495819	.1905671	.1556439	.4280207	66
25	.0779317	.9318432	.3584015	.2026227	.1671250	.4345256	65
26	.0872036	.9023558	.3665819	.2146552	.1787939	.4401084	64
27	.0971256	.8727278	.3741122	.2265743	.1906187	.4446770	63
28	.1077095	.8436043	.3809827	.2384996	.2025721	.4485524	62
29	.1189646	.8144150	.3871810	.2502359	.2146179	.4514376	61
30	.1308998	.7853982	.3926990	.2617994	.2267254	.4534498	60
31	.1435218	.7565900	.3975304	.2731543	.2388603	.4546051	59
32	.1568363	.7280265	.4016691	.2842664	.2509907	.4549215	58
33	.1708476	.6997373	.4051110	.2951004	.2630822	.4544147	57
34	.1855580	.6717595	.4078540	.3056249	.2751010	.4531075	56
35	.2009685	.6441236	.4098967	.3165358	.2870128	.4510202	55
36	.2170789	.6168597	.4112396	.3256181	.2987832	.4481748	54
37	.2338865	.5899977	.4118853	.3350265	.3103779	.4445962	53
38	.2513883	.5635661	.4118369	.3440050	.3217622	.4403062	52
39	.2695787	.5375920	.4110997	.3525260	.3329020	.4353334	51
40	.2884511	.5121006	.4096806	.3605649	.3437628	.4297035	50
41	.3079975	.4871170	.4075877	.3680945	.3543106	.4234443	49
42	.3282075	.4626638	.4048309	.3750942	.3645114	.4165892	48
43	.3490699	.4387632	.4014216	.3815415	.3747317	.4091532	47
44	.3705725	.4154354	.3973728	.3874163	.3837385	.4011812	46
45	.3926991	.3926991	.3926991	.3926991	.3926991	.3926991	45
		$x \sin^2 x$		$x \cos^2 x$		$x \sin x \cos x$	x

TABLE

TABLE OF DATA IN CONNECTION WITH MASONRY

Name.	Date.	Engineer.	No. of Spans.	Span 2c
				Ft.
1. Trezzo, over Adda, Italy	1380+	Order of Visconti	(?)	251
2. Cabin John, Wash., D. C. Aq.	1853-9	Meigs	1	220
3. Jaremcze, Pruth, Austrian State Ry.	1892-3	Huss(?)	1	213
4. Grosvenor, over Dee, Chester, Eng	1833	Hartley	...	200
5. Gour Noir, France Ry.	1888+	Draux	1	196.8
6. Vieille Brioude, over Allier, France	.....	Romans	2	183.8
7. Ballochmyle, Ayr, Scotland Ry.	.....	Miller	...	180
8. Munderkingen, Danube, Met. Hinges (3)	1893	Bois	1	164
9. Vieille-Brioude, France	1354	Greiner & Estone	2	160
10. Wheeling, W. Va	1893	Hoge & White	1	159
11. London, New, Thames	1831	Rennie (George)	5	152
12. Gloucester, England	.....	Telford	.....	150
13. Elyria, Ohio	1886	Kinney	1	150
14. Turin, Italy	.....	Mosca	.....	148
15. Putney, over Thames	1886	Bazalgette	5	144
16. Alma, Paris	.....	Darcel	.....	141
17. Pont-y-tu-Prydd, over Taff, So. Wales	1755	Edwards	1	140
18. Coulouvrenière, Concrete, Geneva	1896	Bois	2	131.2
19. Aberdeen, over Den Burn	1801+	Telford	.....	130
20. Neuilly, Seine, France	1773 / 1774	Perronet	5	128.2
21. Mantes, Seine, France	1757 / 1765	Hupeau	3	128.2
22. Maidenhead, Eng. Ry.	1837	Brunel	2	128
23. Vingeanne Viaduct near Oisilly, France. Ry.	.....	.....	7	127
24. Bourbonnais, France Ry.	.....	Vaudray	.....	124
25. Waterloo, Thames	1816	Rennie	9	120
26. Tongueland, over Dee	1801+	Telford	1	118
27. Cresheim, Phila., Fairmount Park....Sewer	1893	Webster	1	116
28. Napoleon, Paris Ry.	.....	Couche	.....	116
29. Grand-Maitre, Fontainebleau	1869	Belgrand	.....	115.8
30. Nantes, over Loire, France	.....	.....	.....	115.2
31. Mantes, over Seine, France	1757 / 1765	Hupeau	3	115.4
32. Murr, Marbach	1887	.....	1	114.8(?)
33. Alcantara, over the Tagus, Spain	105	Trajan	6	110 (?)
34. Tees at Winstone	1762	Robinson	1	108.8
35. Orleans, over Loire, France	1750 / 1760	Hupeau	9	106.9
36. Blackfriars, over Thames	1760 / 1770	Mylne	9	100
37. Etherow River Ry.	.....	Haskoll	4	100
38. Bishop-Auckland	1388	.....	.....	100
39. Wellington, over Aire, Leeds	1819	Rennie	1	100
40. Rialto, Venice	1588 / 1591	Antonio da Ponte	1	98.5
41. Grand, Saône Charrey	1888	.....	5	98.4
42. Trinity, Florence	1566	Ammanati	3	95.8
43. Louis XIV	.....	Perronet	.....	94
44. Enz, near Hofen	1885	Liebbrand	1	91.8
45. Stoneleigh, Warwickshire	.....	Rennie	1	92
46. Dean, near Edinburgh	.....	Telford	.....	90
47. Tay at Dunkeld	1809	Telford	7	90
48. Licking Aqueduct, C. & O. Canal	.....	Fisk	.....	90
49. Dorlaston	.....	.....	.....	87

* This table has been compiled with much labor, and unfortunately many dates and be fairly correct. Authorities consulted did not agree in dates, names of the engineers, nor note have been omitted, as data concerning them were not at hand.

XXX.*

ARCHES CONSTRUCTED AT VARIOUS PERIODS.

	Rise, k.	Depth of Key, t_0.	Skew t_s.	Curve.	Spandrels.	Remarks.
	Ft.	Ft.	Ft.			
1	......	4.0	......	C.	......	Destroyed by Carmagnola 1416
2	57.0	4.2	6.2	C.	Arched void.	
3	59.6	6.89	10.2	C.	Lateral arched openings......	7 small arches.
4	42.0	4.6	7.0	C.	......	Lead in joints.
5	52.8	5.6	13.8	C.	Lateral arched openings.	
6	70.6	......	......	....	......	Fell 1822; max. span given.
7	90.5	4.5	6.0	Semi-C.		
8	16.4	3.3	3.6	C.	Longitudinal arched voids.....	Ring has max. thickness at haunches.
9	......	......	......	......	......	Span 151(?) by some.
10	28	4.5	6.0	C.	Longitudinal arched voids.	
11	37.1	4.9	10.0	E.	......Do.	Max. span. given.
12	35	4.5	......	E.		
13	27	3.8	4.5	C.	Gravel and stone...	Amherst sandstone.
14	18	4.9	......	C.		
15	19.3	4.2	4.2+	......	Longitudinal voids....	Other spans 112 and 129.
16	28	4.9	......	E.	......	Béton.
17	35	2.6	......	C.(?)	Lateral voids.	
18	18.2	3.0	3.0	C.	Longitudinal arched voids.	Ring has max. thickness at haunches.
19	43	......	......	......		
20	32	5.1	......	E.		
21	38.5	6.4(?)	......	E.	......	Side spans, 115.4 × 34.9.
22	24.2	5.3	7.0	E.	Longitudinal voids......	Brick in cement.
23	46	......	......	E.	Lateral arched openings.	
24	6.9	2.7	3.60	C.	......	Cut granite.
25	30.5	4.9	8.0	E.	Longitudinal voids	Granite.
26	38	3.6	3.6	C.	Longitudinal arched spandrels.	
27	21.2	3.5	4.0	C.	......	Buff sandstone.
28	14.8	4.0	......	C.		
29	......	......	......	......	Lateral arched openings	
30	34.4	6.4	......	E.		
31	......	......	......	E.	......	Side arches No. 21.
32	10.2	3.9	5.3	C.	Longitudinal arched voids.....	Lead at crown and springing joints.
33	55 (?)	8.3(?)	......	Semi-C.	......	Dimen. for centre spans; one span now standing.
34	......	......	......	......		
35	29.1	......	......	......	......	Dimensions for central arch.
36	......	......	......	E.	Lateral arched voids covered with side walls..	Replaced 1864–9. Dimensions for centre arch.
37	25	4.0	4.0	C.		
38	22	1.8	1.8	C.		
39	15	3.0	7.0	C.	......	Stone near site.
40	23	....	......	......		
41	12.3	3.3	4.9	C.	Spandrels filled with earth.	
42	16 (?)	3.1	....	E.	......	Side spans 87.6.
43	9.8	3.7	......	C.		
44	9.2	3.3	4.9	C.	Lateral voids......	Lead at crown and springing.
45	13	4.6(?)	......	C.		
46	30	3.0	......	C.		
47	30	......	......	C.	......	Dimensions of centre arch.
48	15	2.8	......	C.		
49	13.5	3.5	......	C.		

dimensions could not be obtained. Some of the data may not be exact, but it is believed to dimensions; in such instances only the data givenby one has been stated. Some arches of

TABLE

Name.	Date.	Engineer.	No. of Spans.	Span 2c
				Ft.
50. Viaduct over Crueize, Midland Ry., France.			6	82
51. Elkader, Iowa	1888	Tschirgi	2	84
52. Oise ... Ry.				83
53. Pont Royal, Seine, Paris	1685	Mansard	5	82.3
54. Trilport ... Ry.				81
55. Conemaugh Viaduct ... Penna. Ry.				80
56. Royal Border Viaduct				80
57. Posen Viaduct				80
58. Schuylkill Falls ... Ry.	1890	Nichols		80
59. Orleans, France				79
60. Hutcheson, Glasgow		Stephenson		79
61. Falls ... P. & R. Ry.				78
62. Tay at Perth	1760 / 1771	Smeaton	9	77.0
63. St. Maxence, over Oise, France	1774 / 1785	Perronet	3	76.1
64. Westminster ... Thames	1747	Labalye	13	76.0
65. Albany St. Arch, N. Brunswick ... N. J.	1893	Dean & Westbrook	7	75
66. Earn River, Scotland		Rennie	3	75
67. Allentown, England		Stephenson		75
68. Black Rock Tunnel Bridge ... P. & R. Ry.		Robinson		72
69. Edinburgh		Mylne		72
70. Swatara ... P. & R. Ry.		Osborn		70
71. Brent Viaduct, England ... Ry.	1837	Brunel	8	70
72. Over Lea ... Ry.		Brathwaite		70
73. Wellesley, Limerick		Nimmo	5	70
74. Staines, Thames		Geo. Rennie	3	66
75. Over Ouse, near York ... Ry.		Greene	3	66
76. Bow, over Lea	1837	Walker & Burgess	1	66
77. Houghton River, England ... Ry.		Haskoll		65
78. Teviot-Tweed Bridge	1794 / 1795	Elliot	5?	65
79. Rivanna Aqueduct		Ellet	5	65
80. Coldstream, over Tweed	1771?	Smeaton	5	64
81. Conemaugh Viaduct, New ... Pa. Ry.	1890	Brown	2	60
82. Saumur, over arm of Loire	1756 / 1770	De Voglio	12	60
83. Bewdley, England		Telford		60
84. Chestnut St., Phila		Kneass		60
85. Carrollton Viaduct ... Ry.				58
86. Llanrwst, over Conway, Wales,	1636	Inigo Jones	3	58
87. Glatt at Neuneck			1	55.8
88. Monocacy Viaduct ... C. & O. Canal		Fisk		54.0
89. Stirling, Forth				53
90. Nemours, France		Perronet		53
91. Abattoir St., Paris ... Ry.				53
92. Dôle, over Doubs, France				52
93. Château-Thierry, France		Perronet		51
94. Avon Viaduct		Vignoles		50
95. Filbert St ... Penna. Ry.				50
96. Race St., Phila ... P. & R. Ry.	1893	Wilson	1	50
97. James River Aqueduct, Va		Ellet		50
98. Over Cree, at Newton Stewart		Rennie	5	50
99. Byrd Creek Aqueduct		Ellet	1	50
100. Des Basses-Granges, France				49
101. Pesmes, over Oignon, France		Bertrand		45
102. Over the Esk at Musselburgh		Rennie	5	45
103. Penna. Ry		Steele		44
104. Couturette, Arbois, France				43
105. Moret, over Loing, Concrete ... Aqueduct	1869	Belgrand	many	42.5
106. St. Chamas, France		Roman	1	42
107. Tonoloway Culvert ... C. & O. Canal		Fisk	1	40
108. South St. Bridge ... B. & P. Ry.	1887	Ry	1	40

XXX.—(*Continued.*)

	Rise, k	Depth of Key, t_0.	Skew t_s.	Curve.	Spandrels.	Remarks.
	Ft.	Ft.	Ft.			
50	......	....	......	......	Longitudinal voids.	
51	27.9	3.0	4.0	C.	Broken stone?	Limestone.
52	11.8	4.6	......	C.		
53	......	....	......	......	......	Dimensions for centre arch.
54	28	4.5	......	E.		
55	40	3.0	3.5	C.	......	Sandstone in lime without sand.
56	40	2.7	......	C.	......	Brick in cement.
57	16	4.7	......	C.	......	Brick in cement.
58	26	3.0	3.0	......	Masonry and earth.	
59	26.3	4.0	......	E.		
60	13	3.50	......	C.		
61	25	3.0	......	C.		
62	......	....	......	......		
63	6.5	5.0?	12.0?	Seg.	Continuation of the arch stones.	
64	38	7.6	14.0	C.	Longitudinal arched voids with side walls	2 small side arches.
65	15	2.4	2.4	C.	Masonry and earth	Brick ring. Small skew.
66	24	2.4?	3.2?	E.		
67	11.5	2.5	3.0	C.		
68	16.5	2 8	2.8	C.		
69	36	2.8	......	C.		
70	25	3.5	3.5	C.	......	Brick. Stone facing.
71	17 6	3.0	......	E.	Longitudinal voids.	
72	17.5	3.8	3.8	E.	......	Brick.
73	17.5	2.5	......	E.	Longitudinal voids.	
74	8.3	2.3	5.8?	C.	Longitudinal voids.	
75	19.3	3.5	7.0?	E.	Longitudinal voids	Brick interior.
76	13.8	2.5	4.0	E.	Solid masonry	Aberdeen granite.
77	32.5	2.8	2.8	C.		
78	17	....	......	C.	......	Dimensions for centre arch.
79	15	2.8	4.0	C.	Solid masonry	Aqueduct 21 × 5.
80	......	....	......	......	......	Dimensions for centre arch.
81	30	2.7	2.7	C.	Masonry and loose stone	3 tracks on 2° curve.
82	21	....	......	E.		
83	20	2.2	......	C.		
84	18	2.5	......	C.	......	Brick in cement.
85	29	2.5	2.5	C.	......	Granite.
86	17	1.5	......	Seg.	......	Pointed at crown.
87	11.0?	1.3?	2.5?	C.	Longitudinal arched voids	Lead in joints at crown and springing.
88	9.0	2.5	......	......		
89	10.3	2.8	......	C.		
90	3.8	3.2	......	C.		
91	5.1	3.0	......	C.		
92	17.5	3.8	......	E.		
93	17.0	3.8	......	......		
94	15.0	2.0	......	E.	......	Brick in cement.
95	7	2.0	......	C.	......	Brick in lime-mortar.
96	9.8	3.1	3.1	C.	Masonry and stone	Brick.
97	7	2.0	......	C.		
98	6.6	1.8?	2.5?	C.	......	Side spans 39 and 46.
99	7.0	2.8?	4.2?	C.	Solid masonry	Aqueduct 21 × 5.
100	24.5	4.0	......	C.		
101	3.8	3.8	......	C.		
102	5.6	2.8?	2.8?	C.	......	Side spans 37 and 42.
103	8.0	2.5	......	C.		
104	6.1	3.0	......	C.		
105	......	1.3	......	C.	Spandrels with lateral voids.	
106	21?	3.4	......	C.	......	Roman.
107	15	2.0	......	C.		
108	20	2.7	2.7	C.	Masonry and earth	Granite. N. Conway, N. H.

TABLE XXXI.

DIMENSIONS OF SOME CAST-IRON ARCHES.

Name, etc.	Date.	Engineer.	Span.	Rise.	Spandrel-bracing.
			Ft.	Ft.	
Southwark	1819	Rennie	240.0	24.0	
Sunderland	1796	Burdon	236.0	34.0	Circles.
St. Louis, Paris	1867	Martin	210.0	Very flat	Vertical.
Water-pipe Bridge, Washington, D. C.	1858	Meigs	200.0	20.0	Light.
El-Kantara, Algeria	1864	Martin	188.3	26 2	Ornamental.
Carrousel, Paris (wood arch cased with iron)	1834–6	Polenceau	187.0	16.5	Light ornamental.
Chestnut St., Phila. (spandrel-braced)	1866	Kneass	185.0	20.0	Vertical, with ornaments.
Staines	1802	......	180.0	16.0	
Galton	......	......	180.0	18.0	
Lendal	1862	Page	172.2	......	Light ornamental.
Tewksbury	......	Telford	170.0	17.0	
New N. Bridge, Halifax	1869?	Frazer	160.0	16.0	Vertical.
Tees	1837	Hambley	150.0	16.0	Vertical.
Bonar	1811–2	Telford	150.0	20.0	Braced.
Craigellachie	1812?	Telford	150.0	20.0	
Buildwas	1796	Telford	130.0	30.0	
D'Austerlitz	1806	Lamande	106.0	10.7	
Coalbrookdale	1779	Darby	100.5	50.0	
Bristol	......	......	100.0	15.0	
Nottingham	1871	Tarbottom	100.0	10.0	Vert. ornamental.
Myton	1868	Page & Gordon	100.0	10.5	Ornamental.
New Logan	1890	Bell & Miller	91.0	18.3	Braced.
River Meuse .... Ry.	1847	Geo. Rennie	82.0	10.0	
Vauxhall	1816	Walker	78.0	29.0	
Carrington	1842	Wood	70.0	5.0	
Pont du Louvre	1803	......	57.0	10.7	
Laason	1794	......	43.0	......	
St. Denis	1808	......	39.4	3.2	

TABLE XXXII.

Dimensions of a Few Wrought-iron or Steel Arches.

Name, etc.	Date.	Engineer.	Span.	Rise.	No. of Hinges.
			Ft.	Ft.	
Viaur Viaduct, France.....Ry.	1888?	....	820.0	150.0	3
Luiz I., Douro............Hy.	1877	Seyrig	566.0	146.4	0
Niagara, spandrel-braced. Ry. & Hy.	1897	Buck	550.0	114.0	2
Garabit Viaduct...............	1884	Eiffel	541.3	186.6	2
Pia Maria.................Ry.	1877?	Eiffel	525.0	122.4	2
St. Louis. ....Ry. & Hy.	1874	Eads	520.0	47.0	0
Grunthal Bridge..............	1890	....	513.3	clear 137.8	2
Washington...............Hy.	1888	Hutton	510 0	91.7	2
Paderno Viaduct....Hy. & Ry.	1887?	Rothlis-berger	492.0	122.8	0
Minneapolis, spandrel-braced. ..Hy.	1888	Sewall	456.0	about 90.0	3
Rochester, spandrel-braced....	1890	Buck	416.0	67.0	2
Pont de Szegedin..............	......	....	361.1	28.0	
Stony Creek.....Can. Pac. Ry.	1893	Peterson	336.0	80.0	3
Pons Palatinus................	1882–5	Lauter	334.1	radius 455.1	2
Coblenz......................	1866	Harwich & Sternberg	315.0	31.5?	0
Pesth, spandrel-braced........	1873	Gouin	305.0	8.0	2
Verona...................Hy.	1884	Biadego	291.3	32.8	2
Arcole Bridge, Paris..........	1855	Oudry	262.5	20.?	0
Blaauw-Krautz, spandrel-braced	1884	Max Am Ende	229.6	97.0	0
Pont Morande, Rhone.....Hy.	1889	Tavernier	221.1	14.6	2 & 0
St. Giustina, Southern Tyrol..	1889	Hagen	196.8	36.7	2
Pont de Rouen...............	1889	Cadart	179.1	14.9	2 & 0
Garibaldi, Rome, spandrel-braced, ..	1892	Vescovali	173.7	6.3	0?
Cerveyrette Gorge, France, Hy.	1892	Baldy.	172.2	37.7	0
Marburg......................	......	....	172.3	....	
Cedar Ave., Baltimore, spandrel-braced.	1891	Latrobe	150.0	38.0	3
Railway bridge over canal St. Denis. ..	1858	....	148.3		
Street bridge over canal St. Denis at Villette. ..	1867	....	147.7		
Theiss at Szegedin............	1858	....	138.8		
Ruhr at Mühlheim.............	......	....	118.5		
Cron at St. Denis, first wrought-iron arched bridge.	1808				
Albert over Clyde, latticed spandrels. ..	1870	Bell & Miller	114.0		2
30th Street Bridge, spandrel-braced. ....Ry.	1870	Wilson	64 1	12.0	3

TABLE XXXIII.

DIMENSIONS OF A FEW METAL ROOF-ARCHES.*

Name, etc.	Date.	Engineer.	Span.	Rise.	No. of Hinges.
			Ft.	Ft.	
Liberal Arts Building, Columbian Exposition..	1892	Shankland	368.0	206.3	3
Roof Main Bldg., Lyons Ex.	1894	....	361.0	108.0	2
Train-shed, Phila., Pa. Ry..	1893	Brown	300.0	108.5	3
Station, Phila., P. & R. Ry.	1892	Wilson	259.0	clear 88.0	3
Jersey City, Phila. Ry......	1891	Brown	252.7	89.8	3
St. Pancras Ry. Station.....	1868	Barlow	240.0	clear 124.8	0
Cologne, train-shed........	1892	....	209.0	78.7	2
Dome, Horticultural Bldg., Col. Ex.	1892	Jenney	181.6	91.0	0
Drill-hall, 22d Reg., N. Y..	1889	Williamson	176.0	clear 62.0	tie-rod 0
" 12th " " ..	1888	Schneider	171.3	55.6	tie-rod 1
" 1st " Chicago	1894	Shankland	155.5	77.5	3
Dancing-hall, Saltair Beach.	1893	Kletting	118.7	54.0	3
Machinery Hall, Columbian Exposition. ..	1802	Shankland	clear 115.2	clear 93.5	3
Cleveland Arcade..........	1890	Eisenmann	50.0	23.0	3

* These dimensions are sometimes given "in clear" and sometimes centre to centre of pins. When possible we have designated "in clear" by placing the word clear above the dimension figures.

INDEX.

SHORT-TITLE CATALOGUE

OF THE

PUBLICATIONS

OF

JOHN WILEY & SONS,

NEW YORK.

LONDON: CHAPMAN & HALL, LIMITED.

ARRANGED UNDER SUBJECTS.

Descriptive circulars sent on application.
Books marked with an asterisk are sold at *net* prices only.
All books are bound in cloth unless otherwise stated.

AGRICULTURE.

CATTLE FEEDING—DAIRY PRACTICE—DISEASES OF ANIMALS—GARDENING, ETC.

Armsby's Manual of Cattle Feeding	12mo,	$1 75
Downing's Fruit and Fruit Trees	8vo,	5 00
Grotenfelt's The Principles of Modern Dairy Practice. (Woll.)	12mo,	2 00
Kemp's Landscape Gardening	12mo,	2 50
Lloyd's Science of Agriculture	8vo,	4 00
Loudon's Gardening for Ladies. (Downing.)	12mo,	1 50
Steel's Treatise on the Diseases of the Dog	8vo,	3 50
" Treatise on the Diseases of the Ox	8vo,	6 00
Stockbridge's Rocks and Soils	8vo,	2 50
Woll's Handbook for Farmers and Dairymen	12mo,	1 50

ARCHITECTURE.

BUILDING—CARPENTRY—STAIRS—VENTILATION, ETC.

Berg's Buildings and Structures of American Railroads	4to,	7 50
Birkmire's American Theatres—Planning and Construction	8vo,	3 00
" Architectural Iron and Steel	8vo,	3 50
Birkmire's Compound Riveted Girders	8vo,	2 00
" Skeleton Construction in Buildings	8vo,	3 00

Carpenter's Heating and Ventilating of Buildings	8vo,	$3 00
Downing, Cottages	8vo,	2 50
and Wightwick's Hints to Architects	8vo,	2 00
Freitag's Architectural Engineering	8vo,	2 50
Gerhard's Sanitary House Inspection	16mo,	1 00
" Theatre Fires and Panics	12mo,	1 50
Hatfield's American House Carpenter	8vo,	5 00
Holly's Carpenter and Joiner	18mo,	75
Kidder's Architect and Builder's Pocket-book	Morocco flap,	4 00
Merrill's Stones for Building and Decoration	8vo,	5 00
Monckton's Stair Building—Wood, Iron, and Stone	4to,	4 00
Stevens' House Painting	18mo,	75
Worcester's Small Hospitals—Establishment and Maintenance, including Atkinson's Suggestions for Hospital Architecture	12mo,	1 25
World's Columbian Exposition of 1893	4to,	2 50

ARMY, NAVY, Etc.

MILITARY ENGINEERING—ORDNANCE—PORT CHARGES, ETC.

Bourne's Screw Propellers	4to,	5 00
Bruff's Ordnance and Gunnery	8vo,	6 00
Bucknill's Submarine Mines and Torpedoes	8vo,	4 00
Chase's Screw Propellers	8vo,	3 00
Cooke's Naval Ordnance	8vo,	12 50
Cronkhite's Gunnery for Non-com. Officers	18mo, morocco,	2 00
De Brack's Cavalry Outpost Duties. (Carr.)	18mo, morocco,	2 00
Dietz's Soldier's First Aid	12mo, morocco,	1 25
* Dredge's Modern French Artillery	4to, half morocco,	20 00
" Record of the Transportation Exhibits Building, World's Columbian Exposition of 1893	4to, half morocco,	15 00
Dyer's Light Artillery	12mo,	3 00
Hoff's Naval Tactics	8vo,	1 50
Hunter's Port Charges	8vo, half morocco,	13 00
Ingalls's Ballistic Tables	8vo,	1 50
" Handbook of Problems in Direct Fire	8vo,	4 00
Mahan's Advanced Guard	18mo,	1 50
" Permanent Fortifications. (Mercur.)	8vo, half morocco,	7 50

Mercur's Attack of Fortified Places..........12mo, $2 00
" Elements of the Art of War..........8vo, 4 00
Metcalfe's Ordnance and Gunnery..........12mo, with Atlas, 5 00
Phelps's Practical Marine Surveying..........8vo, 2 50
Powell's Army Officer's Examiner..........12mo, 4 00
Reed's Signal Service.......... 50
Sharpe's Subsisting Armies..........18mo, morocco, 1 50
Strauss and Alger's Naval Ordnance and Gunnery..........
Todd and Whall's Practical Seamanship..........8vo, 7 50
Very's Navies of the World..........8vo, half morocco, 3 50
Wheeler's Siege Operations..........8vo, 2 00
Winthrop's Abridgment of Military Law..........12mo, 2 50
Woodhull's Notes on Military Hygiene..........12mo, morocco, 2 50
Young's Simple Elements of Navigation..12mo, morocco flaps, 2 50

ASSAYING.

SMELTING—ORE DRESSING—ALLOYS, ETC.

Fletcher's Quant. Assaying with the Blowpipe..12mo, morocco, 1 50
Furman's Practical Assaying..........8vo, 3 00
Kunhardt's Ore Dressing..........8vo, 1 50
* Mitchell's Practical Assaying. (Crookes.)..........8vo, 10 00
O'Driscoll's Treatment of Gold Ores..........8vo, 2 00
Ricketts and Miller's Notes on Assaying..........8vo, 3 00
Thurston's Alloys, Brasses, and Bronzes..........8vo, 2 50
Wilson's Cyanide Processes..........12mo, 1 50

ASTRONOMY.

PRACTICAL, THEORETICAL, AND DESCRIPTIVE.

Craig's Azimuth..........4to, 3 50
Doolittle's Practical Astronomy..........8vo, 4 00
Gore's Elements of Geodesy..........8vo, 2 50
Michie and Harlow's Practical Astronomy..........8vo, 3 00
White's Theoretical and Descriptive Astronomy..........12mo, 2 00

BOTANY.

GARDENING FOR LADIES, ETC.

Baldwin's Orchids of New England..........8vo, $1 50
Loudon's Gardening for Ladies. (Downing.)..........12mo, 1 50

Thomé's Structural Botany........................18mo, $2 25

Westermaier's General Botany. (Schneider.)..............8vo, 2 00

BRIDGES, ROOFS, Etc.

CANTILEVER—DRAW—HIGHWAY—SUSPENSION.

(*See also* ENGINEERING, p. 6.)

Boller's Highway Bridges..............................8vo, 2 00

* " The Thames River Bridge..................4to, paper, 5 00

Burr's Stresses in Bridges..............................8vo, 3 50

Crehore's Mechanics of the Girder........................8vo, 5 00

Dredge's Thames Bridges............................7 parts, -,-

Du Bois's Stresses in Framed Structures....................4to, 10 00

Foster's Wooden Trestle Bridges..........................4to, 5 00

Greene's Arches in Wood, etc.............................8vo, 2 50

" Bridge Trusses..................................8vo, 2 50

" Roof Trusses...................................8vo, 1 25

Howe's Treatise on Arches..............................8vo,

Johnson's Modern Framed Structures......................4to, 10 00

Merriman & Jacoby's Text-book of Roofs and Bridges.
Part I., Stresses..................................8vo, 2 50

Merriman & Jacoby's Text-book of Roofs and Bridges.
Part II., Graphic Statics............................8vo, 2 50

Merriman & Jacoby's Text-book of Roofs and Bridges.
Part III., Bridge Design.............................8vo, 5 00

Merriman & Jacoby's Text-book of Roofs and Bridges.
Part IV., Continuous, Draw, Cantilever, Suspension, and
Arched Bridges..........................(*In preparation*).

*Morison's The Memphis Bridge....................Oblong 4to, 10 00

Waddell's Iron Highway Bridges..........................8vo, 4 00

Wood's Construction of Bridges and Roofs..................8vo, 2 00

Wright's Designing of Draw Spans.........................8vo, 2 50

CHEMISTRY.

QUALITATIVE—QUANTITATIVE—ORGANIC—INORGANIC, ETC.

Adriance's Laboratory Calculations.......................12mo, 1 25

Allen's Tables for Iron Analysis..........................8vo, 3 00

Austen's Notes for Chemical Students.....................12mo, 1 50

Bolton's Student's Guide in Quantitative Analysis............8vo, 1 50

Classen's Analysis by Electrolysis. (Herrick.)........8vo,		$3 00
Crafts's Qualitative Analysis. (Schaeffer.)............12mo,		1 50
Drechsel's Chemical Reactions. (Merrill.)..............12mo,		1 25
Fresenius's Quantitative Chemical Analysis. (Allen.)......8vo,		6 00
" Qualitative Chemical Analysis. (Johnson.)....8vo,		4 00
Gill's Gas and Fuel Analysis.............................12mo,		1 25
Hammarsten's Physiological Chemistry. (Mandel.).......8vo,		4 00
Helm's Principles of Mathematical Chemistry. (Morgan).12mo,		1 50
Kolbe's Inorganic Chemistry..............................12mo,		1 50
Mandel's Bio-chemical Laboratory.........................12mo,		1 50
Mason's Water-supply.....................................8vo,		5 00
Miller's Chemical Physics.................................8vo,		2 00
Mixter's Elementary Text-book of Chemistry.............12mo,		1 50
Morgan's The Theory of Solutions and its Results.......12mo,		1 00
Nichols's Water Supply (Chemical and Sanitary)..........8vo,		2 50
O'Brine's Laboratory Guide to Chemical Analysis.........8vo,		2 00
Perkins's Qualitative Analysis...........................12mo,		1 00
Pinner's Organic Chemistry. (Austen.)..................12mo,		1 50
Ricketts and Russell's Notes on Inorganic Chemistry (Non-metallic)........................ Oblong 8vo, morocco,		75
Schimpf's Volumetric Analysis............................12mo,		2 50
Spencer's Sugar Manufacturer's Handbook. 12mo, morocco flaps,		2 00
Stockbridge's Rocks and Soils.............................8vo,		2 50
Troilius's Chemistry of Iron..............................8vo,		2 00
Wiechmann's Chemical Lecture Notes.....................12mo,		3 00
" Sugar Analysis.......................................8vo,		2 50
Wulling's Inorganic Phar. and Med. Chemistry...........12mo,		2 00

DRAWING.

ELEMENTARY—GEOMETRICAL—TOPOGRAPHICAL.

Hill's Shades and Shadows and Perspective................8vo,	2 00
MacCord's Descriptive Geometry...........................8vo,	3 00
" Kinematics...8vo,	5 00
" Mechanical Drawing..................................8vo,	4 00
Mahan's Industrial Drawing. (Thompson.).......2 vols., 8vo,	3 50
Reed's Topographical Drawing. (H. A.)....................4to,	5 00
Smith's Topographical Drawing. (Macmillan.)..............8vo,	2 50
Warren's Descriptive Geometry...................2 vols., 8vo,	3 50

Warren's Drafting Instruments........................12mo, 1 25
" Free-hand Drawing 12mo, $1 00
" Higher Linear Perspective 8vo, 3 50
" Linear Perspective.........................12mo, 1 00
" Machine Construction.................2 vols., 8vo, 7 50
" Plane Problems..............................12mo, 1 25
" Primary Geometry...........................12mo, 75
" Problems and Theorems.....................8vo, 2 50
" Projection Drawing..........................12mo, 1 50
" Shades and Shadows..........................8vo, 3 00
" Stereotomy—Stone Cutting...................8vo, 2 50
Whelpley's Letter Engraving 12mo, 2 00

ELECTRICITY AND MAGNETISM.

Illumination—Batteries—Physics.

Anthony and Brackett's Text-book of Physics (Magie). ...8vo, 4 00
Barker's Deep-sea Soundings............................8vo, 2 00
Benjamin's Voltaic Cell.................................8vo, 3 00
Cosmic Law of Thermal Repulsion....................18mo, 75
Crehore and Squier's Experiments with a New Polarizing Photo-Chronograph....................................8vo, 3 00
* Dredge's Electric Illuminations....2 vols., 4to, half morocco, 25 00
" " " Vol. II..................4to, 7 50
Gilbert's De magnete. (Mottelay.)......................8vo, 2 50
Holman's Precision of Measurements....................8vo, 2 00
Michie's Wave Motion Relating to Sound and Light,.......8vo, 4 00
Morgan's, The Theory of Solutions and its Results.......12mo,
Niaudet's Electric Batteries. (Fishback.)..............12mo, 2 50
Reagan's Steam and Electrical Locomotives.............12mo 2 00
Thurston's Stationary Steam Engines for Electric Lighting Purposes...12mo, 1 50
Tillman's Heat...8vo, 1 50

ENGINEERING.

Civil—Mechanical—Sanitary, Etc.

(*See also* Bridges, p. 4; Hydraulics, p. 8; Materials of Engineering, p. 9; Mechanics and Machinery, p. 11; Steam Engines and Boilers, p. 14.)

Baker's Masonry Construction..............................8vo, 5 00

Baker's Surveying Instruments........................12mo, 3 00
Black's U. S. Public Works...............................4to, $5 00
Butts's Engineer's Field-book..................12mo, morocco, 2 50
Byrne's Highway Construction...........................8vo, 5 00
Carpenter's Experimental Engineering 8vo, 6 00
Church's Mechanics of Engineering—Solids and Fluids....8vo, 6 00
" Notes and Examples in Mechanics...............8vo, 2 00
Crandall's Earthwork Tables 8vo, 1 50
Crandall's The Transition Curve..............12mo, morocco, 1 50
* Dredge's Penn. Railroad Construction, etc... Folio, half mor., 20 00
* Drinker's Tunnelling....................4to, half morocco, 25 00
Eissler's Explosives—Nitroglycerine and Dynamite........8vo, 4 00
Gerhard's Sanitary House Inspection......................16mo, 1 00
Godwin's Railroad Engineer's Field-book. 12mo, pocket-bk. form, 2 50
Gore's Elements of Geodesy....... ....8vo, 2 50
Howard's Transition Curve Field-book.....12mo, morocco flap, 1 50
Howe's Retaining Walls (New Edition.).................12mo, 1 25
Hudson's Excavation Tables. Vol. II.................... 8vo, 1 00
Hutton's Mechanical Engineering of Power Plants........8vo, 5 00
Johnson's Materials of Construction......................8vo, 6 00
Johnson's Stadia Reduction Diagram..Sheet, $22\frac{1}{2} \times 28\frac{1}{2}$ inches, 50
" Theory and Practice of Surveying..............8vo, 4 00
Kent's Mechanical Engineer's Pocket-book.....12mo, morocco, 5 00
Kiersted's Sewage Disposal..... ....12mo, 1 25
Kirkwood's Lead Pipe for Service Pipe....................8vo, 1 50
Mahan's Civil Engineering. (Wood.)....................8vo, 5 00
Merriman and Brook's Handbook for Surveyors....12mo, mor., 2 00
Merriman's Geodetic Surveying...........................8vo, 2 00
" Retaining Walls and Masonry Dams...........8vo, 2 00
Mosely's Mechanical Engineering. (Mahan.)..............8vo, 5 00
Nagle's Manual for Railroad Engineers........12mo, morocco,
Patton's Civil Engineering...................................8vo, 7 50
" Foundations..8vo, 5 00
Rockwell's Roads and Pavements in France........ ....12mo, 1 25
Ruffner's Non-tidal Rivers 8vo, 1 25
Searles's Field Engineering...............12mo, morocco flaps, 3 00
Searles's Railroad Spiral ....12mo, morocco flaps, 1 50

Siebert and Biggin's Modern Stone Cutting and Masonry...8vo, 1 50
Smith's Cable Tramways...4to, $2 50
" Wire Manufacture and Uses...4to, 3 00
Spalding's Roads and Pavements...12mo, 2 00
" Hydraulic Cement...12mo,
Thurston's Materials of Construction...8vo, 5 00
*Trautwine's Civil Engineer's Pocket-book...12mo, mor. flaps, 5 00
* " Cross-section...Sheet, 25
* " Excavations and Embankments...8vo, 2 00
* " Laying Out Curves...12mo, morocco, 2 50
Warren's Stereotomy—Stone Cutting...8vo, 2 50
Webb's Engineering Instruments...12mo, morocco, 1 00
Wegmann's Construction of Masonry Dams...4to, 5 00
Wellington's Location of Railways...8vo, 5 00
Wheeler's Civil Engineering...8vo, 4 00
Wolff's Windmill as a Prime Mover...8vo, 3 00

HYDRAULICS.

WATER-WHEELS—WINDMILLS—SERVICE PIPE—DRAINAGE, ETC.

(*See also* ENGINEERING, p. 6.)

Bazin's Experiments upon the Contraction of the Liquid Vein (Trautwine)...8vo, 2 00
Bovey's Treatise on Hydraulics...8vo, 4 00
Coffin's Graphical Solution of Hydraulic Problems 12mo, mor., 2 50
Ferrel's Treatise on the Winds, Cyclones, and Tornadoes...8vo, 4 00
Ganguillet & Kutter's Flow of Water. (Hering & Trautwine).8vo, 4 00
Hazen's Filtration of Public Water Supply...8vo, 2 00
Kiersted's Sewage Disposal...12mo, 1 25
Kirkwood's Lead Pipe for Service Pipe...8vo, 1 50
Mason's Water Supply...8vo, 5 00
Merriman's Treatise on Hydraulics...8vo, 4 00
Nichols's Water Supply (Chemical and Sanitary)...8vo, 2 50
Ruffner's Improvement for Non-tidal Rivers...8vo, 1 25
Wegmann's Water Supply of the City of New York...4to, 10 00
Weisbach's Hydraulics. (Du Bois.)...8vo, 5 00
Wilson's Irrigation Engineering...8vo, 4 00
Wolff's Windmill as a Prime Mover...8vo, 3 00
Wood's Theory of Turbines...8vo, 2 50

MANUFACTURES.

ANILINE—BOILERS—EXPLOSIVES—IRON—SUGAR—WATCHES—WOOLLENS, ETC.

Allen's Tables for Iron Analysis	8vo,	$3 00
Beaumont's Woollen and Worsted Manufacture	12mo,	1 50
Bolland's Encyclopædia of Founding Terms	12mo,	3 00
" The Iron Founder	12mo,	2 50
" " " " Supplement	12mo,	2 50
Booth's Clock and Watch Maker's Manual	12mo,	2 00
Bouvier's Handbook on Oil Painting	12mo,	2 00
Eissler's Explosives, Nitroglycerine and Dynamite	8vo,	4 00
Ford's Boiler Making for Boiler Makers	18mo,	1 00
Metcalfe's Cost of Manufactures	8vo,	5 00
Metcalf's Steel—A Manual for Steel Users	12mo,	2 00
Reimann's Aniline Colors. (Crookes.)	8vo,	2 50
* Reisig's Guide to Piece Dyeing	8vo,	25 00
Spencer's Sugar Manufacturer's Handbook	12mo, mor. flap,	2 00
Svedelius's Handbook for Charcoal Burners	12mo,	1 50
The Lathe and Its Uses	8vo,	6 00
Thurston's Manual of Steam Boilers	8vo,	5 00
West's American Foundry Practice	12mo,	2 50
" Moulder's Text-book	12mo,	2 50
Wiechmann's Sugar Analysis	8vo,	2 50
Woodbury's Fire Protection of Mills	8vo,	2 50

MATERIALS OF ENGINEERING.

STRENGTH—ELASTICITY—RESISTANCE, ETC.

(*See also* ENGINEERING, p. 6.)

Baker's Masonry Construction	8vo,	5 00
Beardslee and Kent's Strength of Wrought Iron	8vo,	1 50
Bovey's Strength of Materials	8vo,	7 50
Burr's Elasticity and Resistance of Materials	8vo,	5 00
Byrne's Highway Construction	8vo,	5 00
Carpenter's Testing Machines and Methods of Testing Materials		
Church's Mechanic's of Engineering—Solids and Fluids	8vo,	6 00
Du Bois's Stresses in Framed Structures	4to,	10 00
Hatfield's Transverse Strains	8vo,	5 00
Johnson's Materials of Construction	8vo,	6 00

Lanza's Applied Mechanics	8vo,	$7 50
" Strength of Wooden Columns	8vo, paper,	50
Merrill's Stones for Building and Decoration	8vo,	5 00
Merriman's Mechanics of Materials	8vo,	4 00
Patton's Treatise on Foundations	8vo,	5 00
Rockwell's Roads and Pavements in France	12mo,	1 25
Spalding's Roads and Pavements	12mo,	2 00
" Hydraulic Cement	12mo,	2 00
Thurston's Materials of Construction	8vo,	5 00
Thurston's Materials of Engineering	3 vols., 8vo,	8 00
Vol. I., Non-metallic	8vo,	2 00
Vol. II., Iron and Steel	8vo,	3 50
Vol. III., Alloys, Brasses, and Bronzes	8vo,	2 50
Weyrauch's Strength of Iron and Steel. (Du Bois.)	8vo,	1 50
Wood's Resistance of Materials	8vo,	2 00

MATHEMATICS.

CALCULUS—GEOMETRY—TRIGONOMETRY, ETC.

Baker's Elliptic Functions	8vo,	1 50
Ballard's Pyramid Problem	8vo,	1 50
Barnard's Pyramid Problem	8vo,	1 50
Bass's Differential Calculus	12mo,	4 00
Brigg's Plane Analytical Geometry	12mo,	1 00
Chapman's Theory of Equations	12mo,	1 50
Chessin's Elements of the Theory of Functions		
Compton's Logarithmic Computations	12mo,	1 50
Craig's Linear Differential Equations	8vo,	5 00
Davis's Introduction to the Logic of Algebra	8vo,	1 50
Halsted's Elements of Geometry	8vo,	1 75
" Synthetic Geometry	8vo,	1 50
Johnson's Curve Tracing	12mo,	1 00
" Differential Equations—Ordinary and Partial	8vo,	3 50
" Integral Calculus	12mo,	1 50
" Least Squares	12mo,	1 50
Ludlow's Logarithmic and Other Tables. (Bass.)	8vo,	2 00
" Trigonometry with Tables. (Bass.)	8vo,	3 00
Mahan's Descriptive Geometry (Stone Cutting)	8vo,	1 50
Merriman and Woodward's Higher Mathematics	8vo,	5 00
Merriman's Method of Least Squares	8vo,	2 00

Parker's Quadrature of the Circle 8vo, $2 50

Rice and Johnson's Differential and Integral Calculus,

2 vols. in 1, 12mo, 2 50

" Differential Calculus....................8vo, 3 50

" Abridgment of Differential Calculus....8vo, 1 50

Searles's Elements of Geometry. 8vo, 1 50

Totten's Metrology..................................8vo, 2 50

Warren's Descriptive Geometry...............2 vols., 8vo, 3 50

" Drafting Instruments..........................12mo, 1 25

" Free-hand Drawing.............................12mo, 1 00

" Higher Linear Perspective....................8vo, 3 50

" Linear Perspective..............................12mo, 1 00

" Primary Geometry...............................12mo, 75

" Plane Problems..................................12mo, 1 25

" Plane Problems..................................12mo, 1 25

" Problems and Theorems........................8vo, 2 50

" Projection Drawing.............................12mo, 1 50

Wood's Co-ordinate Geometry.......................8vo, 2 00

" Trigonometry......................................12mo, 1 00

Woolf's Descriptive Geometry....................Royal 8vo, 3 00

MECHANICS—MACHINERY.

TEXT-BOOKS AND PRACTICAL WORKS.

(*See also* ENGINEERING, p. 6.)

Baldwin's Steam Heating for Buildings.................12mo, 2 50

Benjamin's Wrinkles and Recipes.......................12mo, 2 00

Carpenter's Testing Machines and Methods of Testing Materials..8vo,

Chordal's Letters to Mechanics.........................12mo, 2 00

Church's Mechanics of Engineering.....................8vo, 6 00

" Notes and Examples in Mechanics..............8vo, 2 00

Crehore's Mechanics of the Girder.....................8vo, 5 00

Cromwell's Belts and Pulleys............................12mo, 1 50

" Toothed Gearing.................................12mo, 1 50

Compton's First Lessons in Metal Working...........12mo, 1 50

Dana's Elementary Mechanics..........................12mo, 1 50

Dingey's Machinery Pattern Making...................12mo, 2 00

Dredge's Trans. Exhibits Building, World Exposition, 4to, half morocco, $15 00

Du Bois's Mechanics. Vol. I., Kinematics 8vo, 3 50

" " Vol. II., Statics 8vo, 4 00

" " Vol. III., Kinetics 8vo, 3 50

Fitzgerald's Boston Machinist 18mo, 1 00

Flather's Dynamometers 12mo, 2 00

" Rope Driving 12mo, 2 00

Hall's Car Lubrication 12mo, 1 00

Holly's Saw Filing 18mo, 75

Lanza's Applied Mechanics 8vo, 7 50

MacCord's Kinematics 8vo, 5 00

Merriman's Mechanics of Materials 8vo, 4 00

Metcalfe's Cost of Manufactures 8vo, 5 00

Michie's Analytical Mechanics 8vo, 4 00

Mosely's Mechanical Engineering. (Mahan.) 8vo, 5 00

Richards's Compressed Air 12mo, 1 50

Robinson's Principles of Mechanism 8vo, 3 00

Smith's Press-working of Metals 8vo, 3 00

The Lathe and Its Uses 8vo, 6 00

Thurston's Friction and Lost Work 8vo, 3 00

" The Animal as a Machine 12mo, 1 00

Warren's Machine Construction ....2 vols., 8vo, 7 50

Weisbach's Hydraulics and Hydraulic Motors. (Du Bois.)..8vo, 5 00

" Mechanics of Engineering. Vol. III., Part I., Sec. I. (Klein.) 8vo, 5 00

Weisbach's Mechanics of Engineering. Vol. III., Part I., Sec. II. (Klein.) 8vo, 5 00

Weisbach's Steam Engines. (Du Bois.) 8vo, 5 00

Wood's Analytical Mechanics 8vo, 3 00

" Elementary Mechanics 12mo, 1 25

" " " Supplement and Key 1 25

METALLURGY.

Iron—Gold—Silver—Alloys, Etc.

Allen's Tables for Iron Analysis 8vo, 3 00

Egleston's Gold and Mercury 8vo, 7 50

Egleston's Metallurgy of Silver	8vo,	$7 50
* Kerl's Metallurgy—Copper and Iron	8vo,	15 00
* " " Steel, Fuel, etc.	8vo,	15 00
Kunhardt's Ore Dressing in Europe	8vo,	1 50
Metcalf Steel—A Manual for Steel Users	12mo,	2 00
O'Driscoll's Treatment of Gold Ores	8vo,	2 00
Thurston's Iron and Steel	8vo,	3 50
" Alloys	8vo,	2 50
Wilson's Cyanide Processes	12mo,	1 50

MINERALOGY AND MINING.

Mine Accidents—Ventilation—Ore Dressing, Etc.

Beard's Ventilation of Mines	12mo,	2 50
Boyd's Resources of South Western Virginia	8vo,	3 00
" Map of South Western Virginia	Pocket-book form,	2 00
Brush and Penfield's Determinative Mineralogy	8vo,	3 50
Chester's Catalogue of Minerals	8vo,	1 25
" Dictionary of the Names of Minerals	8vo,	3 00
Dana's American Localities of Minerals	8vo,	1 00
" Descriptive Mineralogy. (E. S.)	8vo, half morocco,	12 50
" Mineralogy and Petrography. (J. D.)	12mo,	2 00
" Minerals and How to Study Them. (E. S.)	12mo,	1 50
" Text-book of Mineralogy. (E. S.)	8vo,	3 50
*Drinker's Tunnelling, Explosives, Compounds, and Rock Drills.	4to, half morocco,	25 00
Egleston's Catalogue of Minerals and Synonyms	8vo,	2 50
Eissler's Explosives—Nitroglycerine and Dynamite	8vo,	4 00
Goodyear's Coal Mines of the Western Coast	12mo,	2 50
Hussak's Rock-forming Minerals. (Smith.)	8vo,	2 00
Ihlseng's Manual of Mining	8vo,	4 00
Kunhardt's Ore Dressing in Europe	8vo,	1 50
O'Driscoll's Treatment of Gold Ores	8vo,	2 00
Rosenbusch's Microscopical Physiography of Minerals and Rocks. (Iddings.)	8vo,	5 00
Sawyer's Accidents in Mines	8vo,	7 00
Stockbridge's Rocks and Soils	8vo,	2 50

Williams's Lithology	8vo,	$3 00
Wilson's Mine Ventilation	16mo,	1 25

STEAM AND ELECTRICAL ENGINES, BOILERS, Etc.

STATIONARY—MARINE—LOCOMOTIVE—GAS ENGINES, ETC.

(*See also* ENGINEERING, p. 6.)

Baldwin's Steam Heating for Buildings	12mo,	2 50
Clerk's Gas Engine	12mo,	4 00
Ford's Boiler Making for Boiler Makers	18mo,	1 00
Hemenway's Indicator Practice	12mo,	2 00
Hoadley's Warm-blast Furnace	8vo,	1 50
Kneass's Practice and Theory of the Injector	8vo,	1 50
MacCord's Slide Valve	8vo,	
* Maw's Marine Engines	Folio, half morocco,	18 00
Meyer's Modern Locomotive Construction	4to,	10 00
Peabody and Miller's Steam Boilers	8vo,	
Peabody's Tables of Saturated Steam	8vo,	1 00
" Thermodynamics of the Steam Engine	8vo,	5 00
" Valve Gears for the Steam-Engine	8vo,	2 50
Pray's Twenty Years with the Indicator	Royal 8vo,	2 50
Pupin and Osterberg's Thermodynamics	12mo,	1 25
Reagan's Steam and Electrical Locomotives	12mo,	2 00
Röntgen's Thermodynamics. (Du Bois.)	8vo,	5 00
Sinclair's Locomotive Running	12mo,	2 00
Thurston's Boiler Explosion	12mo,	1 50
" Engine and Boiler Trials	8vo,	5 00
" Manual of the Steam Engine. Part I., Structure and Theory	8vo,	7 50
" Manual of the Steam Engine. Part II., Design, Construction, and Operation	8vo,	7 50
	2 parts,	12 00
" Philosophy of the Steam Engine	12mo,	75
" Reflection on the Motive Power of Heat. (Carnot.)	12mo,	2 00
" Stationary Steam Engines	12mo,	1 50
" Steam-boiler Construction and Operation	8vo,	5 00

Spangler's Valve Gears	8vo,	$2 50
Trowbridge's Stationary Steam Engines	4to, boards,	2 50
Weisbach's Steam Engine. (Du Bois.)	8vo,	5 00
Whitham's Constructive Steam Engineering	8vo,	10 00
" Steam-engine Design	8vo,	6 00
Wilson's Steam Boilers. (Flather.)	12mo,	2 50
Wood's Thermodynamics, Heat Motors, etc.	8vo,	4 00

TABLES, WEIGHTS, AND MEASURES.

FOR ACTUARIES, CHEMISTS, ENGINEERS, MECHANICS—METRIC TABLES, ETC.

Adriance's Laboratory Calculations	12mo,	1 25
Allen's Tables for Iron Analysis	8vo,	3 00
Bixby's Graphical Computing Tables	Sheet,	25
Compton's Logarithms	12mo,	1 50
Crandall's Railway and Earthwork Tables	8vo,	1 50
Egleston's Weights and Measures	18mo,	75
Fisher's Table of Cubic Yards	Cardboard,	25
Hudson's Excavation Tables. Vol. II	8vo,	1 00
Johnson's Stadia and Earthwork Tables	8vo,	1 25
Ludlow's Logarithmic and Other Tables. (Bass.)	12mo,	2 00
Thurston's Conversion Tables	8vo,	1 00
Totten's Metrology	8vo,	2 50

VENTILATION.

STEAM HEATING—HOUSE INSPECTION—MINE VENTILATION.

Baldwin's Steam Heating	12mo,	2 50
Beard's Ventilation of Mines	12mo,	2 50
Carpenter's Heating and Ventilating of Buildings	8vo,	3 00
Gerhard's Sanitary House Inspection	Square 16mo,	1 00
Mott's The Air We Breathe, and Ventilation	16mo,	1 00
Reid's Ventilation of American Dwellings	12mo,	1 50
Wilson's Mine Ventilation	16mo,	1 25

MISCELLANEOUS PUBLICATIONS.

Alcott's Gems, Sentiment, Language	Gilt edges,	5 00
Bailey's The New Tale of a Tub	8vo,	75

Ballard's Solution of the Pyramid Problem	8vo,	$1 50
Barnard's The Metrological System of the Great Pyramid	8vo,	1 50
Emmon's Geological Guide-book of the Rocky Mountains	8vo,	1 50
Ferrel's Treatise on the Winds	8vo,	4 00
Mott's The Fallacy of the Present Theory of Sound	Sq. 16mo,	1 00
Perkins's Cornell University	Oblong 4to,	1 50
Ricketts's History of Rensselaer Polytechnic Institute	8vo,	3 00
Rotherham's The New Testament Critically Emphathized.	12mo,	1 50
Totten's An Important Question in Metrology	8vo,	2 50
Whitehouse's Lake Mœris	Paper,	25
* Wiley's Yosemite, Alaska, and Yellowstone	4to,	3 00

HEBREW AND CHALDEE TEXT-BOOKS.

FOR SCHOOLS AND THEOLOGICAL SEMINARIES.

Gesenius's Hebrew and Chaldee Lexicon to Old Testament. (Tregelles.)	Small 4to, half morocco,	5 00
Green's Elementary Hebrew Grammar	12mo,	1 25
" Grammar of the Hebrew Language (New Edition)	8vo,	3 00
" Hebrew Chrestomathy	8vo,	2 00
Letteris's Hebrew Bible (Massoretic Notes in English).	8vo, arabesque,	2 25
Luzzato's Grammar of the Biblical Chaldaic Language and the Talmud Babli Idioms	12mo,	1 50

MEDICAL.

Bull's Maternal Management in Health and Disease	12mo,	1 00
Hammarsten's Physiological Chemistry. (Mandel.)	8vo,	4 00
Mott's Composition, Digestibility, and Nutritive Value of Food.	Large mounted chart,	1 25
Steel's Treatise on the Diseases of the Ox	8vo,	6 00
" Treatise on the Diseases of the Dog	8vo,	3 50
Worcester's Small Hospitals—Establishment and Maintenance, including Atkinson's Suggestions for Hospital Architecture	12mo,	1 25

www.ingramcontent.com/pod-product-compliance
Lightning Source LLC
Chambersburg PA
CBHW021354210326
41599CB00011B/864

9783743687035